Fungal Biology

Series Editors

Fungal biology has an integral role to play in the development of the biotechnology and biomedical sectors. It has become a subject of increasing importance as new fungi and their associated biomolecules are identified. The interaction between fungi and their environment is central to many natural processes that occur in the biosphere. The hosts and habitats of these eukaryotic microorganisms are very diverse; fungi are present in every ecosystem on Earth. The fungal kingdom is equally diverse, consisting of seven different known phyla. Yet detailed knowledge is limited to relatively few species. The relationship between fungi and humans has been characterized by the juxtaposed viewpoints of fungi as infectious agents of much dread and their exploitation as highly versatile systems for a range of economically important biotechnological applications. Understanding the biology of different fungi in diverse ecosystems as well as their interactions with living and non-living is essential to underpin effective and innovative technological developments. This series will provide a detailed compendium of methods and information used to investigate different aspects of mycology, including fungal biology and biochemistry, genetics, phylogenetics, genomics, proteomics, molecular enzymology, and biotechnological applications in a manner that reflects the many recent developments of relevance to researchers and scientists investigating the Kingdom Fungi. Rapid screening techniques based on screening specific regions in the DNA of fungi have been used in species comparison and identification, and are now being extended across fungal phyla. The majorities of fungi are multicellular eukaryotic systems and therefore may be excellent model systems by which to answer fundamental biological questions. A greater understanding of the cell biology of these versatile eukaryotes will underpin efforts to engineer certain fungal species to provide novel cell factories for production of proteins for pharmaceutical applications. Renewed interest in all aspects of the biology and biotechnology of fungi may also enable the development of "one pot" microbial cell factories to meet consumer energy needs in the 21st century. To realize this potential and to truly understand the diversity and biology of these eukaryotes, continued development of scientific tools and techniques is essential. As a professional reference, this series will be very helpful to all people who work with fungi and should be useful both to academic institutions and research teams, as well as to teachers, and graduate and postgraduate students with its information on the continuous developments in fungal biology with the publication of each volume.

More information about this series at http://www.springer.com/series/11224

Ram Prasad
Editor

Fungal Nanotechnology

Applications in Agriculture, Industry, and Medicine

Springer

Editor
Ram Prasad
Amity Institute of Microbial Technology
Amity University
Noida, Uttar Pradesh, India

ISSN 2198-7777 ISSN 2198-7785 (electronic)
Fungal Biology
ISBN 978-3-319-68423-9 ISBN 978-3-319-68424-6 (eBook)
DOI 10.1007/978-3-319-68424-6

Library of Congress Control Number: 2017957707

Printed on acid-free paper

This Springer imprint is published by Springer Nature
The registered company is Springer International Publishing AG
The registered company address is: Gewerbestrasse 11, 6330 Cham, Switzerland

Preface

The study of fungal biology has developed into a valuable science in the last century as it has provided control over a number of infectious diseases. In this trend, nanotechnology has emerged as a potential candidate for similar applications. Biogenic tailored nanoparticles from fungi are gaining consideration due to their cost-effective, sustainable, resource-efficient, simple, and eco-friendly nature. In this book entitled *Fungal Nanotechnology*, the editor has accrued numerous advanced approaches for studying the fungal system for the benefit of humankind. The book covers the synthesis of nanoparticles by fungi, the mechanism involved in such biosynthesis, and a unique template for synthesis of tailored nanoparticles targeted at therapeutics, diagnostics, agriculture, and industries.

This book should be immensely useful to microbiologists, nanotechnologists, researchers, technocrats, scientists of fungal biology, and to those who are interested in fungal nanotechnology. I am honored that the leading scientists who have extensive, in-depth experience and expertise in fungal system and nanotechnology took the time and effort to develop these outstanding chapters. Each chapter is written by internationally recognized researchers/scientists so the reader is given an up-to-date and detailed account of our knowledge of the nanobiotechnology and innumerable applications of fungi.

We are indebted to the many people who helped to bring this book to light. I wish to thank series editors Dr. Vijai Kumar Gupta and Dr. Maria G. Tuohy; Eric Stannard, Senior Editor, Botany, Springer; and Jeffrey Taub, Project Coordinator, Springer, for generous assistance, constant support, and patience in initializing the volume. Special thanks go to my lovely wife Dr. Avita Maurya for her constant support and motivations in putting everything together. Dr. Prasad in particular is very thankful to Professor Ajit Varma, Amity University, for the kind support and constant encouragement. Special thanks are due to my esteemed friend and well-wisher Mr. Manjit Kumar and all faculty colleagues of AIMT, Amity University.

Noida, Uttar Pradesh, India — Ram Prasad

Contents

Contributors

Emad Abada Botany and Microbiology Department, Faculty of Science, Helwan University, Ain Helwan, Cairo, Egypt

Biology Department, Faculty of Science, Jazan University, Jazan, King Saudi Arabia (KSA)

Kamel A. Abd-Elsalam Agricultural Research Center, Plant Pathology Research Institute, Giza, Egypt

Zarraq Al-Faifi Biology Department, Faculty of Science, Jazan University, Jazan, King Saudi Arabia (KSA)

Sandra Pérez Álvarez Departamento de Biotecnología Agrícola, Instituto Politécnico Nacional, CIIDIR-IPN, Unidad Sinaloa, Guasave, Sinaloa, Mexico

Rajender Boddula CAS Key Laboratory for Nanosystem and Hierarchical Fabrication, National Center for Nanoscience and Technology, Beijing, People's Republic of China

Juan Bueno Fundación Centro de Investigación de Bioprospección y Biotecnología de la Biodiversidad (BIOLABB), Colombia, South America

Ernestina Castro-Longoria Departmento de Microbiología, Centro de Investigación Científica y de Educación Superior de Ensenada (CICESE), Ensenada, Baja California, Mexico

Diana Garibo-Ruiz Departmento de Microbiología, Centro de Investigación Científica y de Educación Superior de Ensenada (CICESE), Ensenada, Baja California, Mexico

Amaury Méndez Guerrero Empresa Bioagricultura Aplicada AMG, Calle Santos Degollado entre Guillermo Nelson y Cuauhtemoc, Colonia Centro, Guasave, Sinaloa, Mexico

Ayat F. Hashim Food Industries and Nutrition Division, National Research Center, Giza, Egypt

Sairengpuii Hnamte Department of Microbiology, School of Life Sciences, Pondicherry University, Kalapet, Puducherry, India

Ahmed Hussien Central Administration of Plant Quarantine, Ministry of Agriculture and Land Reclamation, Giza, Egypt

Anal K. Jha University Department of Chemistry, T.M. Bhagalpur University, Bhagalpur, India

Aryabhatta Centre for Nanoscience and Nanotechnology, Aryabhatta Knowledge University, Patna, India

Babita Jha Aryabhatta Centre for Nanoscience and Nanotechnology, Aryabhatta Knowledge University, Patna, India

Suphiya Khan Department of Bioscience and Biotechnology, Banasthali Vidyapith, Banasthali Tonk, Rajasthan, India

Muhsin Konuk Üsküdar University, School of Engineering and Natural Sciences, Molecular Biology and Genetics Department, Istanbul, Turkey

Safiye Elif Korcan Uşak University, Vocational School of Health Services, Department of Medical Services and Techniques, Uşak, Turkey

Niraj Kumari Aryabhatta Centre for Nanoscience and Nanotechnology, Aryabhatta Knowledge University, Patna, India

Sonu Kumari Department of Bioscience and Biotechnology, Banasthali Vidyapith, Banasthali Tonk, Rajasthan, India

Sandy Martínez-Castro Residencia Médica en Cirugía General, Universidad Autónoma de Baja California, Mexicali, Baja California, Mexico

Mohamed Osman Botany and Microbiology Department, Faculty of Science, Helwan University, Ain Helwan, Cairo, Egypt

Karel Ismar Acosta Pérez Universidad de Las Tunas, Las Tunas, Cuba

Kamal Prasad Aryabhatta Centre for Nanoscience and Nanotechnology, Aryabhatta Knowledge University, Patna, India

University Department of Physics, T.M. Bhagalpur University, Bhagalpur, India

Mugdha Rao Aryabhatta Centre for Nanoscience and Nanotechnology, Aryabhatta Knowledge University, Patna, India

Aditya Saran CAS Key Laboratory of Standardization and Measurement for Nanotechnology, National Center for Nanoscience and Technology, Beijing, People's Republic of China

Krishan K. Selwal Department of Biotechnology, Deenbandhu Chhotu Ram University of Science & Technology, Murthal, Sonepat, Haryana, India

Manjit K. Selwal Department of Biotechnology, Deenbandhu Chhotu Ram University of Science & Technology, Murthal, Sonepat, Haryana, India

Sharda Ranjan Sharan Department of Physics, Deoghar College, S K M University, Dumka, India

Busi Siddhardha Department of Microbiology, School of Life Sciences, Pondicherry University, Kalapet, Puducherry, India

Marco Antonio Magallanes Tapia Departamento de Biotecnología Agrícola, Instituto Politécnico Nacional, CIIDIR-IPN, Unidad Sinaloa, Guasave, Sinaloa, Mexico

V. Venkateswara Sarma Department of Biotechnology, School of Life Sciences, Pondicherry University, Kalapet, Puducherry, India

Madan L. Verma Center for Chemistry and Biotechnology, Deakin University, Melbourne, Victoria, Australia

Khamis Youssef Agricultural Research Center, Plant Pathology Research Institute, Giza, Egypt

Chapter 1
Fungal Nanotechnology: A Pandora to Agricultural Science and Engineering

Mugdha Rao, Babita Jha, Anal K. Jha, and Kamal Prasad

Abstract This chapter highlights the current status and the awaiting panorama of fungal nanotechnology in the compass of agricultural science and engineering. The existent advances, potential applications, and challenges of myconanotechnology in agri-food sector have been discussed. It summarizes some of the most promising applications of mycogenic nanomaterials in agriculture that involves nanoformulations for increased crop yield, smart field systems with precision farming, and early disease detection measures along with crop improvement through mycomimetic models. Another aspect captivates their use in food packaging materials that possess extremely high gas barriers and antimicrobial properties and nanosensors which can detect microorganisms. There are tremendous potentials of myconanotechnology in agriculture wherein most of the research projects are in their nascent stage, and it will surely bang all doors of agri-food sector with strong intents and purposes. It conclusively focuses on possible benefits of employing myco-fabricated nano-products and their novel application potentialities.

M. Rao • B. Jha
Aryabhatta Centre for Nanoscience and Nanotechnology, Aryabhatta Knowledge University, Patna 800001, India

A.K. Jha
University Department of Chemistry, T.M. Bhagalpur University, Bhagalpur 812007, India

Aryabhatta Centre for Nanoscience and Nanotechnology, Aryabhatta Knowledge University, Patna 800001, India

K. Prasad (✉)
Aryabhatta Centre for Nanoscience and Nanotechnology, Aryabhatta Knowledge University, Patna 800001, India

University Department of Physics, T.M. Bhagalpur University, Bhagalpur 812007, India
e-mail: k.prasad65@gmail.com

R. Prasad (ed.), *Fungal Nanotechnology*, Fungal Biology,
DOI 10.1007/978-3-319-68424-6_1

1.1 Introduction

The awareness about nanotechnology has gained a pace from past a decade which basically deals with the fabrication of matter at nanoscale (1–100 nm). This technology is applicable to multi-sectional area due to the features of nanoparticles (NPs) being extremely small and large surface-to-volume ratio which cumulatively imparts unique properties to them. One of the subbranches of nanotechnology is fungal nanotechnology, which has been coined as 'myconanotechnology' by Rai et al. (2009) to elucidate fungi-mediated nanofabrication. Nanofabrication is also possible through physical (Chen and Yeh 2002; Abou El-Nour et al. 2010) and chemical methods (Wanga et al. 2005; Raza et al. 2016). The significance of these methods has declined over time as biogenic nanosynthesis manifested by bacteria (Jha et al. 2010), actinomycetes (Sowani et al. 2015), fungi (Jha et al. 2008; Korbekandi et al. 2014; Prasad 2016), algae (Patel et al. 2015; Aziz et al. 2015), plants, and plant parts (Jha et al. 2009a; Ahmed et al. 2016; Prasad 2014) has proved to be superior in many aspects. But among all biogenic sources for nanoparticle (NP) synthesis, fungi can be advantageous due to easy maintenance (Honary et al. 2013a), ease in scaling up (Yadav et al. 2015), enriched with extracellular enzyme (Ingle et al. 2008), and supremely a solvent-less synthesis (Duran and Seabra 2012). Also the fungi have preeminent advantage in terms of cellular level organization and metabolic fluxes for nanofabrication. These reasons persuaded the exploration of myconanotechnology for heterogeneous applications (Youtie et al. 2008; Jha et al. 2009a; Gade et al. 2010) and rise in number of publications as well. Overall, the biogenic method is cost-effective. Green biogenic fabrication protocols paved the way to synthesize numerous nanomaterials like silver, gold, platinum, zinc oxide, titanium dioxide, etc. using fungi.

With the growing population, a hefty demand on agriculture/food (agri-food) sector has become a challenge to the society. The upcoming technology like nanotechnologies may act as jackpot to unburden the heaviness imposed on this area. The advancement of technology provided a better insight of nano-dimensional properties; thus new avenues were explored for the penetration of mycosynthesized nano-products in sectors like agriculture and food. Despite huge funding in the nanotechnology research in almost all expansions, the area that has just scratched the periphery of the sphere is agriculture sector, wherein food nanoscience has started blooming. According to Duncan (2011), packaging in agri-food seems to be the highly prompting area. Majority of the research is still at its budding stage and has yet to be commercialized.

It indeed brings laurels to nanotechnology, as it promises to revolutionize the agriculture and food sector with new tools for crop improvement, smart delivery systems for controlled and targeted release of agrochemicals, precision farming, and early detection of diseases, nano-antimicrobials in food pathogen detection, and nanosensors in smart packaging (Prasad et al. 2014, 2017). It works as a catalyst in the agrarian cycle as depicted in Fig. 1.1. Technical innovations in the agri-sector is the demand of time to meet the global challenges like population growth,

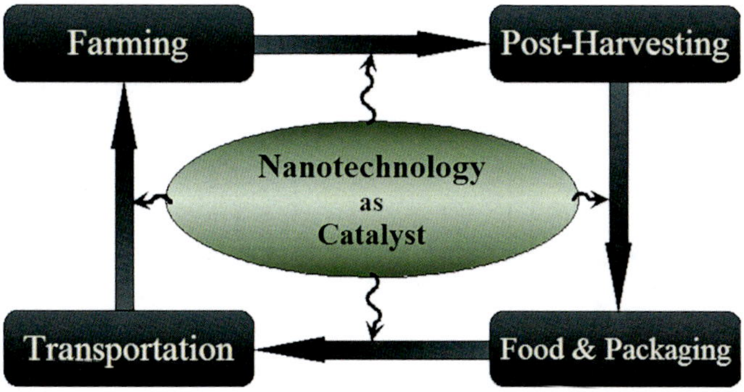

Fig. 1.1 The schematic of agrarian cycle

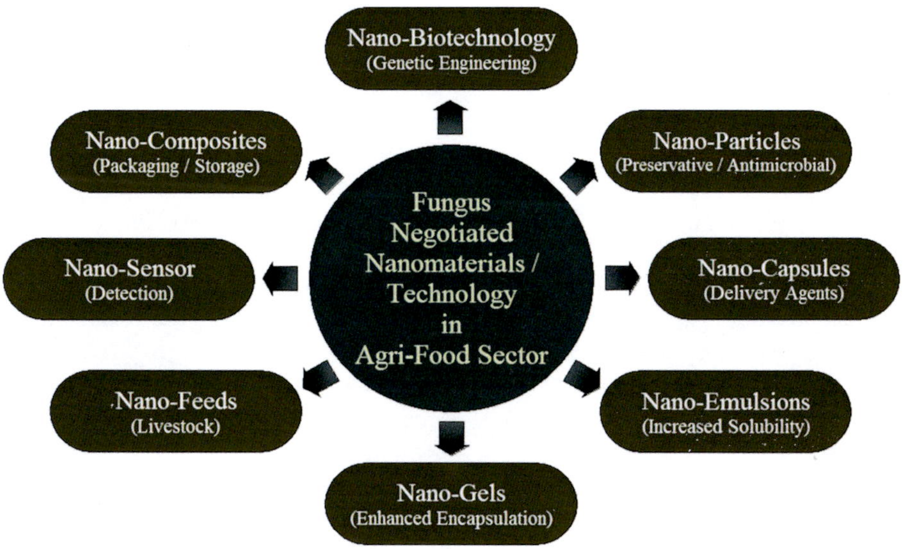

Fig. 1.2 Putative utilization of fungus-negotiated nanomaterials/technology in agri-food sector

depleting soil fertility, pests' resistance to pesticides, climatic changes, and bounded availability of resources. Groundwork on role of fungal nanotechnology in agricultural science and engineering is advancing and probably would be the possible solutions to all problems in agriculture and food sector. Figure 1.2 portrays the potentiality of fungus-negotiated nanomaterials/technology in agri-food sector for diverse pursuits.

This chapter estimates the possible areas and avenues for future myconanotechnology-based agriculture and food applications. It also briefly describes the underlying mechanism of synthesis of nanomaterials by fungi and takes a preliminary look at research efforts, opportunities, and potential applications of fungi in the area of agricultural science and engineering. The goal of this chapter is to look upstream and have an insight of the fungal nano-agri-food envision. With the success of nanotechnology, we will soon be observing the nano-products entering the agri-food supply chain from research laboratories to the fields and finally to our kitchen and environment.

1.2 Myconanofabrication

Fungi are eukaryotic organisms that inhabits in a capacious range of natural environment (Mehra and Winge 1991) with nearly 1.5 million anticipated species, out of which 70,000 species have been characterized (Quester et al. 2013). Being decomposers, these organisms are bestowed with the quality to release enzymes into their surrounding system, thereby becoming capable to decompose materials into simpler molecules.

1.2.1 Fungi as an Efficient Nanofactory

It's been more than a decade that fungal cultures are used for nanoparticle fabrication, for instance, extracellular synthesis of silver nanoparticles (Ag NPs) and CdS nanoparticles, respectively, by *Verticillium* sp. (Mukherjee et al. 2001a) and *Fusarium oxysporum* (Ahmad et al. 2002). There are numerous reasons for fungi to be the choicest candidate for nanofabrication. Fungi can be grown with ease, maintained in the laboratory, and handily subcultured and are renewable source for nanofabrication (Honary et al. 2013a; Soni and Prakash 2012). Fungal nanofabrication can be scaled up easily without experiencing cell disruption during agitation in the bioreactor (Yadav et al. 2015). It is well documented that fungal cell wall provides mechanical strength to overcome osmotic pressure and environmental fluctuations as well (Durán and Nombela 2004; Yadav et al. 2015). Recently, Yadav et al. (2015) reviewed that most of the fungal system supports extracellular synthesis. These organisms are cost-effective for nanoparticle (NP) synthesis due to minimum downstream processing steps in the purification process with an added advantage of nanoparticles' high yield supported by the release of bountiful extracellular enzyme (Rai et al. 2009; Ingle et al. 2008; Birla et al. 2009; Kumar et al. 2007; Quester et al. 2013; Yadav et al. 2015; Gade et al. 2008; Zhang et al. 2011; Honary et al. 2013b; Prasad et al. 2016). Fungal cells when charged with metal ions promote its speedy reduction by the secreted enzymes innoxiously to obtain highly monodispersed nanoparticles with well-defined shape and size (Kashyap et al. 2012). Above all fungi have well-defined nucleus and hence can be genetically manipulated to secrete plentiful of enzymes required for nano-conversions; thereby, huge quantity of nanoparticles could be obtained with reduced particle size (Jha and Prasad 2016).

1.2.2 Myconanosynthesis Procedure

There are primarily a couple of approaches for fabrication of nanoparticles:

(a) Bottom-up, which firstly involves unification of atoms and molecules by self-association forming the base followed by the final product. As an outcome, the well-defined nanoparticles of various sizes and shapes are formed (Mittal et al. 2013) and
(b) Top-down, herein an appropriate preparatory material undergoes reduction either through physical (machining, templating, or lithographic) or chemical techniques (Gade et al. 2010).

According to the literature, the bottom-up approach is reasonably better than top-down as there is a fair chance to obtain nanostructures coerced by Gibbs free energy with thermodynamically controlled chemical compositions (Thakkar et al. 2010; Behari 2010). Myconanofabrication drives the reduction of compounds to their respective nanoparticles through their enzymes and metabolites as a reducing agent (Zhang et al. 2011). Kashyap et al. (2012) mentioned in their review that mycosynthesis follows bottom-up approach for nanofabrication.

1.2.3 Contemplated Mechanism

Till date fungal nanoparticle synthesis mechanism is vague but on intent study by various researchers (Rai et al. 2011; Jha and Prasad 2016; Yadav et al. 2015; Mukherjee et al. 2008; Durán et al. 2011; Ingle et al. 2008) has been hypothetically presented on the basis of microorganisms' ability to withstand high metal concentration which might have developed defense mechanisms, and as a consequence of this, nanoparticle synthesis seemingly occurs. Rai et al. (2011) described fungal mycosynthesis as a three-step process which involves trapping of metal ion, bioreduction, and finally synthesis. Besides, they also explained that nitrate reductase enzyme catalyzes extracellular nanoparticle synthesis. Myconanofabrication is an outcome of reaction between fungal biomass and salt solution, which occurs either extracellular or intracellular, and the former one predominates. Table 1.1 represents the various fungal species-derived nanoparticles implicated/potential in agri-food sector. The mechanism involved in fungal nanoparticle synthesis is mainly twofold (a) extracellular and (b) intracellular.

Jha and Prasad (2016), in their profoundly studied mechanism of myconanosynthesis, have deciphered that fungi undergo stress and reciprocate as a cellular defense at three different levels upon exposure to metal/metal oxide solution. Ion stress triggers bioreduction wherein the nanoparticles are formed on the cell wall surfaces. In previous studies it has been mentioned that fungal cell produces enzymes and metabolites to circumvent metal toxicity and convert it into a least toxic product such as nanoparticles (Kashyap et al. 2012). *Fusarium* is being

Table 1.1 List of potential/applicable myco-fabricated nanoparticles in agriculture

Fungi/yeast	Confinement	References
Silver (Ag) nanoparticles		
Verticillium sp.	Intracellular	Mukherjee et al. (2001a)
Fusarium oxysporum	Extracellular	Ahmad et al. (2003)
Phoma sp.	Extracellular	Chen et al. (2003)
Yeast strain MKY3	Intracellular	Kowshik et al. (2003)
Verticillium sp.	Intracellular	Senapati et al. (2004)
Fusarium oxysporum	Extracellular	Duran et al. (2005)
Aspergillus fumigatus	Extracellular	Bhainsa and D'souza (2006)
Phanerochaete chrysosporium	Extracellular	Vigneshwaran et al. (2006)
Aspergillus flavus	Extracellular	Vigneshwaran et al. (2007a)
Fusarium oxysporum	Extracellular	Mohammadian et al. (2007)
Pleurotus sajor-caju	Extracellular	Vigneshwaran et al. (2007b)
Aspergillus niger	Extracellular	Gade et al. (2008)
Fusarium acuminatum	Extracellular	Ingle et al. (2008)
Fusarium semitectum	Extracellular	Basavaraja et al. (2008)
Penicillium sp.	Extracellular and intracellular	Sadowski et al. (2008)
Trichoderma asperellum	Extracellular	Mukherjee et al. (2008)
Alternaria alternata	Extracellular	Gajbhiye et al. (2009)
Phytophthora infestans	Extracellular	Thirumurugan et al. (2009)
Pleurotus sajor-caju	Extracellular	Nithya and Ragunathan (2009)
Volvariella volvacea	Extracellular	Philip (2009)
Cladosporium cladosporioides	Extracellular	Balaji et al. (2009)
Coriolus versicolor	Extracellular	Sanghi and Verma (2009)
Penicillium brevicompactum	Extracellular	Shaligram et al. (2009)
Fusarium solani	Extracellular	Ingle et al. (2009)
Penicillium fellutanum	Extracellular	Kathiresan et al. (2009)
Penicillium sp.	Extracellular	Maliszewska et al. (2009)
Phoma glomerata	Extracellular	Birla et al. (2009)
Penicillium sp.	Extracellular	Hemanth et al. (2010)
Aspergillus niger	Extracellular	Jaidev and Narasimha (2010)
Aspergillus clavatus	Extracellular	Verma et al. (2010)
Fusarium oxysporum	Extracellular	Pandiarajan et al. (2010)
Aspergillus flavus	Extracellular	Jain et al. (2010)
Aspergillus fumigatus	Extracellular	Navazi et al. (2010)
Bipolaris nodulosa	Extracellular	Saha et al. (2010)
Fusarium solani	Extracellular	El-Rafie et al. (2012)
Aspergillus clavatus	Extracellular	Saravanan and Nanda (2010)
Aspergillus sp.	Extracellular	Saravanan (2010)
Rhizopus stolonifer	Extracellular	Binupriya et al. (2010a)
Trichoderma viride	Extracellular	Fayaz et al. (2010)
Pestalotia sp.	Extracellular	Raheman et al. (2011)

(continued)

Table 1.1 (continued)

Fungi/yeast	Confinement	References
Cochliobolus lunatus	Extracellular	Salunkhe et al. (2011)
Nigrospora oryzae	Extracellular and intracellular	Saha et al. (2011)
Phoma sorghina	Extracellular	Gade et al. (2011)
Trichoderma reesei	Extracellular	Vahabi et al. (2011)
Neurospora crassa	Extracellular and intracellular	Castro-Longoria et al. (2011)
Aspergillus fumigatus	Extracellular	Alani et al. (2012)
Aspergillus terreus	Extracellular	Li et al. (2012)
Saccharomyces boulardii	Extracellular	Kaler et al. (2013)
Cylindrocladium floridanum	Extracellular	Narayanan et al. (2013)
Epicoccum nigrum	Extracellular	Qian et al. (2013)
Penicillium citrinum	Extracellular	Honary et al. (2013b)
Neurospora intermedia	Extracellular	Hamedi et al. (2014)
Penicillium nalgiovense	Extracellular	Maliszewska et al. (2014)
Macrophomina phaseolina	Extracellular	Chowdhury et al. (2014)
Penicillium chrysogenum	Extracellular	Pereira et al. (2014)
Aspergillus oryzae		
Trichoderma harzianum	Extracellular	Sundaravadivelan and Padmanabhan (2014)
Fusarium oxysporum	Extracellular	Gholami-Shabani et al. (2014)
Saccharomyces cerevisiae	Extracellular	Korbekandi et al. (2016)
Guignardia mangiferae	Extracellular	Balakumaran et al. (2015)
Fusarium oxysporum	Extracellular	Salaheldin et al. (2016)
Aspergillus terreus	Extracellular	Ammar and El-Desouky (2016)
Penicillium expansum		
Monascus purpureus	Extracellular	El-Baz et al. (2016)
Rhizopus stolonifer	Extracellular	AbdelRahim et al. (2017)
Gold (Au) nanoparticles		
Verticillium sp.	Intracellular	Mukherjee et al. (2001b)
Fusarium oxysporum	Extracellular	Mukherjee et al. (2002)
Colletotrichum sp.	Extracellular	Shankar et al. (2003)
Fusarium oxysporum	Extracellular	Shankar et al. (2004)
Trichothecium sp.	Extracellular and intracellular	Ahmad et al. (2005)
Verticillium luteoalbum	Intracellular	Gericke and Pinches (2006)
Aspergillus niger	Extracellular	Xie et al. (2007)
Fusarium semitectum	Extracellular	Sawle et al. (2008)
Helminthosporium	Extracellular	Kumar et al. (2008)
Volvariella volvacea	Extracellular	Philip (2009)
Yarrowia lipolytica	Extracellular	Pimprikar et al. (2009)
Penicillium sp.	Intracellular	Zhang et al. (2009a)

(continued)

Table 1.1 (continued)

Fungi/yeast	Confinement	References
Rhizopus oryzae	Extracellular	Das et al. (2009)
		Das and Marsili (2010)
Coriolus versicolor	Extracellular and intracellular	Sanghi and Verma (2010)
Coriolus versicolor	Extracellular and intracellular	Sanghi and Verma (2010)
Rhizopus stolonifer	Extracellular	Binupriya et al. (2010a)
Aspergillus oryzae var. viridis	Extracellular	Binupriya et al. (2010b)
Penicillium chrysogenum	Intracellular	Sheikhloo and Salouti (2011)
Sclerotium rolfsii	Extracellular	Narayanan and Sakthivel (2011)
Penicillium sp.	Extracellular and intracellular	Du et al. (2011)
Aspergillus clavatus	Intracellular	Verma et al. (2011)
Fusarium oxysporum	Extracellular	Anitha and Palanivelu (2011)
Neurospora crassa	Extracellular and intracellular	Castro-Longoria et al. (2011)
Phanerochaete chrysosporium	Extracellular	Sanghi et al. (2011)
Alternaria alternata	Extracellular	Sarkar et al. (2012)
Aspergillus sp.	Extracellular	Gupta and Bector (2013)
Penicillium sp.	Extracellular	Honary et al. (2013a)
Cylindrocladium floridanum	Extracellular	Narayanan and Sakthivel (2013)
Nigrospora oryzae	Extracellular	Kar et al. (2014)
Trichoderma viride	Extracellular	Mishra et al. (2014)
Hypocrea lixii		
Fusarium accuminatum	Extracellular	Tidke et al. (2014)
Aspergillus niger	Extracellular	Chakravarty et al. (2015)
Aspergillus terreus	Extracellular	Balakumaran et al. (2016)
Rhizopus oryzae	Extracellular	Kitching et al. (2016)
Penicillium aculeatum	Extracellular	Barabadi et al. (2017)
Titanium dioxide (TiO_2) nanoparticles		
Fusarium oxysporum	Extracellular	Bansal et al. (2005)
Saccharomyces cerevisiae	Extracellular	Jha et al. (2009b)
Aspergillus flavus	Extracellular	Rajakumar et al. (2012)
Zinc oxide (ZnO) nanoparticles		
Candida albicans	Extracellular	Shamsuzzaman et al. (2013)
Aspergillus fumigatus	Extracellular	Raliya and Tarafdar (2013)
Alternaria alternata	Extracellular	Sarkar et al. (2014)
Aspergillus aeneus	Extracellular	Jain et al. (2012)
Aspergillus fumigatus	Extracellular	Raliya et al. (2016a)
Cadmium sulfide (CdS) nanoparticles		
Candida glabrata	Extracellular	Domeron et al. (1989)

(continued)

Table 1.1 (continued)

Fungi/yeast	Confinement	References
Schizosaccharomyces pombe	Intracellular	Williams et al. (1996)
		Kowshik et al. (2002)
Fusarium oxysporum	Extracellular	Ahmad et al. (2002)
Schizosaccharomyces pombe	Intracellular	Krumov et al. (2007)
Saccharomyces cerevisiae	Extracellular	Prasad and Jha (2010)
Phanerochaete chrysosporium	Extracellular	Chen et al. (2014)
Iron/magnetite (Fe_3O_4) nanoparticles		
Fusarium oxysporum	Extracellular	Bharde et al. (2006)
Aspergillus oryzae	Extracellular	Tarafdar and Raliya (2013)
MgO nanoparticles		
Aspergillus tubingensis	Extracellular	Raliya et al. (2014a)
Aspergillus flavus	Extracellular	Raliya et al. (2014b)
Silica (SiO_2) nanoparticles		
Fusarium oxysporum	Extracellular	Bansal et al. (2005)

majorly explored for nanoparticle synthesis due to its richness in membrane-bound and cytosolic enzymes like cellulase, nitrate reductase, oxidoreductase (Duran et al. 2005), and oxidase (Jha and Prasad 2016). These enzymes are credited due to their flexibility as it confers switch on and off mode upon pH change, as oxidase catalyzes a reaction at lower pH and reductase at higher pH (Jha and Prasad 2010). The role of nitrate reductase in extracellular synthesis is mostly acknowledged (Ahmad et al. 2003; Duran et al. 2005; Ingle et al. 2008; Li et al. 2012). Ahmad et al. (2003) were the first to propose mechanism for Ag NP synthesis by *Fusarium oxysporum* by NADPH-dependent reductase. Accordingly both anthraquinone and NADPH catalyze the silver ion (Ag^+) to neutral silver (Ag^0). In another study, -NADPH upon reduction to –NADP generates electrons that are transferred to Ag^+ ions which in turn forms Ag^0, and the process is accomplished by an electron shuttle hydroxyquinoline (Kumar et al. 2007). The lower fungi like *Aspergillus* and *Penicillium* species release numerous hydroxy/methoxy derivatives of benzoquinones and toluquinones in response to metallic stress (Jha and Prasad 2016). As proposed by Jha and Prasad (2010), these metabolites expeditiously undergo a redox reaction to generate nanoparticles. In other words silver ion acts as a substrate that binds to enzyme (reductase) wherein NADPH gets reduced to NADH releasing electrons which brings about the conversion. The hidden fact is that enzymes are basically proteins. Investigators (Das et al. 2009; Mukherjee et al. 2008; Jain et al. 2010; Gade et al. 2008) have hypothesized the involvement of polypeptide/proteins in bioreduction of metal ions to their corresponding nanoparticles and their role as encapsulating agent. Surprisingly the mechanisms proposed up to now are essentially for silver and gold nanoparticles. A few investigators have come up with metal oxide mechanism (Bansal et al. 2004; Durán and Seabra 2012; Jha and Prasad 2016).

The intracellular synthesis embarks with chelation of metal ions over fungal cell surface. This could be the result of electrostatic interaction between the metal ions and the presence of positively charged groups in enzymes/proteins at the cell wall (Rai et al. 2009; Kashyap et al. 2012). Jha and Prasad (2016) mentioned in their report about extracellular mucilaginous materials having superb toxic metal-binding capabilities, and these metal ions are transported in the cell with the help of transporters/channels (proteins). This creates a metabolic chaos inside the cell and to bypass this situation cells gives off desired nanoparticles. There is a possibility of nanoparticle diffusion out of the cell due to concentration gradient only if the rate at which the substrate enters the cell is in equilibrium with the product formed. Otherwise nanoparticles may get aggregated, and cell disruption may occur due to excess accumulation of particles inside.

1.3 Myconanotechnology for the Agricultural Sector: From Research to the Field

The present scenario of agricultural science and engineering is beyond "cultivation of fields." Conventional farming limitations could not meet the expanding demands of increased productivity and ecosystem restoration. This has paved the way for "conservation farming" (Hobbs et al. 2008) and "organic farming" (Kirchmann and Thorvaldsson 2000). These methods proved to be the helpless stars as they could neither accomplish exorbitant productivity nor ensure environment restoration to its pristine state. In this scenario nanotechnology emerged as the technological platform for the convolution of agri-food system, and few are exemplified in Table 1.2. Agricultural science and engineering field have gained momentum with the burgeoning applications of myconanotechnology in agrarian science. Mycogenic nanomaterials such as nanowires, nanostructures, nanoformulations, nanosensors, and quantum dots are being utilized to suppress plant pathogen and target delivery and for interactive agrochemicals like pesticide and insecticides (Kashyap et al. 2012). The advent of nanotechnology gave access to the development of nanostructures and nanodevices with prospective novel applications in the agricultural sector (Scott and Chen 2003). This accelerating pace of potential agricultural applications has envisioned that ingenious engineering of novel nano-biomaterials would have innovative nanotechnology advances for the benefit of a sustainable society (Scott and Chen 2013; Singhal et al. 2017). The complexities of farm production systems demand nanomaterials with flexible dimensions which could efficiently perform task in a thermodynamically open system (Mukhopadhyay 2014). The mycogenic products are utilized for multitude applications in agriculture which are highly interdisciplinary making its categorization difficult. This section attempts to sum up some of the advances and in a coarse grain analysis categorize it in following subsections.

Table 1.2 Nanomaterials with potential applications in agri-sector

Nanoparticles/ nanomaterials	Potential applications	References
Carbon nanotubes	As growth stimulator in seed germination and plant growth	Khodakovskaya et al. (2009)
		Liu et al. (2009)
		Mondal et al. (2011)
		Tripathi et al. (2011)
		Villagarcia et al. (2012)
		Tiwari et al. (2014)
Nanosilica/ nanoaluminosilicate	Antimicrobial agent/ pesticides/delivery agent for pesticides and chemicals	Liu et al. (2006)
		Torney et al. (2007)
		Barik et al. (2008)
		Goswami et al. (2010)
		Manikandan and Subramanian (2014)
Silica silver	Fungicide/pesticide	Park et al. (2006)
Pt black and carbon nanotube in SR IAA microsensor	In the study of endogenous auxin transport in root epidermal cells	Mclamore et al. (2010)
Gold	Water hygiene management	Das et al. (2009)
Silver	As pesticides/insecticide/ fungicide	Kim et al. (2009)
		Jo et al. (2009)
		Gajbhiye et al. (2009)
		Aguilar-Méndez et al. (2010)
		Lamsal et al. (2011)
		Jayaseelan et al. (2011)
		Velmurugan et al. (2013)
		Aziz et al. (2015, 2016)
		Bhattacharyya et al. (2016a)
Nanostructured alumina	Insecticidal	Stadler et al. (2010)
Nano-copper	Antimicrobial	Cioffi et al. (2004)
		Esteban-Tejeda et al. (2009)
Polyethylene glycol-coated	Insecticidal	Yang et al. (2009)
Titanium dioxide	As fertilizers	Raliya et al. (2015a)
Magnesium oxide	As fertilizers	Raliya et al. (2014b)
Iron	In agriculture	Tarafdar and Raliya (2013)
		Nadi et al. (2013)
$SiO_2 + TiO_2$	As fertilizers	Lu et al. (2002)
Sulfur NPs	Antifungal	Rao and Paria (2013)
Urea-modified hydroxyapatite	As fertilizers	Kottegoda et al. (2011)
Rare earth oxide NPs (CeO_2, La_2O_3, Gd_2O_3, Yb_2O_3)	As fertilizers (root growth)	Ma et al. (2010)

1.3.1 Nanoformulations for Agrochemicals

Nanoparticles/nanomaterials can be directly enforced in the field or could be amenably mutated as formulations for effective action. The agrochemical active ingredients are encapsulated, emulsified, coated, or adhered with inorganic nano-carriers to enhance their activity with minimized wastage. The abovementioned efficacious ingredients could be herbicides, pesticides, fertilizers, pheromones, or plant growth regulators (Bhattacharyya et al. 2016b). These nanoformulations are expedient and prove to be a boon to the agriculture as schematized in Fig. 1.3. Nanocapsulated herbicides offered reduced ecotoxicity (Pérez-de-Luque and Rubiales 2009), while surface-modified hydrophobic nanosilica displayed enhanced pesticidal effect (Barik et al. 2008). Nanoformulations like porous hollow silica nanoparticles depicted shielding protection to photosensitive pesticides like avermectin (Li et al. 2007). However, the major challenges faced by these ideas are in maintaining stability under field conditions and release of the formulations at specific targets in the plant. The idea of nanodevices enabling controlled release of pesticides, fertilizers, and other agrochemicals regulated by time and location could recast the agricultural practices in the near future (Bhattacharyya et al. 2016b). In this direction, two of the most esteemed areas of nanoformulation pertinence are discussed below.

1.3.1.1 Nano-pesticides

Mycosystems serve as an indispensable resource for the fabrication of nanomaterials used for plant disease control. Due to their high metal tolerance and ability to accumulate metals (Prasad et al. 2016), they are extensively exploited for the

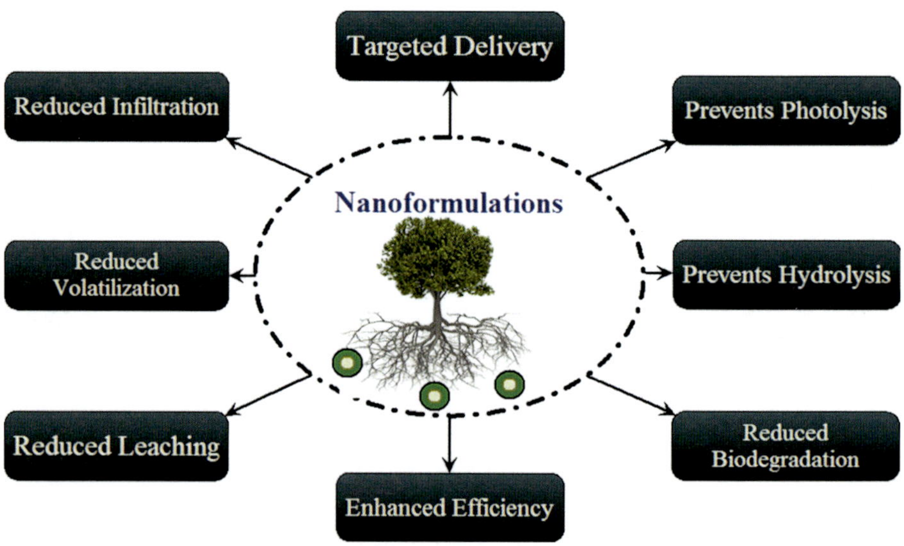

Fig. 1.3 A benefaction to agriculture: nanoformulations

fabrication of metal nanoparticles like silver, gold, platinum, iron, etc. Silver nanoparticles have wide applications as antifungal agents against plant pathogenic fungi (Jo et al. 2009; Lee et al. 2013; Gopinath and Velusamy 2013). Silver nanoparticles synthesized extracellularly by the fungus *Alternaria alternata* were evaluated for antimycotic sweep in association with fluconazole (a fungicide) against phytopathogenic fungi *Phoma glomerata*, *Phoma herbarum*, *Fusarium semitectum*, *Trichoderma* sp., and *Candida albicans*. Maximum inhibition was depicted against *Candida albicans*, and it was confirmed that silver nanoparticles cause significant enhancement of activity of fluconazole (Gajbhiye et al. 2009). Nanoparticles with inorganic nano-carriers like nanosized silica-silver congregate were investigated as a potent biocide for plant diseases (Park et al. 2006). Chitosan and chitin derivatives are incorporated in formulations for delivering pesticides (Yin and Zhang 2010; Ding et al. 2011). *Fusarium oxysporum* has been discovered to be capable of synthesizing silica nanoparticles (Bansal et al. 2005), a functional plant nutrient that augments disease resistance in plants. It has been well documented that fungi are potent sources for synthesis of nanoparticles, like silver and silica, and thus could be easily harvested for indemnity of crops.

1.3.1.2 Nano-fertilizers

Fungus-mediated nanoparticles especially metal oxides delivered through soil are being utilized as nano-nutrient fertilizers. The small size gives them the extra advantage of increased uptake rate as compared to conventional fertilizers which circumvent the problem of eutrophication (Wang et al. 2013). Iron nanoparticles were synthesized by green synthesis approach using fungi *Aspergillus oryzae* TFR9. The synthesized nanoparticles having size between 10 and 24.6 nm were spherical in shape and have wide applications in the field of agriculture, biomedical, and other sectors (Tarafdar and Raliya 2013). Nadi et al. (2013) in a field assay investigated the effect of nano-iron on yield of faba bean (*Vicia faba* L.) at different spraying time and varied concentration. They concluded that highest grain yield was achieved by sprinkling 6 g/l during flowering period. In another venture, titanium dioxide and magnesium oxide nanoparticles have been shown to improve photosynthetic process and efficiency. Magnesium oxide nanoparticles of average size 5.8 nm were synthesized extracellularly from fungus *Aspergillus tubingensis* TFR-3 (Raliya et al. 2014a). Raliya et al. (2014b) in another research have shown that magnesium oxide nanoparticles of size 5.8 nm synthesized from *A. flavus* at 15 mg/l concentration on 2-week-old *Cyamopsis tetragonoloba* could increase the chlorophyll content by 76.1%. Titanium dioxide nanoparticles of 12–15 nm size obtained from *Aspergillus flavus* were seen to enhance chlorophyll content in the mung bean plant leaves by 46.4% (Raliya et al. 2015a). These nanoparticles enhance plant photosynthesis by boosting solar light absorption. ZnO NPs fabricated from $ZnNO_3$ by the fungus *Aspergillus fumigatus* TFR-8 were foliar fed at 10 ppm concentration on cluster bean (*Cyamopsis tetragonoloba* L.) leaves, and its ability to enhance phosphorus-mobilizing enzyme as well as nano-induced gum production was studied (Raliya and Tarafdar 2013). In a recent study,

ZnO NPs synthesized from *Aspergillus fumigatus* TFR-8 were applied as phosphorus fertilizer. Nano-Zn increased the phosphorus uptake by 10.8% in mung bean (Raliya et al. 2016a).

1.3.2 Myconanotechnology in Smart Field Systems and Precision Farming

A smart field system with precision farming is the next-generation farming for "sustainable intensification" (Sekhon 2014). Natural processes of a farm production system with multiple variables are simulated upon scientific intervention for successful implementation in these future generation fields. Engineered nanomaterials in a predesigned and holistic manner restrain the climatic changes and overcome the energy and resource constraints. Nanosensors and nanobiosensors employed in fields effectively sense a wide variety of additives like fertilizers, herbicides, and pesticides and at the same time dispense information about moisture, soil pH, and pathogenic load. Myconanotechnology as an entrant in this area of nanodevices is known to produce quantum dots. Extracellular microbial biosynthesis of face-centered cubic (*fcc*) cadmium sulfide (CdS) quantum dots with average size of 2.56 nm by using white rot fungus *Phanerochaete chrysosporium* has been reported (Chen et al. 2014).

Nanomaterials could be applied as smart delivery systems in various applications in agriculture particularly in pest control and disease management. It could serve as "magical bullets" to target particular plant parts and release the active ingredients. Nanocapsules capable of penetrating through the cuticles and tissues of undesirable plants are used for effective, slow, and constant release of herbicides (Pérez-de-Luque and Rubiales 2009). Agrochemicals like fertilizers and pesticides tagged to nanoparticles lead to precision and controlled discharge of fertilizers and pesticides. This prevents undesirable damage of nontargeted plant parts and reduces the release of chemicals into the environment (Nair et al. 2010). Nanofibrous mats and nanofiber arrays with high surface area and porous structures serve as a perfect medium for site-targeted delivery of diverse macromolecules in plants for varied purposes (Moaveni et al. 2010). *Trichoderma viride*, a fungus used as a biofungicide, has been used in preparation of a nanofibrous mat in combination with chitosan by electrospinning (Spasova et al. 2011). The fibrous mat contained viable spores with the ability to inhibit the growth of test phytopathogens. Fungi could thus be explored for synthesis of vehicles in targeted delivery.

Besides these nanoformulations, early detection of disease-causing plant pathogens is essential; in this area fluorescent probes made with silica nanoparticles have successfully detected plant pathogens (Yao et al. 2009). Mycogenic fabrication of silica nanoparticles is well augmented. Nanoparticles are stable and could be used as diagnostic tools for early disease detection (Sharon et al. 2010). Nanostructured materials like nanowires have been synthesized using fungi that could be useful in miniaturization of electronic devices for use in precision farming and crop protection.

Gold microstructures of 1–2 μm were synthesized by self-assemblage of colloidal gold on living filamentous fungi *Aspergillus nidulans*, *Aspergillus niger*, and *Neurospora crassa* (Sugunan et al. 2007). Porous gold microwires synthesized by heat treatment of highly loaded fungal hyphae of *A. niger* with gold nanoparticles are reported (Rehman et al. 2011). The nanowires could be tuned for inventing diagnostic tools and monitoring devices for detection of pathogens and chemicals in agricultural fields. A nanosensor has been developed that could do early detection of grain spoilage during storage (Neethirajan et al. 2009). This sensor could easily detect the presence of insects or fungus in the stored grain.

Precision farming techniques are based on active surveillance of crop in field conditions through nano-enabled diagnostic techniques and have proved to be a blessing for the agriculture. These diagnostic techniques involve the use of biosensors with components at nanoscale linked to global positioning systems for imparting real-time information about crops' well-being. Remote sensing devices transmit the whole story of crop field, for instance, level of soil nutrient, water stress, pathogenic mass, etc. This information helps in increased productivity with proper agrochemical management and environmental protection. Furthermore nanosensor for the study of plant growth monitoring has been invented. Auxin is a vital hormone for root development and seed germination. The nanosensor reacts with auxin and provides an electrical signal which determines the auxin concentration at a particular point. This data would be very helpful for auxin research and understanding the mechanism of plant root adaptation to environment like marginal soils (McLamore et al. 2010). Smart field systems with precision farming in unison with fungal nanotechnology prove to have tremendous potentials to reconstitute the agricultural sector.

1.3.3 Myconanotechnology and Crop Biotechnology

Researches in the field of agriculture have always focused on improving the efficacy of crop production. Development of plants capable of exhibiting drought resistance, salt tolerance, and excess moisture stress has always been the center of interest. Nanotechnology in amalgamation with biotechnology provides a new tool to pursue such goals through targeted gene delivery for genetic engineering. Mesoporous silica nanoparticles have been utilized as receptacle for delivering DNA into isolated plant cells. These are chemically coated for targeted delivery through the plant cell walls (Torney et al. 2007). 3D DNA crystals have been produced using synthetic DNA sequences that possess the ability to self-assemble into a series of three-dimensional triangle-like patterns. These DNA crystals could be used for crop improvement by conjugating moieties like lipids, nucleic acids, proteins, and carbohydrates to their sticky ends that can attach to another molecule in an organized fashion (Zheng et al. 2009). Mycomimetic models with immense possibilities are under research for synthesis of such nanoscale carriers for crop biotechnology.

1.3.4 Associated Constraints

Agriculture provides prodigious opportunities for myconanotechnologic applications, but its execution is limited, some of which are laid out in Table 1.3. There are several associated constraints in this disposition. The pace of progress of myconanotechnologic intervention in agriculture although with abundance of funding seems to be modest. Research endeavors in the field of agricultural science and engineering have intensively focused on only two broadened areas: future agrochemical formulation and post-harvest food arena. Share of nanotechnology in others remains minuscule. Possible reasons for this delinquency may be the fact that it is fundamentally impossible to control the fate and performance of the nanomaterials added (Mukhopadhyay 2014). The moira of the nanomaterials is diffused from the point source of application. Besides this, the behavior of nanomaterials in contrast to closed systems demands flexibility for a thermodynamically open system. Complicated intrinsic alliance is expected in between nanomaterials and Mother Nature, thereby causing disposal problems, matching nanomedicine in humans and animals. Studies on the toxicological impact of nanoparticles are on the way. Impact of TiO_2 and ZnO nanoparticles translocation in tomato plants and resultant plant response has been examined (Raliya et al. 2015b). The uptake, translocation, and accumulation of 30–80 nm gold nanostructures in watermelon plant have been investigated, and it was observed that nanoparticles' shape, application method, and nature of plant tissues are the decisive factors (Raliya et al. 2016b). Active surveillance in farms with infinite variables is time exhausting and capital intensive. The mode of penetration, transportation, and excretion of nanoparticles by the plants along with reactivity features is an area of extensive research to harvest maximum benefit of nanotechnology in agriculture. The rigor of this challenge demands masterminds from all fields to work in a well-knit fashion.

Table 1.3 Myconanotechnology in agriculture

Fungi species	Nanoparticle	Application	References
Rhizopus oryzae	Au	Water hygiene management	Das et al. (2009)
Alternaria Alternata	Ag	Fungicide	Gajbhiye et al. (2009)
Trichoderma viride	Ag	Vegetable and fruit preservation	Fayaz et al. (2009)
Aspergillus oryzae	Fe	Agriculture	Tarafdar and Raliya (2013)
Aspergillus fumigatus	ZnO	Nano-induced gum production and phosphorus fertilizer	Raliya and Tarafdar (2013)
Aspergillus flavus	MgO	Enhance plant photosynthesis	Raliya et al. (2014b)
Aspergillus flavus	TiO_2	Enhance plant photosynthesis	Raliya et al. (2015a)
Aspergillus fumigatus	ZnO	Phosphorus fertilizer	Raliya et al. (2016a)
Piriformospora indica DSM 11827	Zinc oxide nanorod	Plant gowth	Singhal et al. (2017)

1.4 Myconanotechnology: A Highly Treasured Technology to Food (Processing-Storage-Packaging-Transportation)

Fields are not only the area to be locked up under agricultural nanotechnology; more willingly agricultural products' processing-storage-packaging-transportation (PSPT) is equally exigent aspect that can't be untouched. But above all food safety during PSPT is a humongous disquiet. The worldwide folks are quite aware of edible button mushrooms (*Agaricus bisporus*) a macrofungus, but myconanotechnology seems to be uncommon among all. The common people are not aware of the fact that these edible mushrooms could be packed with low perishability and sold to raise the economic status of their country/place.

Nonetheless nanotechnology has raised its way in, in almost all areas at an absolutely tremendous pace with a low acceptance in the food sector as fresh food products have always being charmed by us. But we cannot overlook the wastage of food due to microbial infection, quoting pathogenic bacteria like *Salmonella* and *E. coli* that surge eggs and beef, lettuce, and vegetables through contamination, respectively (Ray et al. 2013).

This section will emphasize on myconanotechnology in food safety as the primary motto is to reduce the perishability of food. The spotlight is to protect the food at various stages like processing, storage, packaging, and transportation. A discrete number of myconanoparticle have been fabricated as mentioned in Table 1.1 that have been applied/potential in food safety. Food and packaging are basically partitioned as polymer nanocomposites as packaging material (antimicrobial packaging, nanocoated packaging) and nanosensors embedded packaging/intelligent packaging.

1.4.1 Myco-nanocomposites in Food Packaging

The environmental conditions harshly affect the food is a truly commonly authenticated fact. A packaging hybrid system film Durethan KU2-2601 being materialized by Bayer Polymers that incorporates nanosilicates proved to be a barrier to food putrefaction by partially diminishing the access of gases and moisture to the food (Duran and Marcato 2013). A representative scanning electron microscope (SEM) image of nanoparticle-embedded polymer (nanocomposite) is depicted in Fig. 1.4.

Emamifar et al. (2010) in their research have demonstrated the application of low-density polyethylene (LDPE) materials impregnated with silver and zinc oxide nanoparticles hiked the shelf life of crude orange juice near to freezing point (4 °C). There was no lump of nanomaterial in the packaging film, thus making the nanocomposites of exclusively better quality. The ZnO NPs in packaging material have shortened the perishability of orange juice without any change in its phenotypic quality; also Ag NPs are found to have antimicrobial activity on yeast and molds a bit higher than ZnO NPs. The above packaging works for longer time only if raw

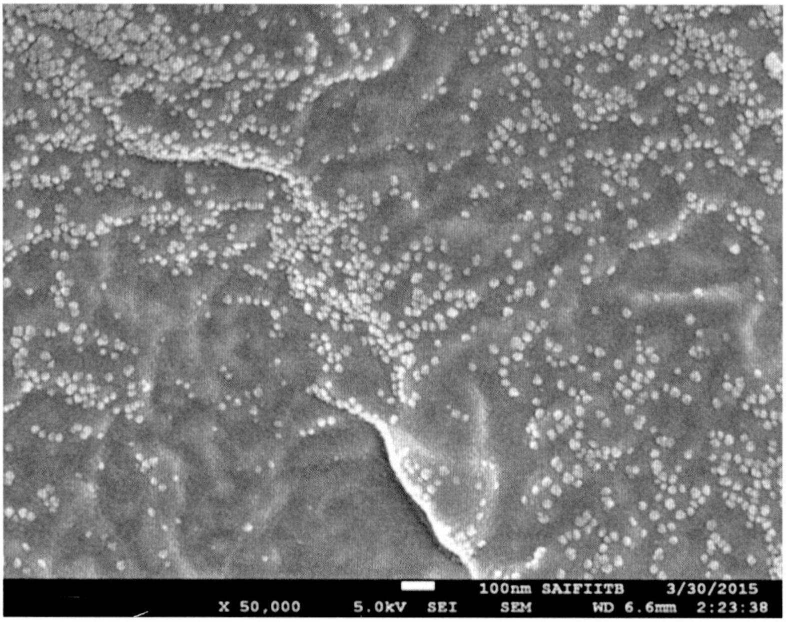

Fig. 1.4 SEM image of a cryo-fractured surface of a representative nanocomposite

orange juices were treated with a mild heat. Few researchers have demonstrated the fungal synthesis of ZnO NPs by *Candida albicans*, *Aspergillus fumigatus*, *and Alternaria alternata* (Shamsuzzaman et al. 2013; Raliya and Tarafdar 2013; Sarkar et al. 2014), and also a huge number of fungal species were explored for Ag NP fabrication (Table 1.1). Zhao et al. (2012) in their work to look at the effect of nanocomposite based packaging in green tea preservation prepared a packaging material by amalgamating polyethylene with Ag NPs, TiO_2 NPs, and attapulgite. This research provided an attractive substitute to maintain the high quality of green tea on storage for enhanced period. Searchingly for safekeeping apples by decrement in the reduction of enzymes (polyphenol oxidase, pyrogallol peroxidase) associated with its deterioration, a packaging composite material as ZnO NP powder embedded onto polyvinylchloride film was developed (Li et al. 2011). In another example diffused TiO_2 NPs copolymerized film EVOH (ethylene vinyl alcohol) was noticed for biocide effect on food poisoning bacteria and yeasts in the presence of light. Also a nanocomposite comprises of TiO_2 NPs and polyethylene showed the antimicrobial effect on leafy vegetable lettuce toward *Escherichia coli* (Duncan 2011).

1.4.2 Potential Myco-bionanosensors

Nanosensors are still burgeoning, but very soon will emerge as handy solutions to the people for self-food analysis. Architecture of general bionanosensor is depicted in Fig. 1.5. A few examples are highlighted in this column. A group of researchers have intimated the hidden potential of bio-conjugated gold nanoparticle (Au NP)-based assay for detection of multiple drug-resistant bacteria (MDRB: *Salmonella*; *E.coli*). But this technique is still at its budding stage. They have found the huge potential of bio-conjugated Au NPs for MDRB diagnostics and their proximate quelling. Herein investigators utilized the plasmonic properties of Au NPs for selective bacterial consigning and photothermal suppression upon conjugation with bacteria-specific biological entity like antibody. The author predicted this bio-conjugated Au NPs as nanoscopic heaters, and they have the capability to vanish MDRB without the use of antibiotics from food samples (Ray et al. 2013). For commercialization of such application at minimal cost, Au NP synthesis has to be scaled up. Recently Kitching et al. (2015) and his group in their comprehensively review conferred practically convenient bioreactors for scale-up of mycogenic Au NPs. Batch air-lift fermentor and fluidized bed bioreactor could be used for scale-up of Au NPs using Rhizopus oligosporus and Trametes versicolor, respectively. They have also mentioned about the less cost-effective scale-up of Au NPs through conventional stirred-type bioreactor.

In many instances *Salmonella enteritidis* arousal in food is found to cause a quick state of affliction (Cogan and Humphrey 2003; Antunes et al. 2016). So a biosensor system was introduced by Alocilja et al. (2013) for the detection a specific IeL gene element of *Salmonella enteritidis*. They have reported the involvement of two nanoparticles in the biosensor setup. The system resembles a sandwich model wherein Au NPs are coated with the target-specific DNA probe which recognizes the target DNA as IeL gene element and magnetite (MNPs) are coated with another specific target DNA to perform easy and flawless separation. The whole sandwiched model is incorporated onto the screen-printed carbon electrode (SPCE) chip. This disposable SPCE biosensor detects (IeL) gene of *Salmonella*. Bharde et al. (2006) in their report have stated the potential of *Fusarium oxysporum* and *Verticillium* sp. to fabricate MNPs. The large-scale mycogenic production of MNPs is not yet reported, but Moon et al. (2010) developed a large-scale production protocol for MNPs using *Thermoanaerobacter*.

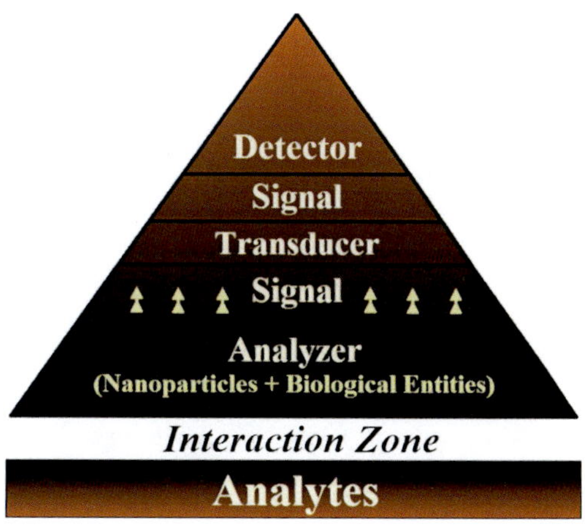

Fig. 1.5 Schematic of a bionanosensor

Fungi release mycotoxins as secondary metabolites required for their survival which in turn contaminate food and animal feed (cereals, stored grains, etc.) but makes our survival difficult. There are various mycotoxins deciphered till date (aflatoxins, zearalenone, deoxynivalenol, cyclopiazonic acid, fumonisins, citrinin, ochratoxin A, patulin, vomitoxins, etc.) (Bullerman and Bianchini 2007; Rai et al. 2015). Among all, aflatoxins (AFs) are of great concern as it poses hepatotoxic, mutagenic, carcinogenic, and teratogenic properties, and their permissible limits in human food and animal feed are 20 ppb and 300 ppb, respectively (Wu et al. 2013). For detection of ochratoxin A (OTA), magnetic nanoparticles were used in the preparation of nanobiosensors to detect the presence of OTA in apples (Fernandez-Baldo et al. 2011). In the research persuaded by Zhang et al. (2009b) on different nanomaterials for the development of nanobiosensors, Au NPs were considered as the productive candidate in the nanobiosensor formation for tracking polyionic drugs such as protamine and heparin. In a study by Liu et al. (2008) to analyze OTA in coffee, gold nanoparticles were used in the immunochromatographic chip. TiO_2 NPs are light sensitive, and employing this property an oxygen-detecting ink has been developed which detects oxygen when exposed to UV light; thus food spoilage due to the presence of oxygen can be understood. However the most important concern is the interaction of these nanoparticles with biological system (FAO/WHO 2010).

1.4.3 Myconano-antimicrobials in Food

Numerous antimicrobials are available till date, and few of them are commercially as well as homely successful (essential oils, organic acids, etc.). There is a possibility that antimicrobials may not withstand rasping industrial steps or microbes may win by developing resistance against these antimicrobials. These disadvantages have paved the way for nano-based antimicrobials and acted as a research arouser. In packaging, materials like Ag and Ag_2O NPs, Cu and CuO NPs, ZnO NPs, CdSe NPs, and TiO_2 NPs are incorporated in the polymer base of the packaging material to defeat food spoilage (Llorens et al. 2012; Sharon et al. 2010; Lu and Bowles 2013; Duncan 2011).

A large number of works have been reported on the synthesis of Ag NPs by fungi and their antimicrobial role in food sciences. In Fig. 1.6, a few potential inventory applications of Ag NPs as antimicrobials in food are represented An et al. 2008; Emamifar et al. 2010; Fayaz et al. 2009; Fernández et al. 2010a, b; Li et al. 2009. In a research by Fayaz et al. (2009), Trichoderma viride fabricated polydispersed Ag NPs were incorporated into sodium alginate and the so-formed antimicrobial film appeared to be a good preservative for fruits (pears) and vegetables (carrots). These edible items were coated with the antimicrobial film after surface sterilization using chlorine dioxide and showed up to 10 days of storage without any loss. There were no changes in the physical appearance and taste thus marked its acceptability (Fayaz et al. 2009).

Duncan (2011) reported numberless application of Ag NPs as antimicrobials. The antimicrobial properties of Ag NPs could be used for preparations of nanocomposite packaging materials in combination with polyethylene and chitosan.

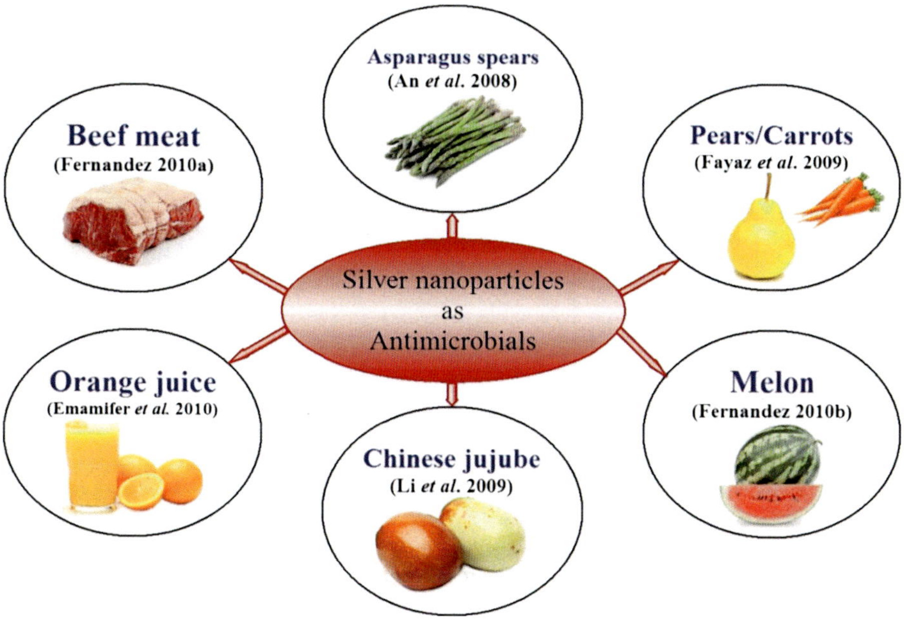

Fig. 1.6 Silver nanoparticles as antimicrobials in food preservation/packaging

These packaging options besides overcoming contamination issues have the added advantage of efficient moldability and gas fencing properties. Besides this, the colloidal Ag NPs also have probable future application as packaging material for protection against various bacterial and fungal infections on being coated upon paper through ultrasonic treatment. An et al. (2008) studied antimicrobial Ag NPs-PVP coating on asparagus, and positive results were obtained as asparagus spears retain its original quality for 25 days at low temperature.

1.5 Conclusion and Future Prospects

Biogenic synthesis of nanomaterials through greener method is advantageous, but its production at commercial scale is a challenge. Fungi are one of the most promising biological entities for scale-up of nanomaterial synthesis. Myconanotechnology will surely bang all doors of agri-food sector with strong intents and purposes. This chapter has summarized some of the most promising applications of mycogenic nanomaterials, including food packaging materials that possess extremely high gas barriers and antimicrobial properties, and nanosensors which can detect microorganisms. Another aspect involves smart field systems, early disease detection by fluorescent probes, crop improvement through mycomimetic model carriers, and crop yield using nanoformulations. There are tremendous potentials of myconanotechnology in agriculture wherein most of the research projects are in their nascent stage.

Myconanotechnology in food may come up with various colorants to energize health drinks at low cost and hence affordable to all, in other words a boon to malnutrition. With the enhanced pace of research in the concerned zone, the possibility of fat and sugar-free fungi-derived nano-modified food will not be a dream. Definitely much aspiration and sweat is needed to bring fungi-derived nanomaterials from lab to industry, but its benefits and advantages will fuel up the efforts. Myconano-functionalized agrochemicals, preharvest and post-harvest crop protection, sensing system, and genetic engineering devices for direct application in farms and fields are set to boost in the near future. Communication of associated risk and privileges of the technology will shape up the public perception and fate of fungal nanotechnology in agri-food science and engineering.

Despite these potential benefits, the most important concern regarding myco-based NPs in the coming future will be the toxicity in human beings, animals, plants, and off course our ecosystem, and these issues can't be overlooked. Much attention is needed on the accumulations of nanomaterials in plants/human systems which can bring about biotransformations.

References

Abdel Rahim K, Sabry Younis Mahmoud SY, Ali AM (2017) Extracellular biosynthesis of silver nanoparticles using *Rhizopus stolonifer*. Saudi J Biol Sci 24:208–216

Abou El-Nour KMM, Eftaiha A, Al-Warthan A, Ammar RAA (2010) Synthesis and applications of silver nanoparticles. Arab J Chem 3:135–140

Aguilar-Méndez MA, Martín-Martínez ES, Ortega-Arroyo L, Cobián-Portillo G, Sánchez-Espíndola E (2010) Synthesis and characterization of silver nanoparticles: effect on phytopathogen *Colletotrichum gloeosporioides*. J Nanopart Res 13:2525–2532

Ahmad A, Mukherjee P, Mandal D, Senapati S, Khan MI, Kumar R, Sastry M (2002) Enzyme-mediated extracellular biosynthesis of CdS nanoparticles by the fungus *Fusarium oxysporum*. J Am Chem Soc 124:12108–12109

Ahmad A, Mukherjee P, Senapati S, Mandal D, Khan MI, Kumar R, Sastry M (2003) Extracellular biosynthesis of silver nanoparticles using the fungus *Fusarium oxysporum*. Colloids Surf B: Biointerfaces 4:313–318

Ahmad A, Senapati S, Khan MI, Kumar R, Sastry M (2005) Extra-/intracellular biosynthesis of gold nanoparticles by an alkalotolerant fungus *Trichothecium* sp. J Biomed Nanotechnol 1:47–53

Ahmed S, Ullah S, Ahmad M, Swami BL, Ikram S (2016) Green synthesis of silver nanoparticles using *Azadirachta indica* aqueous leaf extract. J Radiat Res Appl Sci 9:1–7

Alani F, Moo-Young M, Anderson W (2012) Biosynthesis of silver nanoparticles by a new strain of *Streptomyces* sp. compared with *Aspergillus fumigatus*. World J Microbiol Biotechnol 28:1081–1086

Alocilja EC, Zhang D, Shi C (2013) AuNP-DNA biosensor for rapid detection of Salmonella enterica serovar Enteritidis. Chapter. doi:https://doi.org/10.1021/bk-2013-1143.ch003. ACS Symposium Series, vol. 1143

Ammar HA, El-Desouky TA (2016) Green synthesis of nanosilver particles by *Aspergillus terreus* HA1N and *Penicillium expansum* HA2N and its antifungal activity against mycotoxigenic fungi. J Appl Microbiol 1:89–100

An J, Zhanga M, Wangc S, Tangc J (2008) Physical, chemical and microbiological changes in stored green asparagus spears as affected by coating of silver nanoparticles-PVP. LWT 41:1100–1107

Anitha TS, Palanivelu P (2011) Synthesis and structural characterization of polydisperse silver and multishaped gold nanoparticles using *Fusarium oxysporum* MTCC 284. Dig J Nanomater Biostruc 6:1587–1595

Antunes P, Mourão J, Campos J, Peixe L (2016) Salmonellosis: the role of poultry meat. Clin Microbiol Infect 22:110–121

Aziz N, Faraz M, Pandey R, Sakir M, Fatma T, Varma A, Barman I, Prasad R (2015) Facile algae-derived route to biogenic silver nanoparticles: synthesis, antibacterial and photocatalytic properties. Langmuir 31:11605–11612. https://doi.org/10.1021/acs.langmuir.5b03081

Aziz N, Pandey R, Barman I, Prasad R (2016) Leveraging the attributes of *Mucor hiemalis*-derived silver nanoparticles for a synergistic broad-spectrum antimicrobial platform. Front Microbiol 7:1984. https://doi.org/10.3389/fmicb.2016.01984

Balaji DS, Basavaraja S, Deshpande R, Mahesh DB, Prabhakar BK, Venkataraman A (2009) Extracellular biosynthesis of functionalized silver nanoparticles by strains of *Cladosporium cladosporioide*s fungus. Colloids Surf B: Biointerfaces 1:88–92

Balakumaran MD, Ramachandran R, Kalaichelvan PT (2015) Exploitation of endophytic fungus, *Guignardia mangiferae* for extracellular synthesis of silver nanoparticles and their in vitro biological activities. Microbiol Res 178:9–17

Balakumaran MD, Ramachandran R, Balashanmugam P, Mukeshkumar DJ, Kalaichelvan PT (2016) Mycosynthesis of silver and gold nanoparticles: optimization, characterization and antimicrobial activity against human pathogens. Microbiol Res 182:8–20

Bansal V, Rautray D, Ahmad A, Sastry M (2004) Biosynthesis of zirconia nanoparticles using the fungus *Fusarium oxysporum*. J Mater Chem 14:3303–3305

Bansal V, Rautaray D, Bhadre A, Ahire K, Sanyal A, Ahmad A, Sastry M (2005) Fungus-mediated biosynthesis of silica and titania particles. J Mater Chem 15:2583–2589

Barabadi H, Honary S, Ali Mohammadi M, Ahmadpour E, Rahimi MT, Alizadeh A, Naghibi F, Saravanan M (2017) Green chemical synthesis of gold nanoparticles by using *Penicillium aculeatum* and their scolicidal activity against hydatid cyst protoscolices of *Echinococcus granulosus*. Environ Sci Pollut Res. https://doi.org/10.1007/s11356-016-8291-8

Barik TK, Sahu B, Swain V (2008) Nanosilica-from medicine to pest control. Parasitol Res 103:253–258

Basavaraja S, Balaji SD, Legashetty A, Rasab AH, Venkatraman A (2008) Extracellular biosynthesis of silver nanoparticles using the fungus *Fusarium semitectum*. Mater Res Bull 5:1164–1170

Behari J (2010) Principles of nanoscience: an overview. Indian J Exp Biotechnol 10:1008–1019

Bhainsa KC, D'souza SF (2006) Extracellular biosynthesis of silver nanoparticles using the fungus *Aspergillus fumigatus*. Colloids Surf B: Biointerfaces 1:160–164

Bharde A, Rautaray D, Bansal V, Ahmad A, Sarkar I, Yusuf SM, Sanyal M, Sastry M (2006) Extracellular biosynthesis of magnetite using fungi. Small 2:135–141

Bhattacharyya A, Duraisamy P, Govindarajan M, Buhroo AA, Prasad R (2016a) Nano-biofungicides: emerging trend in insect pest control. In: Prasad R (ed) Advances and applications through fungal nanobiotechnology. Springer International Publishing, Switzerland, pp 307–319

Bhattacharyya A, Prasad R, Buhroo AA, Duraisamy P, Yousuf I, Umadevi M, Bindhu MR, Govindarajan M, Khanday AL (2016b) One-pot fabrication and characterization of silver nanoparticles using *Solanum lycopersicum*: an eco-friendly and potent control tool against Rose Aphid, Macrosiphum rosae. J Nanosci. Article ID 4679410, 7 pages, 2016. https://doi.org/10.1155/2016/4679410

Binupriya AR, Sathishkumar M, Vijayaraghavan K, Yun S-I (2010a) Bioreduction of trivalent aurum to nano-crystalline gold particles by active and inactive cells and cell-free extract of *Aspergillus oryzae* var. viridis. J Hazard Mater 177:539–545

Binupriya AR, Sathishkumar M, Yun S-I (2010b) Biocrystallization of silver and gold ions by inactive cell filtrates of *Rhizopus stolonifer*. Colloids Surf B: Biointerfaces 79:531–534

Birla SS, Tiwari VV, Gade AK, Ingle AP, Yadav AP, Rai MK (2009) Fabrication of silver nanoparticles by *Phoma glomerata* and its combined effect against *Escherichia coli Pseudomonas aeruginosa* and *Staphylococcus aureus*. Lett Appl Microbiol 48:173–179

Bullerman LB, Bianchini A (2007) Stability of mycotoxins during food processing. Int J Food Microbiol 119:140–146

Castro-Longoria E, Vilchis-Nestor AR, Avalos-Borja M (2011) Biosynthesis of silver, gold and bimetallic nanoparticles using the filamentous fungus *Neurospora crassa*. Colloids Surf B: Biointerfaces 83:42–48

Chakravarty I, Pradeepam RJ, Kundu K, Singh PK, Kundu S (2015) Mycofabrication of gold nanoparticles and evaluation of their antioxidant activities. Curr Pharm Biotechnol 8:747–755

Chen YH, Yeh CS (2002) Laser ablation method: use of surfactants to form the dispersed Ag nanoparticles. Colloid Surf A Physicochem Eng Asp 197:133–139

Chen JC, Lin ZH, Ma XX (2003) Evidence of the production of silver nanoparticles via pretreatment of *Phoma* sp.3.2883 with silver nitrate. Lett Appl Microbiol 2:105–108

Chen G, Yi B, Zeng G, Niu Q, Yan M, Chen A, Du J, Huang J, Zhang Q (2014) Facile green extracellular biosynthesis of CdS quantum dots by white rot fungus *Phanerochaete chrysosporium*. Colloids Surf B: Biointerfaces 117:199–205

Chowdhury S, Basu A, Kundu S (2014) Green synthesis of protein capped silver nanoparticles from phytopathogenic fungus *Macrophomina phaseolina* (Tassi) goid with antimicrobial properties against multidrug-resistant bacteria. Nanoscale Res Lett 9:365. 11 pages

Cioffi N, Torsi L, Ditaranto N, Sabbatini L, Zambonin PG (2004) Antifungal activity of polymer-based copper nanocomposite coatings. Appl Phys Lett 85:2417–2419

Cogan TA, Humphrey TJ (2003) The rise and fall of *Salmonella Enteritidis* in the UK. J Appl Microbiol 94:114S–119S

Das SK, Marsili (2010) A green chemical approach for the synthesis of gold nanoparticles: characterization and mechanistic aspect. Rev Environ Sci Biotechnol 9:199–204

Das SK, Das AR, Guha AK (2009) Gold nanoparticles: microbial synthesis and application in water hygiene management. Langmuir 25:8192–8199

Ding X, Richter DL, Matuana LM, Heiden PA (2011) Efficient one – pot synthesis and loading of self-assembled amphiphilic chitosan nanoparticles for low – leaching wood preservation. Carbohydr Polym 86:58–64

Domeron CT, Smith BR, Winge DR (1989) Glutathione-coated cadmium-sulphide crystallites in *Candida glabrata*. J Biol Chem 264:17355–17360

Du L, Xian L, Feng JX (2011) Rapid extra-/intracellular biosynthesis of gold nanoparticles by the fungus *Penicillium* sp. J Nanopart Res 13:921–930

Duncan TV (2011) Applications of nanotechnology in food packaging and food safety: barrier materials, antimicrobials and sensors. J Colloid Interface Sci 363:1–24

Durán N, Marcato PD (2013) Nanobiotechnology perspectives. Role of nanotechnology in the food industry: a review. Int J Food Sci Technol 48:1127–1134

Durán A, Nombela C (2004) Fungal cell wall biogenesis: building a dynamic interface with the environment. Microbiology 150:3099–3103

Durán N, Seabra AB (2012) Metallic oxide nanoparticles: state of the art in biogenic syntheses and their mechanisms. Appl Microbiol Biotechnol 95:275–288

Durán N, Marcato PD, Alves OL, Desouza G, Esposito E (2005) Mechanistic aspects of biosynthesis of silver nanoparticles by several *Fusarium oxysporum* strains. J Nanobiotechnol 3:8–14

Duran N, Marcato PD, Durán M (2011) Mechanistic aspects in the biogenic synthesis of extracellular metal nanoparticles by peptides, bacteria, fungi and plants. Appl Microbiol Biotechnol 90:1609–1624

El-Baz AF, El-Batal AI, Abomosalam FM, Tayel AA, Shetaia YM, Yang ST (2016) Extracellular biosynthesis of anti-Candida silver nanoparticles using *Monascus purpureus*. J Basic Microbiol 5:531–540

El-Rafie MH, Shaheen THI, Mohamed AA, Hebeish A (2012) Bio-synthesis and applications of silver nanoparticles onto cotton fabrics. Cotton Polym 90:915–920

Emamifar A, Kadivar M, Shahedi M, Soleimanian-Zad S (2010) Evaluation of nanocomposite packaging containing Ag and ZnO on shelf life of fresh orange juice. Innovative Food Sci Emerg Technol 11:742–748

Esteban-Tejeda L, Malpartida F, Esteban-Cubillo A, Pecharromán C, Moya JS (2009) Antibacterial and antifungal activity of a soda-lime glass containing copper nanoparticles. Nanotechnology 20:505701

FAO/WHO (2010) Expert meeting on the application of nanotechnologies in the food and agriculture sectors: potential food safety implications. Food and Agriculture Organization of the United Nations and World Health Organization, Rome, Italy

Fayaz AM, Balaji K, Girilal M, Kalaichelvan PT, Venkatesan R (2009) Mycobased synthesis of silver nanoparticles and their incorporation into sodium alginate films for vegetable and fruit preservation. J Agric Food Chem 57:6246–6252

Fayaz AM, Tiwary CS, Kalaichelvan PT, Venkatesan R (2010) Blue orange light emission from biogenic synthesized silver nanoparticles using *Trichoderma viride*. Colloids Surf B: Biointerfaces. 2010 75:175–178

Fernández A, Picouet P, Lloret E (2010a) Reduction of the spoilage-related microflora in absorbent pads by silver nanotechnology during modified atmosphere packaging of beef meat. J Food Prot 73:2263–2269

Fernández A, Picouet P, Lloret E (2010b) Cellulose-silver nanoparticle hybrid materials to control spoilage-related microflora in absorbent pads located in trays of fresh-cut melon. Int J Food Microbiol 142:222–228

Fernández-Baldo MA, Bertolino FA, Fernández G, Messina GA, Sanz MI, Raba J (2011) Determination of Ochratoxin A in apples contaminated with *Aspergillus ochraceus* by using a microfluidic competitive immunosensor with magnetic nanoparticles. Analyst 136:2756–2762

Gade AK, Bonde P, Ingle AP, Marcato PD, Duran N, Rai MK (2008) Exploitation of *Aspergillus niger* for synthesis of silver nanoparticles. J Biobaased Mater Bioenergy 2:243–247

Gade A, Ingle A, Whiteley C, Rai M (2010) Mycogenic metal nanoparticles: progress and applications. Biotechnol Lett 32:593–600

Gade A, Rai M, Kulkarni S (2011) *Phoma sorghina*, a phytopathogen mediated synthesis of unique silver rods. Int J Green Nanotechnol 3:153–159

Gajbhiye M, Kesharwani J, Ingle A, Gade A, Rai M (2009) Fungus mediated synthesis of silver nanoparticles and their activity against pathogenic fungi in combination with fluconazole. Nanomed Nanotechnol Biol Med 5:382–386

Gericke M, Pinches A (2006) Biological synthesis of metal nanoparticles. Hydrometallurgy 83:132–140

Gholami-Shabani M, Akbarzadeh A, Norouzian D, Amini A, Gholami-Shabani Z, Imani A, Chiani M, Riazi G, Shams-Ghahfarokhi M (2014) Antimicrobial activity and physical characterization of silver nanoparticles green synthesized using nitrate reductase from *Fusarium oxysporum*. Appl Biochem Biotechnol 8:4084–4098

Gopinath V, Velusamy P (2013) Extracellular biosynthesis of silver nanoparticles using *Bacillus* sp. GP-23 and evaluation of their antifungal activity towards *Fusarium oxysporum*. Spectrochim Acta A Mol Biomol Spectr 106:170–174

Goswami A, Roy I, Sengupta S, Debnath N (2010) Novel applications of solid and liquid formulations of nanoparticles against insect pests and pathogens. Thin Solid Films 519:1252–1257

Gupta S, Bector S (2013) Biosynthesis of extracellular and intracellular gold nanoparticles by *Aspergillus fumigatus* and *A. flavus*. Antonie Van Leeuwenhoek 103:1113–1123

Hamedi S, Shojaosadati SA, Shokrollahzadeh S, Hashemi-Najafabadi S (2014) Extracellular biosynthesis of silver nanoparticles using a novel and non-pathogenic fungus, *Neurospora intermedia*: controlled synthesis and antibacterial activity. World J Microbiol Biotechnol 30:693–704

Hemanth NKS, Kumar G, Karthik L, Bhaskara Rao KV (2010) Extracellular biosynthesis of silver nanoparticles using the filamentous fungus *Penicillium* sp. Arch Appl Sci Res 2:161–167

Hobbs PR, Sayre K, Gupta R (2008) The role of conservation agriculture in sustainable agriculture. Philos Trans R Soc Lond Ser B Biol Sci 363:543–555

Honary S, Barabadi H, Gharaei-Fathabad E, Naghibi F (2013a) Green synthesis of silver nanoparticles induced by the fungus *Penicillium citrinum*. Trop J Pharm Res 1:7–11

Honary S, Gharaei-Fathabad E, Barabadi H, Naghibi F (2013b) Fungus-mediated synthesis of gold nanoparticles: a novel biological approach to nanoparticle synthesis. J Nanosci Nanotechnol 13:1427–1430

Ingle A, Gade A, Pierrat S, Sonnichsen C, Rai MK (2008) Mycosynthesis of silver nanoparticles using the fungus *Fusarium acuminatum* and its activity against some human pathogenic bacteria. Curr Nanosci 4:141–144
Ingle A, Gade A, Bawaskar M, Rai M (2009) *Fusarium solani*: a novel biological agent for the extracellular synthesis of silver nanoparticles. J Nanopart Res 11:2079–2085
Jaidev LR, Narasimha G (2010) Fungal mediated biosynthesis of silver nanoparticles, characterization and antimicrobial activity. Colloids Surf B: Biointerfaces 2:430–433
Jain N, Bhargava A, Majumdar S, Tarafdar JC, Panwar J (2010) Extracellular biosynthesis and characterization of silver nanoparticles using *Aspergillus flavus* NJP08: a mechanism perspective. Nanoscale 3:635–641
Jain N, Bhargava A, Tarafdar JC, Singh SK, Panwar J (2012) A biomimetic approach towards synthesis of zinc oxide nanoparticles. Appl Microbiol Biotechnol 97:859–869
Jayaseelan C, Rahuman AA, Rajakumar G, Kirthi AV, Santoshkumar T, Marimuthu S, Bagavan A, Kamaraj C, Zahir AA, Elango G (2011) Synthesis of pediculocidal and larvicidal silver nanoparticles by leaf extract from heartleaf moonseed plant, *Tinospora cordifolia* Miers. Parasitol Res 109:185–194
Jha AK, Prasad K (2010) Understanding biosynthesis of metallic/oxide nanoparticles: a biochemical perspective. In: Kumar SA, Thiagarajan S, Wang S-F (eds) Biocompatible nanomaterials synthesis, characterization and applications. NOVA Sci Publ, New York
Jha AK, Prasad K (2016) Understanding mechanism of fungus mediated nanosynthesis: a molecular approach. In: Prasad R (ed) Adv app through fungal nanobiotechnol. Chap 1. p 1–23. doi:https://doi.org/10.1007/978-3-319-42990-8_1
Jha AK, Prasad K, Kulkarni AR (2008) Yeast mediated synthesis of silver nanoparticles. Int J Nanosci Nanotechnol 4:17–21
Jha AK, Prasad K, Kulkarni AR (2009a) Synthesis of TiO_2 nanoparticles using microorganisms. Colloids Surf B: Biointerfaces 71:226–229
Jha AK, Prasad K, Kumar V, Prasad K (2009b) Biosynthesis of silver nanoparticles using *Eclipta* leaf. Biotechnol Prog 25:1476–1479
Jha AK, Prasad K, Kulkarni AR (2010) Synthesis of Gd_2O_3 nanoparticles using *Lactobacillus* sp.: a novel green approach. Int J Green Nanotechnol Phys Chem 2:31–38
Jo YK, Kim BH, Jung GH (2009) Antifungal activity of silver ions and nanoparticles on phytopathogenic fungi. Plant Dis 93:1037–1043
Kaler A, Jain S, Banerjee UC (2013) Green and rapid synthesis of anticancerous silver nanoparticles by *Saccharomyces boulardii* and insight into mechanism of nanoparticle synthesis. Biol Med Res Int 2013.: Article ID 872940:8. https://doi.org/10.1155/2013/872940
Kar PK, Murmu S, Saha S, Tandon V, Acharya K (2014) Anthelmintic efficacy of gold nanoparticles derived from a phytopathogenic fungus, *Nigrospora oryzae*. PLoS One 9:e84693. 9 pages
Kashyap PL, Kumar S, Srivastava AK, Sharma AK (2012) Myconanotechnology in agriculture: a perspective. World J Microbiol Biotechnol 29:191–207
Kathiresan K, Manivannan S, Nabeel AM, Dhivya B (2009) Studies on silver nanoparticles synthesized by a marine fungus *Penicillium fellutanum* isolated from coastal mangrove sediment. Colloids Surf B: Biointerfaces 71:133–137
Khodakovskaya M, Dervishi E, Mahmood M, Xu Y, Li Z, Watanabe F, Biris AS (2009) Carbon nanotubes are able to penetrate plant seed coat and dramatically affect seed germination and plant growth. ACS Nano 3:3221–3227
Kim SW, Kim KS, Lamsal K, Kim YJ, Kim SB, Jung M, Sim SJ, Kim HS, Chang SJ, Kim JK, Lee YS (2009) An in vitro study of the antifungal effect of silver nanoparticles on oak wilt pathogen *Raffaelea* sp. J Microbiol Biotechnol 19:760–764
Kirchmann H, Thorvaldsson G (2000) Challenging targets for future agriculture. Eur J Agron 12:145–161
Kitching M, Ramani M, Marsili E (2015) Fungal biosynthesis of gold nanoparticles: mechanism and scale up. Microb Biotech 8:907–917

Kitching M, Choudhary P, Inguva S, Guo Y, Ramani M, Das SK, Marsili E (2016) Fungal surface protein mediated one-pot synthesis of stable and hemocompatible gold nanoparticles. Enzym Microb Technol 95:76–84

Korbekandi H, Jouneghani RM, Mohseni S, Pourhossein M, Iravani S (2014) Synthesis of silver nanoparticles using biotransformations by *Saccharomyces boulardii*. Green Process Synth 3:271–277

Korbekandi H, Mohseni S, Jouneghani RM, Pourhossein M, Iravani S (2016) Biosynthesis of silver nanoparticles using *Saccharomyces cerevisiae*. Artif Cells Nanomed Biotechnol 44:235–239

Kottegoda N, Munaweera I, Madusanka N, Karunaratne V (2011) A green slow-release fertilizer composition based on urea-modified hydroxyapetite nanoparticles encapsulated wood. Curr Sci 101:73–78

Kowshik M, Deshmukh N, Vogel W, Urban J, Kulkarni SK, Paknikar KM (2002) Microbial synthesis of semiconductor CdS nanoparticles, their characterization, and their use in the fabrication of an ideal diode. Biotechnol Bioeng 78:583–588

Kowshik M, Ashtaputre S, Kharrazi S, Vogel W, Urban J, Kulkarni SK, Paknikar KM (2003) Extracellular synthesis of silver nanoparticles by a silver-tolerant yeast strain MKY3. Nanotechnology 14:95–100

Krumov N, Oder S, Perner-Nochta I, Angelov A, Posten C (2007) Accumulation of CdS nanoparticles by yeasts in a fed-batch bioprocess. J Biotechnol 132:481–486

Kumar SA, Abyaneh MK, Gosavi SW, Kulkarni SK, Pasricha R, Ahmad A, Khan MI (2007) Nitrate reductase-mediated synthesis of silver nanoparticles from $AgNO_3$. Biotechnol Lett 29:439–445

Kumar SA, Peter YA, Nadaeu JL (2008) Facile biosynthesis, separation, conjugation of gold nanoparticles to doxorubicin. Nanotechnology 19:495101

Lamsal K, Kim SW, Jung JH, Kim YS, Kim KS, Lee YS (2011) Inhibition effects of silver nanoparticles against powdery mildews on cucumber and pumpkin. Mycobiology 39:26–32

Lee KJ, Park SH, Govarthanan M, Hwang PH, Seo YS, Cho M, Lee WH, Lee JY, Kannan SK, Oh BT (2013) Synthesis of silver nanoparticles using cow milk and their antifungal activity against phytopathogens. Mater Lett 105:128–131

Li ZZ, Chen JF, Liu F, Liu AQ, Wang Q, Sun HY, Wen LX (2007) Study of UV-shielding properties of novel porous hollow silica nanoparticles carriers for avermectin. Pest Manag Sci 63:241–246

Li H, Li F, Wang L, Sheng J, Xin Z, Zhao L, Xiao H, Zheng Y, Hu Q (2009) Effect of nanopacking on preservation quality of Chinese jujube (*Ziziphusjujuba* Mill. var. inermis (Bunge) Rehd). Food Chem 114:547–552

Li X, Li W, Jiang Y, Ding Y, Yun J, Tang Y, Zhang P (2011) Effect of nano-ZnO-coated active packaging on quality of fresh-cut 'Fuji' apple. Int J Food Sci Technol 46:1947–1955

Li G, He D, Qian Y, Guan B, Gao S, Cui Y, Yokoyama K, Wang L (2012) Fungus-mediated green synthesis of silver nanoparticles using *Aspergillus terreus*. Int J Mol Sci 13:466–476

Liu F, Wen LX, Li ZZ, Yu W, Sun HY, Chen JF (2006) Porous hollow silica nanoparticles as controlled delivery system for water soluble pesticide. Mater Res Bull 41:2268–2275

Liu BH, Tsao ZJ, Wang JJ, Yu FY (2008) Development of a monoclonal antibody against ochratoxin A and its application in enzyme-linked immunosorbent assay and gold nanoparticle immunochromatographic strip. Anal Chem 80:7029–7035

Liu Q, Chen B, Wang Q, Shi X, Xiao Z, Lin J, Fang X (2009) Carbon nanotubes as molecular transporters for walled plant cells. Nano Lett 9:1007–1010

Llorens A, Lloret E, Picouet PA, Trbojevich R, Fernandez (2012) Metallic-based micro and nanocomposites in food contact materials and active food packaging. Trends Food Sci Technol 24:19–29

Lu J, Bowles M (2013) How will nanotechnology affect agricultural supply chains? Int Food Agribusiness Manag Rev 2:21–42

Lu CM, Zhang CY, Wen JQ, Wu GR, Tao MX (2002) Research of the effect of nanometer materials on germination and growth enhancement of glycine max and its mechanism. Soybean Sci 21:168–171

Ma Y, Kuang L, He X, Bai W, Ding Y, Zhang Z, Zhao Y, Chai Z (2010) Effects of rare earth oxide nanoparticles on root elongation of plants. Chemosphere 78:59–78

Maliszewska I, Szewczyk K, Waszak K (2009) Biological synthesis of silver nanoparticles. J Phys Conf Ser 146:2025

Maliszewska I, Juraszek A, Bielska K (2014) Green synthesis and characterization of silver nanoparticles using ascomycota fungi *Penicillium nalgiovense* AJ12. J Clust Sci 25:989–1004

Manikandan A, Subramanian KS (2014) Fabrication and characterisation of nanoporous zeolite based N fertilizer. Afr J Agric Res 9:276–284

McLamore ES, Diggs A, Marzal PC, Shi J, Blakeslee JJ, Peer WA, Murphy AS, Porterfield DM (2010) Non-invasive quantification of endogenous root auxin transport using an integrated flux microsensor technique. Plant J 63:1004–1016

Mehra RK, Winge DR (1991) Metal ion resistance in fungi: molecular mechanisms and their regulated expression. J Cell Biochem 45:30–40

Mishra A, Kumari M, Pandey S, Chaudhry V, Gupta KC, Nautiyal CS (2014) Biocatalytic and antimicrobial activities of gold nanoparticles synthesized by *Trichoderma* sp. Bioresour Technol 166:235–242

Mittal AK, Chisti Y, Banerjee UC (2013) Synthesis of metallic nanoparticles using plant extracts. Biotechnol Adv 31:346–356

Moaveni P, Karimi K, Valojerdi MZ (2010) The nanoparticles in plants: review paper. J Nanostruct Chem 2:59–78

Mohammadian A, Shaojaosadati SA, Rezee MH (2007) *Fusarium oxysporum* mediates photogeneration of silver nanoparticles. Sci Iran 14:323–326

Mondal A, Basu R, Das S, Nandy P (2011) Beneficial role of carbon nanotubes on mustard plant growth: an agricultural prospect. J Nanopart Res 13:4519–4528

Moon JW, Rawn CJ, Rondinone AJ, Love LJ, Roh Y, Everett SM, Lauf RJ, Phelps TJ (2010) Large-scale production of magnetic nanoparticles using bacterial fermentation. J Ind Microbiol Biotechnol 37:1023–1031

Mukherjee P, Ahmad A, Mandal D, Senapati S, Sainkar SR, Khan MI, Parishcha R, Ajaykumar PV, Alam M, Kumar R, Sastry M (2001a) Fungus-mediated synthesis of silver nanoparticles and their immobilization in the mycelial matrix: a novel biological approach to nanoparticle synthesis. Nano Lett 1:515–519

Mukherjee P, Ahmad A, Mandal D, Senapati S, Sainkar SR, Khan MI, Ramani R, Parischa R, Ajayakumar PV, Alam M, Sastry M, Kumar R (2001b) Bioreduction of $AuCl_4-$ ions by the fungus, *Verticillium* sp. and surface trapping of the gold nanoparticles formed. Angew Chem 40:3585–3588

Mukherjee P, Senapati S, Mandal D, Ahmad A, Khan MI, Kumar R, Sastry M (2002) Extracellular synthesis of gold nanoparticles by the fungus *Fusarium oxysporum*. Chem Biol Chem 3:461–463

Mukherjee P, Roy M, Mandal BP, Dey GK, Mukherjee PK, Ghatak J, Tyagi AK, Kale SP (2008) Green synthesis of highly stabilized nanocrystalline silver particles by a non-pathogenic and agriculturally important fungus *T. asperellum*. Nanotechnology 19:075103

Mukhopadhyay SS (2014) Nanotechnology in agriculture: prospects and constraints. Nanotechnol Sci Appl 7:63–71

Nadi E, Aynehband A, Mojaddam M (2013) Effect of nano-iron chelate fertilizer on grain yield, protein percent and chlorophyll content of Faba bean (*Vicia faba* L.) Int J Biosci 3:267–272

Nair R, Varghese SH, Nair BG, Maekawa T, Yoshida Y, Kumar DS (2010) Nanoparticulate material delivery to plants. Plant Sci 179:154–163

Narayanan KB, Sakthivel N (2011) Facile green synthesis of gold nanostructures by NADPH-dependent enzyme from the extract of *Sclerotium rolfsii*. Colloid Surf A Physicochem Eng Asp 380:156–161

Narayanan KB, Sakthivel N (2013) Mycocrystallization of gold ions by the fungus *Cylindrocladium floridanum*. World J Microbiol Biotechnol 29:2207–2211

Narayanan KB, Park HH, Sakthivel N (2013) Extracellular synthesis of mycogenic silver nanoparticles by *Cylindrocladium floridanum* and its homogeneous catalytic degradation of 4-nitrophenol. Spectrochim Acta A Mol Biomol Spectr 116:485–490

Navazi ZR, Pazouki M, Halek FS (2010) Investigation of culture conditions for biosynthesis of silver nanoparticles using *Aspergillus fumigatus*. Iran J Biotechnol 8:56–61

Neethirajan S, Jayas DS, Sadistap S (2009) Carbon dioxide (CO_2) sensors for the agri-food industry-a review. Food Bioprocess Technol 2:115–121

Nithya R, Ragunathan R (2009) Synthesis of silver nanoparticle using pleurotus *Pleurotus sajor-caju* and its antimicrobial study. Dig J Nanomater Biosci 4:623–629

Pandiarajan G, Govindaraj R, Kumar M, Ganesan V (2010) Biosynthesis of silver nanoparticles using silver nitrate through biotransformation. J Ecobiotechnol 2:13–18

Park HJ, Kim SH, Kim HJ, Choi SH (2006) A new composition of nanosized silica-silver for control of various plant diseases. Plant Pathol J 22:295–302

Patel V, Berthold D, Puranik P, Gantar (2015) Screening of cyanobacteria and microalgae for their ability to synthesize silver nanoparticles. Biotechnol Rep 5:112–119

Pereira L, Dias N, Carvalho J, Fernandes S, Santos C, Lima N (2014) Synthesis, characterization and antifungal activity of chemically and fungal-produced silver nanoparticles against *Trichophyton rubrum*. J Appl Microbiol 16:1601–1613

Pérez-de-Luque A, Rubiales D (2009) Nanotechnology for parasitic plant control. Pest Manag Sci 65:540–545

Philip D (2009) Biosynthesis of Au, Ag and Au-Ag nanoparticles using edible mushroom extract. Spectrochim Acta A 73:374–381

Pimprikar PS, Joshi SS, Kumar AR, Zinjardea SS, Kulkarni SK (2009) Influence of biomass and gold salt concentration on nanoparticle synthesis by the tropical marine yeast *Yarrowia lipolytica* NCIM 3589. Colloids Surf B: Biointerfaces 74:309–316

Prasad R (2014) Synthesis of silver nanoparticles in photosynthetic plants. J Nanopart. Article ID 963961, https://doi.org/10.1155/2014/963961

Prasad R (2016) Advances and applications through fungal nanobiotechnology. Springer, International Publishing Cham (ISBN: 978-3-319-42989-2)

Prasad K, Jha AK (2010) Biosynthesis of CdS nanoparticles: an improved green and rapid procedure. J Colloid Interface Sci 342:68–72

Prasad R, Kumar V, Prasad KS (2014) Nanotechnology in sustainable agriculture: present concerns and future aspects. Afr J Biotechnol 13:705–713

Prasad R, Pandey R, Barman I (2016) Engineering tailored nanoparticles with microbes: *quo vadis*. WIREs Nanomed Nanobiotechnol 8:316–330

Prasad R, Bhattacharyya A, Nguyen QD (2017) Nanotechnology in sustainable agriculture: recent developments, challenges, and perspectives. Front Microbiol 8:1014. https://doi.org/10.3389/fmicb.2017.01014

Qian Y, Yu H, He D, Yang H, Wang W, Wan X, Wang L (2013) Biosynthesis of silver nanoparticles by the endophytic fungus *Epicoccum nigrum* and their activity against pathogenic fungi. Bioprocess Biosyst Eng 36:1613–1619

Quester K, Avalos-Borja M, Castro-Longoria E (2013) Biosynthesis and microscopic study of metallic nanoparticles. Micron 54–55:1–27

Raheman F, Deshmukh S, Ingle A, Gade A, Rai M (2011) Silver nanoparticles: novel antimicrobial agent synthesized from an endophytic fungus *Pestalotia* sp. isolated from leaves of *Syzygium cumini* (L). Nano Biomed Eng 3:174–178

Rai M, Yadav A, Bridge PD, Gade A (2009) Myconanotechnology: a new and emerging science. In: Rai M, Bridge PD (eds) Appl mycology. Chapter 14. p 258. doi:https://doi.org/10.1079/9781845935344.0258

Rai M, Gade A, Yadav A (2011) Biogenic nanoparticles: an introduction to what they are, how they are synthesized and their applications. In: Rai M, Durán N (eds) Metal nanoparticles microbiology. Springer, Berlin. https://doi.org/10.1007/978-3-642-18312-6_1

Rai M, Jogee PS, Ingle AP (2015) Emerging nanotechnology for detection of mycotoxins in food and feed. Int J Food Sci Nutr 1-8. DOI: 10.3109/09637486.2015.1034251

Rajakumar G, Rahumana AA, Roopan SM, Khanna VG, Elangoa G, Kamaraj C, Zahir AA, Velayutham K (2012) Fungus-mediated biosynthesis and characterization of TiO_2 nanoparticles and their activity against pathogenic bacteria. Spectrochim Acta A Mol Biomol 91:23–29

Raliya R, Tarafdar JC (2013) ZnO nanoparticle biosynthesis and its effect on phosphorous-mobilizing enzyme secretion and gum contents in clusterbean (*Cyamopsis tetragonoloba* L.) Agric Res 2:48–57

Raliya R, Tarafdar JC, Choudhary K, Mal P, Raturi A, Gautam R, Singh SK (2014a) Synthesis of MgO nanoparticles using *Aspergillus Tubingensis* TFR-3. J Bionanosci 8:34–38

Raliya R, Tarafdar JC, Singh SK, Gautam R, Choudhary K, Maurino VG, Saharan V (2014b) MgO nanoparticles biosynthesis and its effect on chlorophyll contents in the leaves of Clusterbean (*Cyamopsis tetragonoloba* L.) Adv Sci Eng Med 6:538–545

Raliya R, Biswas P, Tarafdar JC (2015a) TiO_2 nanoparticle biosynthesis and its physiological effect on mung bean (*Vigna radiata* L.) Biotechnol Rep 5:22–26

Raliya R, Nair R, Chavalmane S, Wang WN, Biswas P (2015b) Mechanistic evaluation of translocation and physiological impact of titanium dioxide and zinc oxide nanoparticles on the tomato (*Solanum lycopersicum* L.) plant. Metallomics 7:158494

Raliya R, Franke C, Chavalmane S, Nair R, Reed N, Biswas P (2016a) Quantitative understanding of nanoparticles uptake in watermelon plants. Front Plant Sci 7:1288. https://doi.org/10.3389/fpls.2016.01288

Raliya R, Tarafdar JC, Biswas P (2016b) Enhancing the mobilization of native phosphorous in mung bean rhizosphere using ZnO nanoparticles synthesized by soil fungi. J Agric Food Chem 64:3111–3118

Rao KJ, Paria S (2013) Use of sulfur nanoparticles as a green pesticide on *Fusarium solani* and *Venturia inaequalis* phytopathogens. RSC Adv 3:10471–10478

Ray PC, Khan SA, Fan Z, Senapati D (2013) Gold nanotechnology for targeted detection and killing of multiple drug resistant bacteria from food samples. doi:https://doi.org/10.1021/bk-2013-1143.ch001. ACS Symposium Series, vol. 1143

Raza MA, Kanwal Z, Rauf A, Sabri AN, Riaz S, Naseem S (2016) Size- and shape-dependent antibacterial studies of silver nanoparticles synthesized by wet chemical routes. Nano 6:74. https://doi.org/10.3390/nano6040074

Rehman A, Majeed MI, Ihsan A, Hussain SZ, Rehman S, Ghauri MA, Khalid ZM, Hussain I (2011) Living fungal hyphae-templated porous gold microwires using nanoparticles as building blocks. J Nanopart Res 13:6747–6754

Sadowski Z, Maliszewska IH, Grochowalska B, Polowczyk I, Koźlecki T (2008) Synthesis of silver nanoparticles using microorganisms. Mater Sci Pol 26:419–424

Saha S, Sarkar J, Chattopadhyay D, Patra S, Chakraborty A, Acharya K (2010) Production of silver nanoparticles by a phytopathogenic fungus *Bipolaris nodulosa* and its antimicrobial activity. Dig J Nanomater Bios 5:887–895

Saha S, Chattopadhyay D, Acharya K (2011) Preparation of silver nanoparticles by bio-reduction using *Nigrospora oryzae* culture filtrate and its antimicrobial activity. Dig J Nanomater Bios 6:1519–1528

Salaheldin TA, Husseiny SM, Al-Enizi AM, Elzatahry A, Cowley AH (2016) Evaluation of the cytotoxic behavior of fungal extracellular synthesized Ag nanoparticles using confocal laser scanning microscope. Int J Mol Sci 3:329

Salunkhe RB, Patil SV, Salunke BK, Patil CD, Sonawane AM (2011) Studies on silver accumulation and nanoparticle synthesis by *Cochliobolus lunatus*. Appl Biochem Biotechnol 165:221–234

Sanghi R, Verma P (2009) Biomimetic synthesis and characterization of protein capped silver nanoparticles. Bioresour Technol 100:501–504

Sanghi R, Verma P (2010) pH dependant fungal proteins in the 'green' synthesis of gold nanoparticles. Adv Mater Lett 1:193–199

Sanghi R, Verma P, Puri S (2011) Enzymatic formation of gold nanoparticles using *Phanerochaete chrysosporium*. Adv Chem Eng Sci 1:154–162

Saravanan M (2010) Biosynthesis and in vitro studies of silver bionanoparticles synthesized from *Aspergillus* species and its antimicrobial activity against multi drug resistant clinical isolates. World Acad Sci Eng Technol 68:728–731

Saravanan M, Nanda A (2010) Extracellular synthesis of silver bionanoparticles from *Aspergillus clavatus* and its antimicrobial activity against MRSA and MRSE. Colloids Surf B: Biointerfaces 77:214–218

Sarkar J, Ray S, Chattopadhyay D, Laskar A, Acharya K (2012) Mycogenesis of gold nanoparticles using a phytopathogen *Alternaria alternata*. Bioprocess Biosyst Eng 35:637–643

Sarkar J, Ghosh M, Mukherjee A, Chattopadhyay D, Acharya K (2014) Biosynthesis and safety evaluation of ZnO nanoparticles. Bioprocess Biosyst Eng 37:165–171

Sawle BD, Salimath B, Deshpande R, Bedre MD, Prabhakar BK, Venkataraman A (2008) Biosynthesis and stabilization of Au and Au-Ag alloy nanoparticles by fungus, *Fusarium semitectum*. Sci Technol Adv Mater 9:1–6

Scott N, Chen H (2003) Nanoscale science and engineering for agriculture and food systems. Report: Cooperative State Research, Education and Extension Service, United States Department of Agriculture, Natl. Planning Workshop, November 18–19, 2002 Washington, DC

Scott N, Chen H (2013) Nanoscale science and engineering for agriculture and food systems. Ind Biotechnol 9:17–18

Sekhon BP (2014) Nanotechnology in agri-food production: an overview. Nanotechnol Sci Appl 7:31–53

Senapati S, Mandal D, Ahmad A, Khan MI, Sastry M, Kumar R (2004) Fungus mediated synthesis of silver nanoparticles: a novel biological approach. Indian J Phys A 78:101–105

Shaligram NS, Bule M, Bhambure R, Singhal RS, Singh SK, Szakac SG, Pandey A (2009) Biosynthesis of silver nanoparticles using aqueous extract from the compactin producing fungal strain. Process Biochem 44:939–943

Shamsuzzaman, Mashrai A, Khanam H, Aljawfi RN (2013) Biological synthesis of ZnO nanoparticles using *C. albicans* and studying their catalytic performance in the synthesis of steroidal pyrazolines. Arab J Chem. https://doi.org/10.1016/j.arabjc.2013.05.004

Shankar SS, Ahmad A, Pasricha R, Sastry M (2003) Bioreduction of chloroaurate ions by Geranium leaves and its endophytic fungus yields gold nanoparticles of different shapes. J Mater Chem 13:1822–1826

Shankar SS, Rai A, Ankamwar B, Singh A, Ahmad A, Sastry M (2004) Biological synthesis of triangular gold nanoprisms. Nat Mater 3:482–488

Sharon M, Choudhary AK, Kumar R (2010) Nanotechnology in agricultural diseases and food safety. J Phytol 2:83–92

Sheikhloo Z, Salouti M (2011) Intracellular biosynthesis of gold nanoparticles by the fungus *Penicillium Chrysogenum*. Int J Nanosci Nanotechnol 7:102–105

Singhal U, Khanuja M, Prasad R, Varma A (2017) Impact of synergistic association of ZnO-nanorods and symbiotic fungus *Piriformospora indica* DSM 11827 on *Brassica oleracea* var. botrytis (Broccoli). Front Microbiol 8:1909. doi: 10.3389/fmicb.2017.01909

Soni N, Prakash S (2012) Efficacy of fungus mediated silver and gold nanoparticles against Aedes aegypti larvae. Parasitol Res 110:175–184

Sowani H, Mohite P, Munot H, Shouche Y, Bapat T, Kumar AR, Kulkarni M (2015) Green synthesis of gold and silver nanoparticles by an actinomycete *Gordonia amicalis* HS-11: mechanistic aspects and biological application. Process Biochem 51:374–383

Spasova M, Manolova N, Naydenov M, Kuzmanova J, Rashkov I (2011) Electrospun biohybrid materials for plant biocontrol containing chitosan and *Trichoderma viride* spores. J Bioact Compat Polym 26:48–55

Stadler T, Buteler M, Weaver DK (2010) Novel use of nano-structured alumina as an insecticide. Pest Manag Sci 66:577–579

Sugunan A, Melin P, Schnürer J, Hilborn JG, Dutta J (2007) Nutrition-driven assembly of colloidal nanoparticles: growing fungi assemble gold nanoparticles as microwires. Adv Mater 19:77–81

Sundaravadivelan C, Padmanabhan MN (2014) Effect of mycosynthesized silver nanoparticles from filtrate of *Trichoderma harzianum* against larvae and pupa of dengue vector *Aedes aegypti* L. Environ Sci Pollut Res Int 6:4624–4633

Tarafdar JC, Raliya R (2013) Rapid, low-cost, and ecofriendly approach for iron nanoparticle synthesis using *Aspergillus oryzae* TFR9. J Nanopart 2013. Article ID 141274, 4

Thakkar KN, Mhatre SS, Parikh RY (2010) Biological synthesis of metallic nanoparticles. Nanomed Nanotechnol Biol Med 6:257–262

Thirumurugan G, Shaheedha SM, Dhanaraju MD (2009) In vitro evaluation of anti-bacterial activity of silver nanoparticles synthesised by using *Phytophthora infestans*. Int J Chem Tech R 1:714–716

Tidke PR, Gupta I, Gade A, Rai M (2014) Fungus- mediated synthesis of gold nanoparticles and standardization of parameters for its biosynthesis. IEEE Trans Nanobioscience 13:397–402

Tiwari DK, Schubert ND, Cendejas LMV, Villegas J, Montoya LC, García SEB (2014) Interfacing carbon nanotubes (CNT) with plants: enhancement of growth, water and ionic nutrient uptake in maize (*Zea mays*) and implications for nanoagriculture. Appl Nanosci 4:577–591

Torney F, Trewyn BG, Lin VS-Y, Wang K (2007) Mesoporous silica nanoparticles deliver DNA and chemicals into plants. Nat Nanotechnol 2:295–300

Tripathi S, Sonkar SK, Sarkar S (2011) Growth stimulation of gram (*Cicer arietinum*) plant by water soluble carbon nanotubes. Nanoscale 3:1176–1181

Vahabi K, Ali Mansoori G, Karimi S (2011) Biosynthesis of silver nanoparticles by fungus *Trichoderma reesei* (a route for large scale production of AgNPs). Insci J 1:65–79

Velmurugan P, Lee SM, Iydroose M, Lee KJ, Oh BT (2013) Pine cone-mediated green synthesis of silver nanoparticles and their antibacterial activity against agricultural pathogens. Appl Microbiol Biotechnol 97:361–368

Verma VC, Kharwar RN, Gange AC (2010) Biosynthesis of antimicrobial silver nanoparticles by the endophytic fungus *Aspergillus clavatus*. Nanomedicine 5:33–40

Verma VC, Singh SK, Solanki R, Prakash S (2011) Biofabrication of anisotropic gold nanotriangles using extract of endophytic *Aspergillus clavatus* as a dual functional reductant and stabilizer. Nanoscale Res Lett 6:16–22

Vigneshwaran N, Kathe AA, Varadarajan PV, Nachane RP, Balasubramanya RH (2006) Biomimetics of silver nanoparticles by white rot fungus, *Phanerochaete chrysosporium*. Colloids Surf B: Biointerfaces 53:55–59

Vigneshwaran N, Ashtaputre NM, Varadarajan PV, Nachane RP, Paralikar KM, Balasubramanya RH (2007a) Biological synthesis of silver nanoparticles using the fungus *Aspergillus flavus*. Mater Lett 61:1413–1418

Vigneshwaran N, Kathe AA, Varadarajan PV, Nachane RP, Balasubramanya RH (2007b) Silver-protein (Core-Shell) nanoparticle production using spent mushroom substrate. Langmuir 23:7113–7117

Villagarcia H, Dervishi E, Silva K, Biris AS, Khodakovskaya MV (2012) Surface chemistry of carbon nanotubes impacts the growth and expression of water channel protein in tomato plants. Small 8:2328–2334

Wang WN, Tarafdar JC, Biswas P (2013) Nanoparticle synthesis and delivery by an aerosol route for watermelon plant foliar uptake. J Nanopart Res 15:1417

Wanga H, Qiaoa X, Chena J, Ding S (2005) Preparation of silver nanoparticles by chemical reduction method. Colloid Surf A Physicochem Eng Asp 256:111–115

Williams P, Keshavarz-Moore E, Dunnill P (1996) Production of cadmium sulphide microcrystallites in batch cultivation by *Schizosaccharomyces pombe*. J Biotechnol 48:25967

Wu X, Chen J, Park B, Huang YW, Zhao Y (2013) The use of silver nanorod array-based surface-enhanced raman scattering sensor for food safety applications. Chapter. doi:https://doi.org/10.1021/bk-2013-1143.ch005. ACS Symp Series, vol. 1143

Xie J, Lee JY, Ting YP (2007) High-yield synthesis of complex gold nanostructures in a fungal system. J Phys Chem C111:16858–16865

Yadav A, Kon K, Kratosova G, Duran N, Ingle AP, Rai M (2015) Fungi as an efficient mycosystem for the synthesis of metal nanoparticles: progress and key aspects of research. Biotechnol Lett 37:2099–2120

Yang FL, Li XG, Zhu F, Lei CL (2009) Structural characterization of nanoparticles loaded with garlic essential oil and their insecticidal activity against *Tribolium castaneum* (Herbst) (Coleoptera: Tenebrionidae). J Agric Food Chem 57:10156–10162

Yao KS, Li SJ, Tzeng KC, Cheng TC, Chang CY, Chiu CY, Liao CY, Hsu JJ, Lin ZP (2009) Fluorescence silica nanoprobe as a biomarker for rapid detection of plant pathogens. Adv Mater Res 79–82:513–516

Yin XD, Zhang ZY (2010) Preparation, properties and characteristics of azadirachtin/carboxymethyl chitosan/ phosphorylation chitosan nanoparticles. Acta Agric Jiangxi 01:69–73

Youtie J, Shapira P, Porter AL (2008) Nanotechnology publications and citations by leading countries and blocs. J Nanopart Res 10:981–986

Zhang X, Guo Q, Cui D (2009a) Recent advances in nanotechnology applied to biosensors. Sensors 9:1033–1053

Zhang X, He X, Wang K, Wang Y, Li H, Tan W (2009b) Biosynthesis of size-controlled gold nanoparticles using fungus, *Penicillium* sp. J Nanosci Nanotechnol 9:5738–5744

Zhang X, Yan S, Tyagi RD, Surampalli RY (2011) Synthesis of nanoparticles by microorganisms and their application in enhancing microbiological reaction rates. Chemosphere 82:489–494

Zhao L, Li F, Chen G, Fang Y, An X, Zheng Y, Xin Z, Zhang M, Yang Y, Hu Q (2012) Effect of nanocomposite-based packaging on preservation quality of green tea. Int J Food Sci Technol 47:572–557

Zheng J, Birktoft JJ, Chen Y, Wang T, Sha R, Constantinou PE, Ginell SL, Mao C, Seman NC (2009) From molecular to macroscopic via the rational design of a self-assembled 3D DNA crystal. Nat Lett 461:74–77

Chapter 2
Agriculture Applications of Entomopathogenic Fungi Using Nanotechnology

Sandra Pérez Álvarez, Marco Antonio Magallanes Tapia, Karel Ismar Acosta Pérez, and Amaury Méndez Guerrero

Abstract Globally, a great percent of crops are affected or loses due to plant diseases, and commonly, pathogen control is through chemical (pesticides). For sustainable agriculture and environment safety, the reduction of pesticides is absolutely necessary, so the use of entomopathogen such as fungi, bacteria, virus, protozoa, and nematodes is important in the natural regulation of pest population. Nanotechnology is one of the main technologies of the twenty-first century that promises to enhance agricultural practices through the improvement of management and conservation of crops. Delivery of pesticides, herbicides, biocontrol agents, and sensors for the control of plant diseases are some of the topics solved by nanotechnology for agriculture. Some aspect about the potential of nanotechnology in crop protection, nano-entomopathogenic fungi, and chitosan nanoparticles are discussed in this chapter.

S.P. Álvarez (✉) • M.A.M. Tapia
Departamento de Biotecnología Agrícola, Instituto Politécnico Nacional, CIIDIR-IPN, Unidad Sinaloa, Blvd. Juan de Dios Bátiz Paredes No 250, Guasave, Sinaloa, Mexico, C.P. 81101
e-mail: perezalvarezsandra2015@gmail.com

K.I.A. Pérez
Universidad de Las Tunas, Avenida Carlos J. Finlay s/n Reparto Santos, Las Tunas, Cuba, C.P. 75200
e-mail: karel0978@gmail.com; karelap@ult.edu.cu

A.M. Guerrero
Empresa Bioagricultura Aplicada AMG, Calle Santos Degollado entre Guillermo Nelson y Cuauhtemoc, Colonia Centro, Guasave, Sinaloa, Mexico, C.P. 81000
e-mail: amaurymg13@gmail.com

R. Prasad (ed.), *Fungal Nanotechnology*, Fungal Biology,
DOI 10.1007/978-3-319-68424-6_2

2.1 Introduction: Entomopathogenic Fungi as Part of Integrated Pest Management (IPM) in Modern Agriculture

Nanotechnology for modern agriculture is being used for insect pest control with metal particles that can be used to prepare formulations of pesticides, insecticides, and repellent (Esteban-Tejeda et al. 2010; Zahir et al. 2012). Another application of nanotechnology in agriculture is the gene (deoxyribonucleic acid – DNA) transfer by the nanoparticles, so some desire characteristics can be transferred into the plant tissues for host plant defense against the pest insects (Torney 2009).

Entomopathogenic fungi are microorganisms that specifically infect and often kill insects and other arthropods. Most are nonpathogenic to plants and relatively nontoxic to humans and animals. Even with their many advantages over other biological and chemical products, entomopathogenic fungi remain underutilized. Modern agriculture is a necessary combination of different tool to increase food production, and for this purpose entomopathogenic fungus is poised to become a significant component of IPM (Skinner et al. 2014).

2.1.1 Entomopathogenic Fungi as Biological Control Agents

The entomopathogenic fungi have proved to be effective as a tool in IPM systems as microbial insecticides, due to their biological characteristics and mode of action. Its action and effectiveness are usually independent of the host's eating habits, infecting the insect through the integument, a characteristic that makes its use feasible in the control of sucking insects, eggs, and pupae of all taxonomic orders (González et al. 2012; Barrios et al. 2016).

The development of biopesticides is fundamentally based on obtaining pure isolates in culture media from the control agent, and its selection must be made mainly in terms of its virulence and other essential characteristics such as specificity in action, tolerance to environmental factors, ease of production, durability in storage, and high safety in mammals (Méndez 2012).

These characteristics presented in the development of biopesticides have made it possible for these biological control agents to be used worldwide. Some products obtained from entomopathogenic fungi are shown in Table 2.1 (Faria and Wraight 2007). In Table 2.1 a total of 171 products are showed and only 97 of them are currently on the market. South America and Central America use 50% of these products. In North America, the use of products based on entomopathogenic fungi is about 21% of which 13% is produced and used in Mexico which demonstrates the rise in the use of biopesticides. The rest of the percentage is distributed in the rest of the world (Faria and Wraight 2007).

Table 2.1 Commercial mycoinsecticides and geographical zone

Fungi species	Products	South America	Central America	North America	Europe	Asia	Africa	Oceania
Aschersonia aleyrodis	1					1		
Beauveria bassiana	58	22	5	17	5	7	2	
Beauveria brongniartii	7	1			4	1	1	
Conidiobolus thromboides	2	1					1	
Helminthosporium thompsonii	3			1		2		
Isaria fumosorosea	10	4		4	1	1		
Isaria sp.	1					1		
Lagenidium giganteum	1			1				
Lecanicillium longisporum	2	1			1			
Lecanicillium muscarium	3				1	2		
Lecanicillium sp.	11	4	1	1	2	3		
Metarhizium acridum	3						1	2
Metarhizium anisopliae	57	32	6	8	6	3		2
Metarhizium brunneum	1			1			1	2
Nomuraea rileyi	1	1						
Sporothrix insectorum	3	3						
Mezclas (2 o más)	7	4		2	1			
Total	171	73	12	35	21	21	5	4
	100%	42.7%	7.1%	20.6%	12.4%	12.4%	2.9%	2.4%

2.1.2 *Main Methods of Production of Biopesticides Based on Entomopathogenic Fungi*

The development of simple and reliable entomopathogenic fungus production methods is mainly based on processes such as liquid fermentation submerged for the production of biomass, solid state fermentation for the production of aerial conidia, and the biphasic method (Sahayaraj et al. 2008).

The latter method is considered the most viable in the mass production of entomopathogenic fungi, combining the high biomass production of the fungus in liquid medium submerged with the production of stable and hydrophobic conidia in solid medium, achieving yields in the order of 10^9 conidia g^{-1} substrate in a shorter time and natural products and low cost (Leite et al. 2003, cited by Santoro et al. 2005).

2.1.3 Production of Entomopathogenic Fungi by Biphasic Method

This method is being well developed for massive production methods of *B. bassiana* (Bradley et al. 1992) and *Metarhizium flavoviride* Gams y Rozsypal (Jenkins and Thomas 1996). Jenkins et al. (1998) have established the following advantages for the biphasic method:

- The competitiveness of the fungus is favored, since it reduces the risk of colonization of the substrate by contaminating microorganisms.
- The liquid phase can act as a filter for contaminations that may be present in the original culture.
- Uniform growth of the fungus is ensured throughout the substrate, with good coverage of all solid particles.
- Substrate colonization and conidia formation are faster, reducing incubation time and saving space.

2.1.4 Production of Fungal Biomass in Liquid Medium

To produce the fungus in an industrial way, standardized methodologies are required, which use liquid media. Few species develop well with this process, producing conidia, such as *B. bassiana*, while others produce high yields of blastospores and mycelium, such as *M. anisopliae* and *N. rileyi*, which are very suitable for a second stage on a solid substrate, like rice (Jarrold et al. 2007).

The liquid culture can be made with relatively inexpensive and easily obtainable substrates that combine a source of carbon, a source of nitrogen, and some microelements (Kao 2008). In the case of *N. rileyi*, Caro et al. (2005) informed that the combination of carbon and yeast extract sources in liquid medium was effective for the production of biomass, which can be very useful in a biphasic process of mass production of the fungus.

The addition of sterile and moist air to the entomopathogenic fungi production system generally increases yields and can accelerate the sporulation process and help extract excess heat from the system (Bradley et al. 1992). This aspect has a great importance since all mitosporic fungi are aerobic and require oxygen for growth and sporulation. Most low-cost production systems do not incorporate forced aeration but rather take advantage of the passive exchange between the breeding unit and the environment (Jenkins et al. 1998).

2.1.5 Production of Conidia in Solid Medium

Once the culture in liquid medium reaches the exponential growth phase, it must be transfer to the solid medium (Jenkins et al. 1998). Substrate selection depends on several factors that included local availability, costs, and the preference of the fungus to be cultivated (Desphande et al. 2003).

There is a great diversity of solid substrates available for the mass production of fungi for biological control among which can be mentioned cereal grains and other agricultural products or by-products (Jenkins et al. 1998; Kao 2008).

The production units can be of many different types such as glass jars, plastic containers, metal or plastic trays, and polypropylene bags that can be sterilized, the latter being the most used in the majority of production processes worldwide (Desphande and Tour 2001).

Special attention should be paid to the optimization of the various parameters of the production process, such as temperature, substrate moisture content, light conditions, and incubation time, among others (Santoro et al. 2005).

At the end of the production process, the substrate colonized by the fungus must be dried, and its moisture content should be below 15%. Similarly, the way in which the conidia extraction will be performed must be studied, and in these processes, the greatest automation must be used (Jenkins et al. 1998; Cherry et al. 1999).

2.2 Potential of Nanotechnology in Crop Protection

Nanotechnology is a new technology that plays a vital role in different fields of science such as medicine, engineering, electronic, pharmaceuticals, agriculture, and food industry (Prasad 2014; Prasad et al. 2014, 2016, 2017). Nanoparticles have unique targeted characteristics with high strength, conductance of electricity and extra chemical reactivity (Nykypanchuk et al. 2008). The nanoparticles have a size of 10^{-9} in diameter with typical chemical, physical, and biological properties (Bhattacharyya et al. 2010; Sabbour 2013; Prasad et al. 2016). Scientists these days are trying to synthesize organic, inorganic, and hybrid nanoparticles with physical, optical, and biological properties. From agricultural point of view, nanotechnology has become a helping tool for phytopathologists in detection treatment of plant diseases by the use of different nanoparticles, detection of pests by the use of nanosensors, enhanced capability of plants for nutrients absorption, and so on (Chaudhry et al. 2008; Rai and Ingle 2012; Prasad et al. 2014, 2017; Ismail et al. 2017; Sangeetha et al. 2017).

2.2.1 *Agriculture and Diseases in Crops*

Agriculture represents one of the activities more important in the world due to its economic value, nutrition, and jobs generated. However, it is not exempt from damage that causes considerable decrease of agricultural production. Plant diseases study the yield losses they cause; plants become sick when their normal functions are altered by environmental conditions or by pathogens that infect them, causing symptoms such as wilts, root rots, leaf spots, cankers, mosaics, yellowing, and blight, among others. Agents causing disease in plants usually are grouped on the basis of disorders resulting in the plants organs affected; nevertheless, the most useful way of classifying refers to the agents, which are divided into noninfectious (abiotic) and infectious (biotic) including such pathogens as bacteria, fungi, nematodes, phytoplasma, and virus (Vidhyasekaran 2004; Agrios 2005).

2.2.1.1 Disorders Abiotic Noninfectious

Most environmental factors that affect plants hinder their normal physiological processes. These noninfectious disorders are conditions in which no primary parasite is involved, and they may be due by the lack of an essential substance such as water or nutrients that causes an excess of a toxic substance in the soil or in the air and also to abnormal conditions such as an atmospheric pollution, carbon dioxide, climate change, humidity, light, oxygen, or temperature (high and low). Much economic loss of crops is brought about by adverse environmental factors (Weir and Cresswell 1993; Shurtleff and Averre 1997; Agrios 2005).

2.2.1.2 Bacteria as a Disease of Plants

Bacteria are very small microscopic organisms (0.2–1.2 μm in diameter and 0.4–14 μm in length), which are usually constituted by prokaryotic cell wall that can be subdivided into Gram positive and Gram negative. These pathogens may be in the form of rod (bacillus), spherical, ellipsoid, spiral, or filamentous. This pathogen group can cause a wide variety of disease symptoms in plants such as chlorosis, wilting, stunting, rots, necrosis, galls, and scabs. The type and severity of symptoms are related with several factors including the host, climatic conditions, the amount of inoculum, and the method of infection (Sigee 1993; Saddler 2002).

2.2.1.3 Fungi Plant Pathogens

This group of phytopathogens is large and diverse. They occur in most taxonomy groups and can grow easily in culture and survive as saprophytes in the soil or dead vegetative material, while the active fungus pathogen may produce spores to disseminate the disease. Only few organisms of this group are obligate parasites, such as rusts and powdery and downy mildews, obtaining their nutrition directly from plant tissue in development. The symptoms on leaves are produced by a shot-hole effect when the lesions become necrotic and the dead area falls away, while extensive leaf necrosis is often due to blight. In respect to the stems, cankers or limited lesions of anthracnose can be observed and in root and collar may lead to damping off in seedlings or wilting or sudden death of mature plants. The term mildew is referred when the pathogen produces a visible growth over the host surface (Webster 1988; Ulloa and Hanlin 2000; Agrios 2005).

2.2.1.4 Plant-Parasitic Nematodes

Nematodes are the largest group of non-segmented organisms on earth. They have bilateral symmetry with a physiological system similar to the higher animals and measure from 0.2 to 10 mm in length and 0.01–0.5 mm in width. They have a cylindrical and thin form, except those genera in which the females in their advanced stage of development adopt forms of lemon, circular, or kidney (Perry and Wright 1998; Bird and Kaloshian 2003). The parasite nematodes are fed by their stylet, which causes a secretion of enzymes to occur inside the cells, in order to degrade the cellular components and to be able to extract them from the protoplasm. The form of feeding carried out by these organisms causes the damaged cell set to cause death of the plant or severely weaken the root growth points such as the buds and root tips, to form lesions, and to degrade the tissue, producing the abnormal growth of the cells and with this the formation of galls in different parts of the plant (Luc et al. 1991; Nickle 1991).

2.2.1.5 Phytoplasma Plant Pathogens

The phytoplasmas, formerly known as mycoplasma-like organisms (MLOs), are unicellular prokaryotes belong to the class Mollicutes. These parasitic organisms do not possess a cell wall and cause diseases of over 300 plant species around the world. The phytoplasmas are very small; their diameter can vary between 200 and 800 nm and possesses a very small chromosome ranging in size from about 680 to 2000 kb, with high gene content and a single tRNA gene isoleucine (McCoy et al. 1989; Kirkpatrick et al. 1994; Marcone et al. 1999; Bertaccini 2007). Insect vector of these phytopathogens is restricted to phloem-feeding leafhopper and plant hopper members of the Cicadellidae, Coccidia, Fuloroidea, and Psylloidea (Tsai 1979; Weintraub and Beanland 2006). The symptomatology caused by the phytoplasmas is very diverse and manifests itself differently in the wide range of plants or is often confused with viral symptoms. Plants infected include generalized stunting, filodia, floral abnormalities, virescence, shoot proliferation, floral necrosis, yellowing of foliage, and decline of vigor often leading to death of the plant (Lee et al. 1996; Zhang et al. 2004; Bertaccini 2007).

2.2.1.6 Virus Diseases

Viruses represent 47% of the diseases that affect plants. These phytopathogens are infective particles considered as obligate intracellular parasites, as they perform their own replication within a host cell. 90% of these biological entities are composed by ribonucleic acid (RNA), whereas 10% by DNA. Viruses are usually encapsulated in a protein or lipoprotein layer and do not form a cellular structure. Plant viruses need the help of insect vectors, such as mites, nematodes, and certain soil-borne fungi to penetrate the cell through the cell wall; however, they can also be transmitted from plant to plant such as vegetative propagation and mechanical transmission through the sap and through seeds and pollen (Hull 2002; Anderson et al. 2004; Madigan et al. 2006). Most viral diseases often shorten the life of infected plants, and severe stunting can occur. Other symptoms include chlorosis or yellowing of leaves veins, mosaic or mottled, flower abortion, necrosis, and deformation in the leaves, such as wrinkling and curling. As a result, crop yields are severely affected (Agrios 2005; Schuman and D'Arcy 2010).

2.2.2 *Impact of Crop Protection*

Crop plants live in a very complex ecosystem in competition with neighboring plants including weeds, which are attacked by bacteria, birds, fungi, insects, mammals, mites, nematodes, spiders, and viruses, among others. One mission of agriculture is protecting crops from pests and diseases, and in this sense, there have been several efforts to provide a measure of global crop losses. Thus, the principle of protection is avoiding the infection of the crops. Many of the measures considered powerful involve the use of pesticides on seeds and in the root system and foliage plants, environmental regulation, selection of growing areas, cultivation practices, and preventive methods against pathogens. The use of chemical insecticides acts as an important control of pests; however, some collateral effects are attributed to their indiscriminate use such as environmental pollution, reduction in the number of natural enemies, resistance to insecticides and of course human poisoning. It is important to mention that global warming and climate change will have serious consequences on diversity and abundance of pests and the extent of losses due to pests, which will impact both crop production and food security (Loebenstein and Thottappilly 2007; Perlatii et al. 2013; Reddy 2015).

2.2.3 *How Nanotechnology Help in Crop Protection?*

During the years, several chemicals have been used to control or kill the insects on agriculture; however, it is necessary to make it more effective on the pests, because its indiscriminate use leads to problems of ecosystem pollution, affecting human health, together with the costs of application for the producer, coupled with the increasing worldwide demand for foods. This problem requires new technological advances of agricultural production to minimizing losses in the crops, and to increase nutrient- and water-use efficiency, new tools are emerging such as nanotechnology. The word "nano" is developed from the Greek word meaning "dwarf." Naturally, the word nanotechnology evolved due to the use of nanometer-sized particles (size of 1–100 nm). The potential of nano- and microparticles has been reaching a prominent position, opening up a wide variety of uses and benefits. These include agricultural productivity enhancement involving nanoporous zeolites for slow release and efficient dosage of water and fertilizer, nanocapsules for herbicide delivery and vector and pest management, and nanosensors for pest detection, not only in lethal action on the target insect but also in all ecosystems, including natural enemies, animals, and the man himself (Panagiotakopulu et al. 1995; Moraru et al. 2003; Bhattacharyya et al. 2010).

Nanopesticides define as any formulation that intentionally includes elements in the nm size range and/or claim novel properties associated with this small-sized range; it would appear that some nanopesticides have already been on the market for several years. Nanopesticides encompass a great variety of products and cannot be considered as a single category. Nanopesticides can consist of organic ingredients (e.g., polymers) and/or inorganic ingredients (e.g., metal oxides) in various forms (e.g., particles and micelles) (Ragaei and Sabry 2014). In this sense, micro- or nanoformulations containing insecticides prepared in powder or colloidal suspensions have been developed where they have several advantages such as synergism, reduction in foliar leaching, specificity, increased stability of the active organic compound, foliar discoloration, and systemic action, among others. These characteristics enable the use of smaller amount of active compound per area, as long as the formulation may provide an optimal concentration delivery for the target insecticide for longer times and reduce number of applications, human exposure to insecticides, and environmental impact. Nevertheless, this technology is also used in natural products such as herbal extracts and entomopathogenic organisms (Tamez-Guerra et al. 2002; Casanova et al. 2005; Owolade et al. 2008; Peteu et al. 2010; Tsuda et al. 2011; Margulis-Goshen and Magdassi 2012; Batista et al. 2014; Sabbour 2015a).

2.3 Nano-entomopathogenic Fungi

The applications of nanotechnology in agriculture specifically the use of nanomaterials will aloud the development of efficient and potential approaches for the management of pest (Rai and Ingle 2012), so it is important that more researchers all over the world work in this important area. The use of nanotechnology together with entomopathogenic fungi as biotic alternate, because they can control effectively a wide range of insect pests (Sabbour and Singer 2013; Sabbour and El Aziz 2015; Sahab et al. 2015), it is a potential tool for modern agriculture.

Nanomaterials-nanoparticles in different forms could be used for effective management of pests and formulations of insecticides and pesticides (Rai and Ingle 2012; Bhattacharyya et al. 2016). Most of these applications are used with metal or some other materials that have been proven to control plant pathogens, insect, and pests through the preparation of formulations such as pesticides, insecticides, and insect repellants (Gajbhiye et al. 2009; Goswami et al. 2010). Some studies showed the use of silver nanoparticles against the fungus *Raffaelea* sp., which was responsible for the mortality of many oak trees in Korea, and researchers found that silver nanoparticles inhibited significantly the growth of fungi (Kim et al. 2009). Esteban-Tejeda et al. (2009) studied copper nanoparticles in soda lime glass powder and results showed efficient antimicrobial activity against Gram-positive, Gram-negative bacteria and fungi.

Other materials like oil in water (nano-emulsions) were useful for the formulations of pesticides, and these could control some insect pests in agriculture (Wang et al. 2007); also essential oil-loaded solid lipid nanoparticles have been study and they can be used for the formulations of nanopesticides (Liu et al. 2006).

Nanosilica, unique nanomaterial, has been used as nanopesticide because nanosilica gets absorbed into the cuticular lipids by physisorption and thereby causes death of insects by physical means when applied on leaves and stem surface (Barik et al. 2008).

The topic of nanotechnology applied to entomopathogenic fungi has not been investigated deeply. Some nanomaterials have been proven with entomopathogenic fungus such as three kinds of multi-walled carbon nanotubes (MWCNT) toward the conidia of entomopathogenic fungus *Isaria fumosorosea* that was examined in an in vitro study by Gorczyca et al. (2015). Results showed that germination of conidia after contact with the MWCNT was not significantly modified, and raw MWCNT significantly stimulates in vitro pathogenicity of *I. fumosorosea* conidia toward insects; this is not observed in the case of carboxylated nanotubes. Also it was observed that carboxylation of MWCNT reduces the bioactivity of this nanomaterial toward the investigated conidia (Gorczyca et al. 2015).

The entomopathogenic species *Metarhizium anisopliae* (Metsch.) Sorok also was studied in relation to mycelium sporulation and pathogenicity toward test insects, and the research demonstrated a result identical to that of *I. fumosorosea* (Gorczyca et al. 2015), such as the possibility of great sporulation in subculturing, especially in the case of short contact periods with nanomaterial and improved pathogenicity compared to the control (Gorczyca et al. 2014).

Some other researchers studying the bioactivity of raw and carboxylated MWCNT for other than fungal live eukaryotic cells also confirmed that any initial reactivity, including raw nanotube toxicity, could be eliminated by carboxylation-type functionalization (Hamilton et al. 2013; Liu et al. 2014).

Recently some studies revealed that intracellular sustained release of any substances loaded with nanoparticles is more effective, and chitosan is an example. Chitosan has been selected as a material for the preparation of nanoparticles in several applications due to its biodegradable and nontoxic properties (Yien et al. 2012), for example, Helan et al. (2013) evaluated LC_{50} and LT_{50} of chitosan nanoparticle coated fungal metabolite (CNPCFM), uncoated fungal metabolite (UFM), and fungal spores (FS) of entomopathogenic fungi *Nomuraea rileyi* (F) Samson on the pest *Spodoptera litura*; as a result the pesticidal activity was improved using chitosan nanoparticles being CNPCFM better according to the pesticidal activity when compared with UFM and FS (Helan et al. 2013).

Some nanoparticles can be obtained from entomopathogenic fungi, for example, silver and gold. *Beauveria bassiana* (Balsamo) Vuillemin was used for the synthesis of silver nanoparticles to obtain a biolarvicide against the different larval instars of dengue vector, *Aedes aegypti*, and results showed that first and second larval instars were controlled at 100% (all dead), and for the third instar, mortality was between 50% and 80%, so it is suggested that the entomopathogenic fungus synthesized silver nanoparticles would be appropriate for environmentally safer and greener approach for vector control strategy through a biological process (Banu and Balasubramanian 2014). Gold nanoparticles are widely used because of the oxidation resistance, biocompatibility, and stability, but their chemical obtained produces toxic residues that damage the environment, so microorganism like entomopathogenic fungi can be used for this purpose such as *Trichoderma viride* Pers. that can be used for the rapid Au^{3+} bioreduction kinetics in cell-free filtrate (Mishra et al. 2014).

Nowadays nanotechnology is used in agriculture as nanopesticides, nanofungicides, and nanoherbicides (Owolade et al. 2008). Companies around the world use nanotechnology in different ways, some made formulations with nanoparticles within the 100–250 nm size range because they can be dissolving in water more effectively than others, and some other companies use nano-emulsions (nanoscale particle), which can be water or oil based and contain uniform suspensions of pesticidal or herbicidal nanoparticles in the range of 200–400 nm; this nano-emulsions have many uses such as preventative measures, treatment, or preservation of the harvested product and some others (Goswami et al. 2010).

Nanotechnology based in the use of entomopathogenic fungi can provide green and eco-friendly alternatives for insect pest management without harming the nature; this is something really necessary these days when the environment is damage continuously.

2.4 Chitosan Nanoparticles

Chitosan (CHT) is a natural, safe, and cheap biopolymer produced from chitin, the major constituent of arthropods exoskeleton and fungi cell walls and the second renewable carbon source after lignocellulosic biomass (Kurita 2006). For industrial production, solid chitin is soaked in 40–50% (w/v) NaOH. This process removes more than 80% of the acetyl residues and converts N-acetyl-D-glucosamine into -1, 4-D-glucosamine. The CHT preparations are heterogeneous as for deacetylation degree, molecular mass, polymerization degree, viscosity, and acid dissociation constant (pKa value), and the term "chitosan" does not describe a unique compound but a group of commercially available copolymers. This heterogeneity can greatly affect the physical properties of CHT, thus governing its biological applications (Kurita 2006).

In addition to its low cost of production, CHT also possesses several favorable biological properties such as biodegradability, biocompatibility, and non-allergenicity. CHT is susceptible to degradation by specific and nonspecific enzymes, and it shows low toxicity to humans (Wang et al. 2006; Park and Kim 2010; Keawchaoon and Yoksan 2011), thus it has been the material of choice for the preparation of nanoparticles in various applications. All these characteristics make CHT very useful for several industries, namely, cosmetology, food, biotechnology, pharmacology, and medicine (Hamed et al. 2016; Choi et al. 2016).

The availability of industrial quantities of CHT in the late 1980s enabled it to be tested in agriculture. CHT has been proven to stimulate plant growth, to protect the safety of edible products, and to induce abiotic and biotic stress tolerance in various horticultural commodities. Recent innovative uses of CHT include synthesis of CHT nanoparticles as a valuable delivery system for fertilizers, herbicides, pesticides, and micronutrients for crop growth promotion by a balanced and sustained nutrition (Malerba and Cerana 2016). Their application in agriculture is relatively recent compared with their use in drug delivery and pharmaceuticals. Smart delivery of nutrients, pesticides, herbicides, biocontrol agents, and sensors for the diagnosis of plant diseases are some of the emerging topics of nanotechnology for agriculture.

All over the world, the conventional practice of using chemical pesticides and other agrochemicals is either waning or being banned due to their toxic effects on human beings and livestock, residual toxicity, pest outbreaks, and adverse impact on beneficial nontarget organism (Enkerli et al. 2004; Sahayaraj et al. 2008).

In the current situation, it is absolutely necessary to limit the use of chemical pesticides to remain in the world market and sustain the competition. Among the safe and effective alternatives considered, entomopathogens such as bacteria, virus, fungi, protozoa, and nematodes even otherwise play a major role in the natural regulation of pest population and, therefore, offer opportunities for gainful employment as augmentative biocontrol agents. Their relative specificity to target pest, safety to nontarget beneficial organisms, and their ability to cause epizootics make them attractive alternatives for sustainable pest management (Ghazal and Houssein 2005; Hafez et al. 1994; Sahayaraj et al. 2011).

Fungi are ubiquitous natural entomopathogens that often cause epizootics in host insects and possess many desirable traits that favor their development as microbial control agents (MCAs). Presently, commercialized microbial pesticides based on entomopathogenic fungi largely occupy niche markets. A greater understanding of fungal ecology is needed to define their roles in nature and evaluate their limitations in biological control. More efficient mass production, formulation, and delivery systems must be devised to supply an ever-increasing market (Lacey et al. 2015).

In this sense, using nanotechnology, the nanoparticle carrier and nano-metabolite variants with entomopathogenic fungi have been reported as an alternative formulation and delivery systems. Thus, further nanoparticles composed of a chitosan nanocarrier and a metabolite from entomopathogenic fungi *N. rileyi* showed higher pesticide activity against the pest *Spodoptera litura* than the corresponding uncoated fungal metabolite and fungal spores (Chandra et al. 2013).

Chitosan nanoparticles incorporated insecticidal protein beauvericin (CSNp-BV) was evaluated against improved pesticide activity with *Spodoptera litura* and revealed that all the life stages were susceptible to the CSNp-BV formulation, and maximum mortality was recorded in early larval instars. CSNp-BV treatment revealed reduced pupal and adult emergence (Arvind et al. 2014).

In other studies, using the nanoparticles of destruxin from *Metarhizium anisopliae* against the stored product insect pests *Schistocerca agregaria*, found that the infections were significantly decreased when treated (Sabbour 2014a). Sabbour (2014b) found that, under laboratory conditions, the LC 50s were significantly decreased when the adult female of grasshopper *Hetiracris littoralis* was treated with nano-destruxin and reached to 153×10^4 spores/ml. While under laboratory conditions, the number of eggs laid/female of *Plodia interpunctella* (Lepidoptera-Pyralidae) was significantly decreased to 17.4 ± 3.8 and 10.6 ± 9.5 eggs/female after destruxin and nano-destruxin applications, as compared to 99.9 ± 7.9 eggs/female in the control after 120 days (Sabbour 2015b). However, in these studies chitosan-coated nanoformulations was not used, feature that could be improvement effect of metabolite from entomopathogenic fungi against the pests.

Recently, a variety of molecular tools and technologies have recently allowed reclassification of numerous species based on phylogeny, as well as matching anamorphs (asexual forms) and teleomorphs (sexual forms) of several entomopathogenic taxa in the Phylum Ascomycota. Although these fungi have been traditionally regarded exclusively as pathogens of arthropods, recent studies have demonstrated that they occupy a great diversity of ecological niches. Entomopathogenic fungi are now known to be plant endophytes, plant disease antagonists, rhizosphere colonizers, and plant growth promoters (Lacey et al. 2015). These newly understood attributes provide possibilities to use fungi in multiple roles.

Unfortunately there is dearth of information regarding the toxicity of chitin/chitosan to beneficial microbes, nematodes, or insects other than those purposely applied as biological control agents alongside chitin products, including saprotrophic/necrotrophic organisms. One study (Palma-Guerrero et al. 2008) did show that chitosan was more toxic to phytopathogenic fungi than to beneficial nematophagous and entomopathogenic fungi.

Further work is still required to optimize the application methods for chitin-based treatments. Notable areas where there is currently a lack of information include the effects on mite, beetle, and fly pests, as well as requirement for an improved understanding of the phytotoxicity resulting from super-optimal application. Also, the use of nanoparticles provides an effective means of protecting entomopathogenic fungi against premature degradation, resulting in prolongation of its effect on the target pest. Therefore, adding chitin-based products to the growing environment may aid beneficial antagonists by stimulating the production and activation of chitinases that can then be used to attack pests. Thus the addition of chitin to the soil around cultivated crops could be to promote the growth of entomopathogenic fungi for control of some insect pests that live or present any life stage in the soil. Such information will be useful for improving the integration of chitin treatments into IPM systems on farms and nurseries.

From the point of view of sustainable agriculture, nanotechnology can help in the development of environmentally friendly agricultural inputs, improving the safety and stability of active agents, enhancing their activity in pest control, and, consequently, increasing their acceptance by producers (Nair et al. 2010; Srilatha 2011; Khot et al. 2012; Agrawal and Rathore 2014; Prasad et al. 2014, 2017; Ismail et al. 2017). There is also a need to streamline the legislative procedures for registering biopesticides to make chitin-based products realistic alternatives to conventional synthetic pesticides (Chandler et al. 2011).

2.5 Conclusions and Future

The application of chemical pesticides has adverse effects on animals, the ecosystem, and humans, and an integrated pest management strategy is insufficient. It is necessary to reconsider the strategies used for the conservation of biodiversity through alternative approaches such as nanotechnology. Nanomaterials have demonstrated their capability to act as an artificial immune system for plants, as they have the potential to better interaction and mode of action at a target site of the plant or in a desired pest due to its tunable controlled release system and larger superficial area, this confer it more selective. Likewise, they can use smaller amount of active compound per area, so there is no need for reapplications. This potential of nanobiotechnology has been demonstrated in crop protection by increasing crop yields.

References

Agrawal S, Rathore P (2014) Nanotechnology pros and cons to agriculture: a review. Int J Curr Microbiol Appl Sci 3:43–55

Agrios GN (2005) Plant pathology, 5th edn. Elsevier, Philadelphia, p 918

Anderson PK, Cunningham AA, Patel NG, Morales FJ, Epstein PR, Daszak P (2004) Emerging infectious diseases of plants: pathogen pollution, climate change and agrotechology drivers. Trends Ecol Evol Sci 19(10):535–544

Arvind Bharani RS, Karthick R, Namasivayam S, Sai Shankar S (2014) Biocompatible chitosan nanoparticles incorporated pesticidal protein Beauvericin (Csnp-Bv) preparation for the improved pesticidal activity against major groundnut defoliator *Spodoptera litura* (Fab.) (Lepidoptera; Noctuidae). Int J ChemTech Res 6(12):5007–5012

Banu AN, Balasubramanian C (2014) Myco-synthesis of silver nanoparticles using *Beauveria bassiana* against dengue vector, *Aedes aegypti* (Diptera: Culicidae). Parasitol Res 113(8):2869–2877

Barik TK, Sahu B, Swain V (2008) Nano-silica-from medicine to pest control. Parasitol Res 103:253–258

Barrios C, Bustillo C, Ocampo K, Alvarado R (2016) Eficacia de hongos entomopatógenos en el control de Leptopharsa gibbicarina (Hemiptera: Tingidae) en palma de aceite. Rev Colomb Entomol 42(1):22–27

Batista DPC, Souza RS, Santos-Magalhães NS, de Sena-Filho JG, Teodoro AV, Grillo LAM, Dornelas CV (2014) Polymer/layered silicate nanocomposite as matrix for bioinsecticide formulation. Macromol Symp 344:14–21

Bertaccini A (2007) Phytoplasmas: diversity, taxonomy, and epidemiology. Front Biosci 12(2):673–689

Bhattacharyya A, Bhaumik A, Rani PU, Mandal S, Epidi TT (2010) Nanoparticles – a recent approach to insect pest control. Afr J Biotechnol 9(24):3489–3493

Bhattacharyya A, Duraisamy P, Govindarajan M, Buhroo AA, Prasad R (2016) Nano-biofungicides: emerging trend in insect pest control. In: Prasad R (ed) Advances and applications through fungal nanobiotechnology. Springer International Publishing, Switzerland, pp 307–319

Bird D, Kaloshian I (2003) Are roots special? Nematodes have their say. Physiol Mol Plant P 62:115–123

Bradley CA, Black WE, Kearns R, Wood P (1992) Role of production technology in mycoinsecticide development. In: Leatham GF (ed) Frontiers in industrial mycology. Chapman and Hall, London, pp 160–173

Caro L, Villamizar L, Espinel C, Cotés A (2005) Efecto del medio de cultivo en la virulencia de Nomuraea rileyi sobre Spodoptera frugiperda (Lepidoptera: Noctuidae). Rev Colomb Entomol 31(1):25–28

Casanova H, Araque P, Ortiz C (2005) Nicotine carboxylate insecticide emulsions: effect of the fatty acid chain length. J Agr Food Chem 53:9949–9953

Chandler D, Bailey AS, Tatchell GM, Davidson G, Greaves J, Grant WP (2011) The development, regulation and use of biopesticides for integrated pest management. Philos T Roy Soc B 366:1987–1998

Chandra JH, Raj LFAA, Namasivayam SKR, Bharani RSA (2013) Improved pesticidal activity of fungal metabolite from *Nomuraea rileyi* with chitosan nanoparticles. Book of Abstracts International Conference on Advanced Nanomaterials and Emerging Engineering Technologies (ICANMEET). Chennai, India, p 387–390. doi: https://doi.org/10.1109/ICANMEET.2013.6609326

Chaudhry Q, Scotter M, Blackburn J, Ross B, Boxall A, Castle L, Watkins R (2008) Applications and implications of nanotechnologies for the food sector. Food Addit Contam 25(3):241–258

Cherry AJ, Jenkins N, Heviefo G, Bateman R, Lomer C (1999) Operational and economic analysis of a West African pilot-scale production plant for aerial conidia of *Metarhizium* spp. for use as a mycoinsecticide against locusts and grasshoppers. Biocontrol Sci Tech 9:35–51

Choi C, Nam JP, Nah JW (2016) Application of chitosan and chitosan derivatives as biomaterials. J Ind Eng Chem 33:1–10

Desphande M, Tour U (2001) Microbial control of pests: entomopathogenic fungi as mycoinsecticides. Disponible en: http:\\www.biotech.biol.ethz.ch/India/forms/pmfai_paper.pdf. (Consulted: 4 of May of 2015)

Desphande M, Chandele A, Nahar P, Hadapad A, Patil G, Ghormade V (2003) Entomopathogenic fungi: mycoinsecticides useful against lepidopteran pests in pulses. Bull OILB/SROP 26(1):27–30

Enkerli J, Widmer F, Keller S (2004) Long term field persistence of *Beauveria brongniartii* strains applied as biocontrol agents' European cockchafer larvae in Switzerland. Biol Control 29:115–123

Esteban-Tejeda L, Malpartida F, Esteban-Cubillo A, Pecharromán C, Moya JS (2009) Antibacterial and antifungal activity of a soda-lime glass containing copper nanoparticles. Nanotechnology 20:505701

Esteban-Tejeda L, Malpartida F, Pecharroman C, Moya JS (2010) High antibacterial and antifungal activity of silver monodispersed nanoparticles embedded in a glassy matrix. Adv Eng Mater 12(7):B292–B297

Faria MR, Wraight SP (2007) Mycoinsecticides and mycoacaricides: a comprehensive list with worldwide coverage and international classification of formulation types. Biol Control 43(3):237–256

Gajbhiye M, Kesharwani J, Ingle A, Gade A, Rai M (2009) Fungus mediated synthesis of silver nanoparticles and its activity against pathogenic fungi in combination of fluconazole. Nanomedicine 5(4):282–286

Ghazal SA, Houssein BA (2005) Effect of actinomycetes strain on biological aspects of Spodoptera litura, Abstracts, 57th International Symposium on Crop Protection, Belgium

González M, Aguilar C, Rodríguez R (2012) Control de insectos-plaga en la agricultura utilizando hongos entomopatógenos:retosy perspectivas. Rev Científica Univ Autónoma Coahuila 4(8):48–57

Gorczyca A, Kasprowicz MJ, Lemek T (2014) Physiological effect of multi-walled carbon nanotubes (MWCNTs) on conidia and development of the entomopathogenic fungus, Metarhizium anisopliae (Metsch.) Sorok. J Environ Sci Heal A Toxic/Hazard Subst Environ Eng. https://doi.org/10.1080/10934529.2014.867217

Gorczyca A, Kasprowicz MJ, Lemek T (2015) Bioactivity of MWCNT in conidia of entomopathogenic fungus *Isaria fumosorosea*. Water Air Soil Poll 226:75–86

Goswami A, Roy I, Sengupta S, Debnath N (2010) Novel applications of solid and liquid formulations of nanoparticles against insect pests and pathogens. Thin Solid Films 519:1252–1257

Hafez M, Zaki FN, Mours A, Sabbsour M (1994) Biological effects of the entomopathogenic fungus Beauveria bassiana on the potato tuber moth. J Islam Acade Sci 7(4):211–214

Hamed I, Ozogul F, Regenstein JM (2016) Industrial applications of crustacean by-products (chitin, chitosan, and chitooligosaccharides): a review. Trends Food Sci Technol 48:40–50

Hamilton RF, Wu Z, Mitra S, Shaw PK, Holian A (2013) Effect of MWCNT size, carboxylation, and purification on in vitro and in vivo toxicity, inflammation and lung pathology. Part Fibre Toxicol. https://doi.org/10.1186/1743-8977-10-57

Helan JCh, Anand LFA, Karthick SRN, Bharani RSA (2013) Improved pesticidal activity of fungal metabolite from Nomureae riZeyi with chitosan nanoparticles proceedings of the "International Conference on Advanced Nanomaterials & Emerging Engineering Technologies" (ICANMEET-20J3). New Delhi, India, 24–26 July, p 387–390

Hull R (2002) Matthews' plant virology. 4th edn. Academic, San Diego, California, p 1001.

Ismail M, Prasad R, Ibrahim AIM, Ahmed ISA (2017) Modern prospects of nanotechnology in plant pathology. In: Prasad R, Kumar M, Kumar V (eds) Nanotechnology. Springer Nature Singapore Pte Ltd., Singapore, pp 305–317

Jarrold SL, Moore D, Potter U, Charnley AK (2007) The contribution of surface waxes to pre-penetration growth of an entomopathogenic fungus on host cuticle. Mycol Res 111:240–249

Jenkins NE, Thomas MB (1996) Effect of formulation and application method on the efficacy of aerial and submerged conidia of *Metarhizium flavoviride* for locust and grasshopper control. Pestic Sci 46:229–306

Jenkins N, Heviefo G, Jürgen L (1998) Development of mass production technology for aerial conidia for use as mycopesticides. Biocontrol News Inf 19(1):21–31

Kao SS (2008) Current status of bio-pesticides development, farmer's acceptance and utilization, and future perspective in Taiwan. Disponible in: http://kmintra.coa.gov.tw/coa/InternetDownloadFileServlet?fileid=124228&actor_info_id=0. (Consulted: 20 of July of 2016)

Keawchaoon L, Yoksan R (2011) Preparation, characterisation and in-vitro release study of carvacrol-loaded chitosan nanoparticles. Colloid Surf B 84:163–171

Khot LR, Sankaran S, Maja JM, Ehsani R, Schuster EW (2012) Applications of nanomaterials in agricultural production and crop protection: a review. Crop Prot 35:64–70

Kim SW, Kim KS, Lamsal K, Kim YJ, Kim SB, Jung M, Sim SJ, Kim HS, Chang SJ, Kim JK, Lee YS (2009) An in vitro study of the antifungal effect of silver nanoparticles on oak wilt pathogen Raffaelea sp. J Microbiol Biotechnol 19:760–764

Kirkpatrick BC, Smart C, Blomquist C, Guerra L, Harrison N, Ahrens U, Lorenz KH, Schneider B, Seemüller E (1994) Identification of MLO-specific PCR primers obtained from 16S/23S rRNA spacer sequences. International Organization for Mycoplasmology (IOM) p 261–262

Kurita K (2006) Chitin and chitosan: functional biopolymers from marine crustaceans. Mar Biotech (NY) 8:203–226

Lacey LA, Grzywacz D, Shapiro-Ilan DI, Frutos R, Brownbridge M, Goettel MS (2015) Insect pathogens as biological control agents: back to the future. J Invertebr Pathol 132:1–41

Lee M, Danielli A, Bertaccini A, Vibio M, Bartoszyk IM (1996) Multiple phytoplasmas detected in two species of Homoptera feeding on pear trees with decline symptoms. Int Organ Mycoplasmol (IOM) 4:199

Liu F, Wen LX, Li ZZ, Yu W, Sun HY, Chen JF (2006) Porous hollow silica nanoparticles as controlled delivery system for water soluble pesticide. Mater Res Bull 41:2268–2275

Liu Z, Dong X, Song L, Zhang H, Liu L, Zhu D et al (2014) Carboxylation of multiwalled carbon nanotube enhanced its biocompatibility with L02 cells through decreased activation of mitochondrial apoptotic pathway. J Biomed Mater Res A. https://doi.org/10.1002/jbm.a.34729

Loebenstein G, Thottappilly G (2007) Agricultural research management. Springer Dordrecht, The Netherlands

Luc M, Sikora RA, Bridge J (1991) Plant parasitic nematodes in subtropical and tropical agriculture. CAB International, Wallingford, p 629

Madigan MT, Martinko JM, Parker J (2006) Biología de los microorganismos. PEARSON Prentice Hall, España, p 1011.

Malerba M, Cerana R (2016) Chitosan effects on plant systems. Int J Mol Sci 17:996

Marcone C, Neimark H, Ragozzino A, Laur U, Seemuller E (1999) Chromosome sizes of phytoplasmas comparing major phylogenetic groups and subgroups. Phytopathology 89:805–810

Margulis-Goshen K, Magdassi S (2012) Nanotechnology: an advanced approach to the development of potent insecticides. In: Ishaaya I, Horowitz AR, Palli SR (eds) Advanced technologies for managing insect pests. Springer, Dordrecht, pp 295–314

McCoy RE, Caudwell A, Chang CJ, Chen TA, Chiykowski LN, Cousin MT, Dale JL, De Leeuw GTN, Golino DA, Hackett KJ, Kirkpatrick BC, Marwitz R, Petzold H, Sinha RC, Sugiura M, Whitcomb RF, Yang IL, Zhu BM, Seemuller E (1989) Plant diseases associated with mycoplasma-like organisms. In: Whitcomb RF, Tully JG (eds) The mycoplasmas: spiroplasmas, acholeplasmas and mycoplasmas of plants and arthropods, vol 5. Academic, San Diego

Méndez A (2012) Potencialidad de aislamientos autóctonos de *Nomuraea rileyi* (Farlow) Samson para el control de *Spodoptera frugiperda* (J.E. Smith) y *Plutella xylostella* (L.). Rev Protección Veg 27(1)

Mishra A, Kumari M, Pandey S, Chaudhry V, Gupta KC, Nautiyal CS (2014) Biocatalytic and antimicrobial activities of gold nanoparticles synthesized by *Trichoderma sp.* Bioresour Technol 166:235–242

Moraru C, Panchapakesan C, Huang Q, Takhistov P, Liu S, Kokini J (2003) Nanotechnology: a new frontier in food science. Food Technol 57(12):24–29

Nair R, Varghese SH, Nair BG, Maekawa T, Yoshida Y, Kumar DS (2010) Nanoparticulate material delivery to plants. Plant Sci 179:154–163

Nickle WR (1991) Manual of agricultural nematology. Marcel Dekker, New York, p 1035

Nykypanchuk D, Maye MM, van der Lelie D, Gang O (2008) DNA-guided crystallization of colloidal nanoparticles. Nature 451:549–552

Owolade OF, Ogunleti DO, Adenekan MO (2008) Titanium dioxide affects disease development and yield of edible cowpea. Electron J Environ Agr Food Chem 7(50):2942–2947

Palma-Guerrero J, Jansson HB, Salinas J, Lopez-Llorca LV (2008) Effect of chitosan on hyphal growth and spore germination of plant pathogenic and biocontrol fungi. J Appl Microbiol 104:541–553

Panagiotakopulu E, Buckland PC, Day PM, Sarpaki AA, Doumas C (1995) Natural insecticides and insect repellents in antiquity: a review of the evidence. J Archaeol Sci 22:705–710

Park BK, Kim MM (2010) Applications of chitin and its derivatives in biological medicine. Int J Mol Sci 11:5152–5164

Perlatti B, de Souza Bergo PL, das Graças Fernandes da Silva MF, Batista Fernandes J, Rossi Forim M (2013) Polymeric nanoparticle-based insecticides: a controlled release purpose for agrochemicals, insecticides – development of safer and more effective technologies. Prof. Stanislav Trdan (ed.). InTech, DOI: https://doi.org/10.5772/53355. http://www.intechopen.com/books/insecticides-development-of-safer-and-more-effective-technologies/polymeric-nanoparticle-based-insecticides-a-controlled-release-purpose-for-agrochemicals

Perry RN, Wright DJ (eds) (1998) The physiology and biochemistry of free-living and plant parasitic nematodes. CAB International, Wallingford, p 438

Peteu SF, Oancea F, Sicuia OA, Constantinescu F, Dinu S (2010) Responsive polymers for crop protection. Polymers 2:229–251

Prasad R (2014) Synthesis of silver nanoparticles in photosynthetic plants. J Nanoparticles, Article ID 963961, https://doi.org/10.1155/2014/963961

Prasad R, Vivek K, Prasad KS (2014) Nanotechnology in sustainable agriculture: present concerns and future aspects. Afr J Biotech 13:705–713

Prasad R, Pandey R, Barman I (2016) Engineering tailored nanoparticles with microbes: quo vadis. WIREs Nanomed Nanobiotechnol 8:316–330. https://doi.org/10.1002/wnan.1363

Prasad R, Bhattacharyya A, Nguyen QD (2017) Nanotechnology in sustainable agriculture: recent developments, challenges, and perspectives. Front Microbiol 8:1014. https://doi.org/10.3389/fmicb.2017.01014

Ragaei M, Sabry AH (2014) Nanotechnology for insect pest control. IJSET 3(2):528–545

Rai M, Ingle A (2012) Role of nanotechnology in agriculture with special reference to management of insect pests. Appl Microbiol Biot 94:287–293

Reddy PP (2015) Climate resilient agriculture for ensuring food security. Springer, New Delhi. https://doi.org/10.1007/978-81-322-2199-9

Sabbour MM (2013) Entomotoxicity assay of Nanoparticle 3-(Zinc oxide ZnO) against *Sitophilus oryzae* under laboratory and store conditions in Egypt. Sci Rep 1(2):51–57

Sabbour MM, Singer SM (2013) Efficacy of nano *Isaria fumosorosea* and *Metarhizium flavoviride* against corn pests under laboratory and field conditions in Egypt. Int J Sci Res 4(7):945–949

Sabbour MM (2014a) Evaluating toxicity of nano-extracted Destruxin from Metarhizium anisopliae against the grasshopper *Hetiracris littoralis* in Egypt. J Egypt Acad Soc Environ Dev 15(2):1–7

Sabbour MM (2014b) Evaluating toxicity of extracted nano-Destruxin against the desert locust *Schistocerc agregaria* in Egypt. J Egypt Acad Soc Environ Dev 15(2):9–17

Sabbour MM (2015a) The toxicity effect of nano fungi *Isaria fumosorosea* and *Metarhizium flavoviride* against the potato tuber moth, Phthorimaea operculella (Zeller). Am J Biol Life Sci 3(5):155–160

Sabbour MM (2015b) Laboratory and store efficacy of nano-extracted Destruxin from Metarhizium anisopliae against Indian meal moth Plodia interpunctella (Lepidoptera-Pyralidae). J Nanosci Nanoeng 1(3):142–147

Sabbour MM, El Aziz SEA (2015) Efficacy of some nano-Imidacloprid against red flour beetle *Tribolium castaneum* and confused flour beetle, *Tribolium confusum* (Coleoptera:Tenebrionidae) under laboratory and store conditions. Bull Env Pharmacol Life Sci 4(7):54–59

Saddler GS (2002) Bacteria and plant disease. In: Waller JM, Lenné J, Waller SJ (eds) Plant pathologist's pocketbook, 3rd edn. CABI Publishing, Wallingford, pp 94–107

Sahab AF, Waly AI, Sabbour MM, Nawar LS (2015) Synthesis, antifungal and insecticidal potential of chitosan (CS)-g-poly (acrylic acid) (PAA) nanoparticles against some seed borne fungi and insects of soybean. Int J ChemTech Res 8(2):589–598

Sahayaraj K, Karthick R, Namasivayam S (2008) Mass production of entomopathogenic fungi using agricultural products and byproducts. Afr J Biotechnol 17:213–218

Sahayaraj K, Karthick R, Namasivayam S (2011) Field evaluation of three entomopathogenic fungi on groundnut pests. Tropicultura 29(3):143–147

Sangeetha J, Thangadurai D, Hospet R, Purushotham P, Karekalammanavar G, Mundaragi AC, David M, Shinge MR, Thimmappa SC, Prasad R, Harish ER (2017) Agricultural nanotechnology: concepts, benefits, and risks. In: Prasad R, Kumar M, Kumar V (eds) Nanotechnology. Springer Nature Singapore Pte Ltd., Singapore, pp 1–17

Santoro PH, Janeiro PM, Silva RZ, Akimi S, Zorzetti J (2005) Produção de esporos de Beauveria bassiana (Bals). Vuill. num proceso bifásico utilizando diferentes meios líquidos. Semin: Ciências Agrárias 26:313–320

Schuman GL, D'Arcy CJ (2010) Essential plant pathology, 2nd edn. APS, St Paul, p 369
Shurtleff MC, Averre CW (1997) The plant disease clinic and field diagnosis of abiotic diseases. APS, St Paul, p 245
Sigee DC (1993) Bacterial plant pathology: cell and molecular aspects. Cambridge University Press, Cambridge, UK
Skinner M, Parker BL, Kim JS (2014) Integrated pets management: current concepts and ecological' perspectives. In: Abrol DP (ed) Role of entomopathogenic fungi in integrated pest management. Academic, San Diego, California, pp 169–191.
Srilatha B (2011) Nanotechnology in agriculture. J Nanomed Nanotechnol 2:123
Tamez-Guerra P, McGuire MR, Behle RW, Shasha BS, Pingell RL (2002) Storage stability of *Anagrapha falcifera* nucleopolyhedrovirus in spray-dried formulations. J Inver Pathol 79:7–16
Torney F (2009) Nanoparticle mediated plant transformation. In: Emerging technologies in plant science research. Interdepartmental Plant Physiology Major Fall Seminar Series Phys, p 696
Tsai JH (1979) Vector transmission of mycoplasmal agents of plant diseases. In: Whitcomb RF, Tully JG (eds) The mycoplasmas: plant and insect mycoplasmas. Academic, New York, pp 226–307
Tsuda N, Ohtsubo T, Fuji M (2011) Preparation of self-bursting microcapsules by interfacial polymerization. Adv Powder Technol. https://doi.org/10.1016/j.apt. 2011.09.005
Ulloa M, Hanlin RT (2000) Illustrated dictionary of micology. APS Press, St Paul, p 200
Vidhyasekaran P (2004) Concise encyclopedia of plant pathology. Food Products Press, New York, USA, p 587.
Wang J, Zen C, Xiao RZ, Xie T, Zhou L, Zhan XR, Wang Sh L (2006) Recent advances of chitosan nanoparticles as drug carriers. Int J Nanomedicine 6:765–774
Wang L, Li Z, Zhang G, Dong J, Eastoe J (2007) Oil-in-water nanoemulsions for pesticide formulations. J Colloid Interf Sci 314:230–235
Webster J (1988) Introduction to fungi, 3rd edn. Cambridge University Press, Cambridge, UK
Weintraub PG, Beanland L (2006) Insect vectors of phytoplasmas. Annu Rev Entomol 51:91–111
Weir RG, Cresswell GC (1993) Plant nutrient disorders 3: vegetable crops. Inkata Press, North Ryde, p 105
Yien L, Mohamad NZ, Sarwar A, Katas H (2012) Antifungal activity of chitosan nanoparticles and correlation with their physical properties. Int J Biomaterials 2012:1–9
Zahir AA, Bagavan A, Kamaraj C, Elango G, Rahuman AA (2012) Efficacy of plant-mediated synthesized silver nanoparticles against *Sitophilus oryzae*. J Biopest 288(Suppl 5):95–102
Zhang J, Hogenhout SA, Nault LR, Hoy CW, Miller SA (2004) Molecular and symptom analyses of phytoplasmas strains from lettuce reveal a diverse population. Phytopathol 94(8):842–849

Chapter 3
Fungi as Ecosynthesizers for Nanoparticles and Their Application in Agriculture

Khamis Youssef, Ayat F. Hashim, Ahmed Hussien, and Kamel A. Abd-Elsalam

Abstract Nanotechnology covers various fields of science and technology including agriculture sector and plant disease management. To produce nanoparticles, several means including chemical and physical methods were cited. However, weaknesses, i.e., chemical toxicity and production of high-energy supplies, make them difficult to be broadly employed. In contrast, nanoparticle fabrication using the biological means is promising as it would overcome the previous problems; hence using them for nanoparticle synthesis has been encouraged. Several fungi (*Aspergillus* spp., *Fusarium* spp., *Alternaria* spp., *Trichoderma* spp., *Verticillium* spp., *Penicillium* spp., etc.) were employed for nanoparticle synthesis with fine size and shape due to their resistance to many harsh conditions. Myconanotechnology can offer eco-friendly and economically alternative ways. In this chapter, several approaches including nanoparticle biosynthesis using fungi as natural factories, characterization of mycosynthesized metal nanoparticles, mycosynthesis mechanism, nanoparticles strategies, and the application of nanoparticles in plant disease management will be discussed.

K. Youssef (✉) • K.A. Abd-Elsalam
Agricultural Research Center, Plant Pathology Research Institute,
9 Gamaa St., Giza 12619, Egypt
e-mail: youssefeladawy@arc.sci.eg

A.F. Hashim
Food Industries and Nutrition Division, National Research Center,
33 Al Behous, Ad Doqi, Giza, Egypt

A. Hussien
Central Administration of Plant Quarantine, Ministry of Agriculture and Land Reclamation,
1 Nadi ElSaid St., Dokki, Giza 12618, Egypt

R. Prasad (ed.), *Fungal Nanotechnology*, Fungal Biology,
DOI 10.1007/978-3-319-68424-6_3

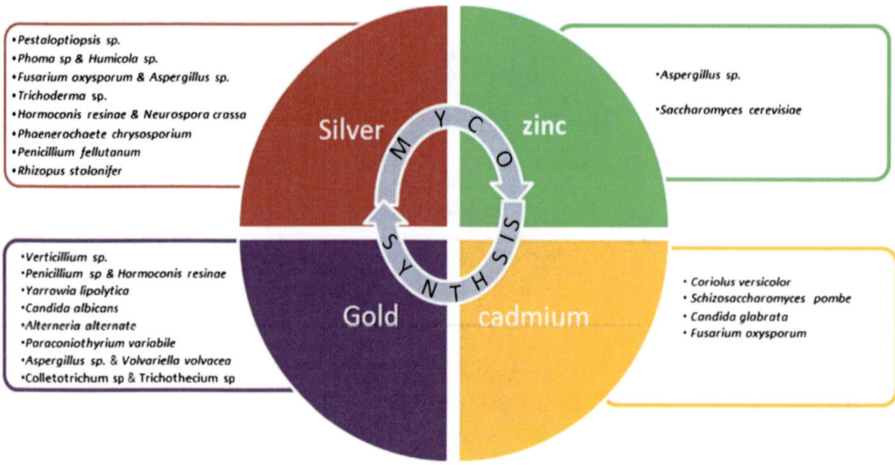

Fig. 3.1 Major fungal species served as ecosynthesizers for nanoparticles

3.1 Introduction: Mycosynthesis of Metal (Silver, Zinc, Gold, and Cadmium) Nanoparticles

A novel interdisciplinary field of science that combines knowledge from nanotechnology and mycology is representing as a new applied science with great possibilities due to extraordinary diversity of fungi. Myconanofabrication (nanoparticle production through fungi) can be defined as the synthesis of metal nanoparticles using fungi. The myconanofactories have already been exploited for the biosynthesis of metal nanoparticles of silver, gold, zinc, CdS, etc. (Fig. 3.1) (Sastry et al. 2003).

3.2 Mycosynthesis of Silver Nanoparticles

A great number of fungal strains can synthesize silver nanoparticles (AgNPs), such as *Pestalotiopsis* sp., *Phoma* sp., *Humicola* sp., *Fusarium oxysporum*, *Aspergillus niger*, *Trichoderma* sp., *Hormoconis resinae*, *Phaenerochaete chrysosporium*, *Penicillium fellutanum*, and *Rhizopus stolonifera*.

3.2.1 Pestalotiopsis *sp.*

It has been revealed that the endophytic fungus *Pestalotiopsis* sp. is capable of fabricating silver nanoparticles extracellularly. The fungal mat of *Pestalotiopsis* sp. was inoculated in PDB medium and was incubated for 2 weeks at 25 °C for optimum growth. Once the growth is observed, different volumes (1, 1.5 and 2 ml) of 1 mM $AgNO_3$ were added to the medium and then kept in a shaker to allow the fungi to intake the nanoparticles. The synthesis of nanoparticles was detected to be 123–195 nm in scanning electron microscope analysis (Vardhana and Kathiravan 2015).

3.2.2 **Phoma** *sp. and* **Humicola** *sp.*

Silver nitrate solution was treated with freeze-dried mycelium of *Phoma* sp. strain 3.2883 for 50 h. Adsorption assays showed that the mycelium had absorbed approximately 13 mg of silver, and TEM micrographs exhibited silver nanoparticles of around 70 nm (Chen et al. 2003). When the thermophilic fungus *Humicola* sp. reacted with Ag^{+} ions, that reduces the precursor solution, and extracellular nanoparticles were synthesized. The cytotoxicity of silver nanoparticles on NIH3T3 mouse embryonic fibroblast cell line and MDA-MB-231 human breast carcinoma cell line was assessed (Syed et al. 2013).

3.2.3 **Fusarium oxysporum** *and* **Aspergillus** *sp.*

Some strains of *Fusarium oxysporum* can achieve the extracellular fabrication of silver nanoparticles with the help of nitrate-dependent reductase enzyme and a shuttle quinone extracellular pathway (Balaji et al. 2009). That has demonstrated the synthesis of silver nanoparticles using *Fusarium oxysporum*. The enzyme nitrate reductase might be responsible for the reduction of silver ions (Duran et al. 2005). Pure AgNPs were synthesized at a size range of 5–15 nm.

The synthesis of silver nanoparticles by *Aspergillus fumigatus* was verified using XRD and TEM analysis. The XRD spectrum showed intense peaks in consent with the Bragg reflections of crystalline silver. TEM micrograph of nanoparticles showed variable shape with majority of them spherical with some triangular form with size range of 5–25 nm (Bhainsa and D'Souza 2006). Endophytic fungus *Aspergillus clavatus* effectively synthesized silver nanoparticles with spherical and hexagonal shape (Verma et al. 2010). Final concentration of 1 mM of $AgNO_3$ was used in fungal extract LCF of *Aspergillus foetidus* (reducing agent) and incubated at 28 ± 2 °C and agitated at 150 rpm in dark. Positive control (biomass only, without the silver nitrate) and negative control (silver nitrate solution only) were also run along with the experimental flask. It was found that the dimension of the synthesized silver nanoparticles was in the range of 20–40 nm in TEM (Roy et al. 2013).

3.2.4 **Trichoderma** *sp.*

Many species of *Trichoderma* can synthesize large amounts of extracellular metabolites and enzymes which can catalytically reduce toxic silver ions to nontoxic nanoparticles when the mycelium is challenged with silver nitrate solution

(Ramanathan et al. 2009). *Trichoderma viride* is reported to behave similarly (Fayaz et al. 2011).

Trichoderma reesei is a useful fungus, and it is well known for its development of extracellular enzyme. Volume of 1 mM of silver nitrate was mixed with 10 *g* of fresh biomass fungus *T. reesei*. Then the reaction mixture was placed on a 100 rpm shaker for 120 h at 28 °C. The results indicated that biosynthesis of AgNPs by *T. reesei* produces particles with the diameters 5–50 nm (Vahabi et al. 2011).

Trichoderma asperellum is a nonpathogenic biocontrol agent and can be used for synthesis of silver nanoparticles which are stable after 6 months of storage. This particle capping is possibly responsible for their stabilization as established by FTIR and Raman spectroscopy studies (Mukherjee et al. 2008).

3.2.5 Hormoconis resinae *and* Neurospora crassa

Treatment of silver nitrate solution with unfamiliar fungus *Hormoconis resinae* will produce within an hour silver nanoparticles. The shape is not uniform and varies from triangle to spherical and the size from 20 to 80 nm (Varshney et al. 2009).

Intracellular and extracellular filamentous fungus *Neurospora crassa* can produce silver nanoparticles. This fungus displays a rapid growth rate, with ability to reduce metallic ions quickly to regularly sized nanoparticles (Longoria et al. 2011).

3.2.6 Phaenerochaete chrysosporium

Silver nanoparticles were synthesized extracellularly when incubating mycelium of a white rot fungus *Phaenerochaete chrysosporium* with silver nitrate solution. Under TEM, pyramidal shape of nanoparticles is detected. The fungus remains non-pathogenic, and its easy control and mass productions contribute to its benefit (Vigneshwaran et al. 2006).

3.2.7 Penicillium fellutanum

Penicillium fellutanum isolated from coastal mangrove sediments is able to produce nanoparticles under controlled parameters of temperature, pH, silver ion concentration, and time of exposure (Kathiresan et al. 2009). This fungus is also able to facilitate rapid production within few minutes by adding substrate to the culture filtrate. High polydispersity is achieved in compacting and producing fungus *Penicilliun brevicompactum*, and nanoparticle's varying size 23–105 nm is obtained under TEM.

3.2.8 Rhizopus stolonifer

Particles synthesized from *Rhizopus stolonifer* are showed by TEM to be actually small and have size ranges from 3 to 20 nm which may verify benefits for particular applications. This extracellular synthesis method yields nanoparticles in 24 h (Banu et al. 2011).

3.3 Mycosynthesis of Zinc Oxide Nanoparticles

The synthesis of zinc oxide nanoparticles using culture filtrate of *Aspergillus* sp. and *Saccharomyces cerevisiae* was reported.

3.3.1 Aspergillus *sp. and* Saccharomyces cerevisiae

The fungus *Aspergillus terreus* was grown aerobically and was agitated in an orbital shaker at 160 rpm for 4 days at 32 °C. The fungal culture was filtered under vacuum. Concentration of 1 mM $ZnSO_4$ salt was added to the filtrate and kept for 2 days at 32 °C in a shaking incubator at 150 rpm. White precipitate deposition started to occur which showed the transformation process. The white aggregate formed was separated from the filtrate by centrifugation at 10,000 rpm for 10 min and lyophilized.

Aspergillus terreus can be used for large-scale synthesis of zinc oxide nanoparticles. FTIR analysis confirmed the functional groups present in the zinc oxide nanoparticles by the peaks in the range of 560–3690 cm^{-1}. In XRD analysis, sharper and stronger diffraction peaks confirmed the synthesis. From SEM analysis, synthesized zinc oxide nanoparticles were in size ranges from 54.8 to 82.6 nm and were found to be spherical in shape (Baskar et al. 2013).

Concentration of 1.0 mM $ZnSO_4$ salt was added to *A. fumigatus* filtrate by adjusting the pH to 6.5 and incubated for 150 rpm at 32 °C for 72 h. Zinc oxide nanoparticles synthesized by *A. fumigatus* JCF that was confirmed with UV have shown a peak at 350 nm. Zinc oxide nanoparticles of spherical structure were observed using SEM (Rajan et al. 2016).

The ZnS nanoparticles in significant system by *Saccharomyces cerevisiae* MTCC 2918 yeast were synthesized in low yield and the size range of 30–40 nm (Mala and Rose 2014).

3.4 Mycosynthesis of Gold Nanoparticles

3.4.1 Verticillium *sp.*

One of the earliest reports of the synthesis of gold nanoparticles by fungi was demonstrated by the fungus *Verticillium* sp. Gericke and Pinches (2006) have confirmed intracellular synthesis of gold nanoparticles. The fungus *Verticillium luteoalbum* was used for the synthesis of gold nanoparticles. The biosynthesized nanoparticle showed variable morphologies including spherical, triangular, hexagonal, and other shapes in TEM micrographs. It was detected that the cells age at the time of exposure to $AuCl^{-4}$ solution did not have any significant effect on the shape of nanoparticles. The pH of the reaction solution has played a significant role in the particle synthesis. TEM images of the particle synthesized at pH 3, 5, 7, and 9 revealed particles with shape including triangles, hexagons, spheres, and rods. The EDS spectrum showed that the Au nanoparticles were mainly composed of Au with trace amounts of C, O, Na, Si, and Al. Thus, the synthesis of gold nanoparticles was related to the incubation temperature. Increased temperature caused faster particle growth rate. Gold nanoparticles were synthesized intracellularly by growing *Verticillium* sp. cells in a well-defined medium and moved to aqueous auric chloride solution; the pale yellow color of the fungal cells changed to vivid purple over 24 h. The gold nanoparticle synthesis was confirmed by UV visible spectra at an absorption peak on 540 nm. The TEM and higher magnification TEM studies showed the presence of gold nanoparticles with an average size of 20–28 nm on the cell wall and the cytoplasm (Mukherjee et al. 2001).

3.4.2 Penicillium *sp. and* Hormoconis resinae

Size-controlled nanoparticles can be synthesized from *Penicillium* sp. where the nanoparticle separation could be simply released from the fungal cell lysate by using the ultrasonication and centrifugation methods (Xiaorong et al. 2009). The gold nanoparticles have been synthesized from *Penicillium brevicompactum*. The cytotoxic nature of nanoparticles has been examined using mouse mayo blast cancer C_2C_{12} cells (Mishra et al. 2011).

The fungus *Hormoconis resinae* was found to produce stable gold nanoparticles that are spherical. This showed economy source of material offers the chance for cost-effective preparation of many gold-based nanostructures (Mishra et al. 2010).

3.4.3 Candida albicans

Volume of 5 ml solution of 10^{-3} M aqueous $HAuCl_4$ is added to different volumes (1–5 ml) of cytosolic extract. The volume was made up to 10 ml by adding the suitable volume of deionized water. The mixture was incubated to complete the reaction for 24 h. The synthesized gold particle morphology depended on the abundance of *C. albicans* in cytosolic extract. TEM micrograph, nanophox particle analysis, and atomic force microscopy showed the size of spherical gold nanoparticles to be in the range of 20–40 nm, and nonspherical gold particles were revealed to be 60–80 nm (Chauhan et al. 2011).

3.4.4 Yarrowia lipolytica

Great amount of biomass of the nonconventional yeast *Yarrowia lipolytica* has been used for the generation of gold nanoparticles. The synthesis occurs in a pH-dependent method in both seawater and freshwater. The mechanism of the causal process is not obviously defined, but the presence of inherent proteases and reductases may be involved (Agnihotri et al. 2009).

3.4.5 Alterneria alternate *and* Paraconiothyrium variabile

Culture filtrate of *Alterneria alternate* produced gold nanoparticles with average diameter of 12 nm and is very stable (Sarkar et al. 2011). Laccase purified from *Paraconiothyrium variabile* is used to reduce $HAuCl_4$ (Faramarzi and Forootanfar 2011). Particle formation increased with decreasing laccase activity assuming that exposure to reducing functional groups may be responsible for the formation of particle.

3.4.6 Aspergillus *sp.*

Endophytic fungus *Aspergillus clavatus* has synthesized gold nanotriangles. This fungus intracellularly reacted with chloroaurate ions which may be found to have possible uses in various areas (Verma et al. 2011). Both live and dead cells of *Aspergillus oryzae* produce gold nanoparticles which are economically significant in the food industry. Purple color formation could be seen in the reaction medium by reduction of trivalent aurum to gold nanoparticles (Binupriya et al. 2010). TEM micrograph showed high polydispersity in the range of 10–400 nm.

The interaction of gold ions with the supernatant of *Aspergillus niger* leads to the production of gold nanoparticles (Bhambure et al. 2009). The extracellular enzyme responsible for the reduction process may also cap the particles and prevent them from aggregation.

3.4.7 Volvariella volvacea

The additional protein presence in the system may favor nanoparticle synthesis as occurred in the instance of extracts from naturally arising edible mushroom *Volvariella volvacea* which synthesizes gold nanoparticles along with silver and gold-silver ones of 20–150 nm. The efficiency improvement is a positive issue, and using mushrooms is more reasonable too (Philip 2009).

3.4.8 Colletotrichum *sp.*

The gold nanoparticles were mostly spherical while some were triangular or hexagonal. An endophytic fungus *Colletotrichum* sp. isolated from *Pelargonium graveolens* showed ability to synthesis of spherical gold nanoparticles (Shankar et al. 2003). The XRD study showed formation of stable gold nanoparticle aggregates while the TEM studies showed synthesis of predominantly spherical nanoparticles and some aggregated into irregular structures with no well-defined morphology. The smaller spherical nanoparticles ranged in size from 8 to 40 nm. It was found that the fungal proteins were responsible for the stabilization of gold nanoparticles.

3.4.9 Trichothecium *sp.*

The fungus *Trichothecium* sp. could synthesize gold nanoparticles by both extra and intracellular method (Ahmad et al. 2005). Rapid extracellular production of gold nanoparticles was obtained when the biomass of fungal was kept in stationary condition, whereas intracellular synthesis of nanoparticles was resulted when the biomass was kept in shaking condition on a rotary shaker. The extracellularly synthesized gold nanoparticles showed individual gold nanoparticles along with few aggregates in TEM photographs. The TEM photographs of the extracellularly synthesized gold nanoparticles showed number of individual gold nanoparticles along with few aggregates. The particle's morphological study described the presence of polygons (especially triangles and hexagons) and some polydisperse spheres and rods. The selected area electron diffraction (SAED) pattern exposed the Scherrer ring pattern which is characteristic of face-centered cubic (fcc) gold structure while the TEM micrographs of the intracellularly synthesized gold nanoparticles displayed formation of small particles with spherical morphology. The SAED pattern of these gold particles exhibited diffuse rings with lattice spacing which were in agreement with that of gold. According to Bragg, reflections of gold XRD data revealed the presence of intense peaks (111, 200, 220 and 311). It was determined that when the reaction conditions are changed, the enzymes and proteins which are released in the stationary phase do not release in shaking condition and hence result in the intracellular and extracellular synthesis of gold nanoparticles.

3.5 Mycosynthesis of Nanocadmium and Cadmium Quantum Dots

3.5.1 Coriolus versicolor

Coriolus versicolor, immobilized on a ceramic bead column, has reduced cadmium to nanoparticles which were quite stable without the addition of any stabilizers. SH-containing proteins, released by the fungus, may be involved in the fabrication of these particles (Sanghi and Verma 2009).

3.5.2 Schizosaccharomyces pombe *and* Candida glabrata

Two different types of yeasts *Schizosaccharomyces pombe* and *Candida glabrata* were able to collect cadmium metal when cultured in a metal-rich batch-fed process (Krumov et al. 2007). As the metal is toxic to cells, detoxification by an intracellular mechanism is adopted by them during which the metal is reduced to CdS nanoparticles of 35 kDa.

3.5.3 Fusarium oxysporum

Q-state cadmium sulfide nanoparticle fabrication has been reported to be produced by *Fusarium oxysporum*. The enzyme sulfate reductase from the fungus was revealed to facilitate the synthesis of nanoparticles, and this enzymatic pathway can be used for the synthesis of a varied range of nanoparticles (Ahmad et al. 2002). The synthesis of CdS quantum dots by the fungus *Fusarium oxysporum* was showed by extracellular enzymatic reduction of sulfate ions. This fungus plays the role of a bioreducing agent (enzyme sulfate reductase) in the reaction of Cd^{2+} ions with sulfate ions and the enzymatic reduction of sulfate ions to sulfide ions for the formation of CdS quantum dots. The semiconductor nanoparticles were monodisperse in size from 5 to 20 nm. The XRD analysis of particles exhibited the nanocrystalline nature of nanoparticles with Bragg reflections characteristic for hexagonal CdS particles.

It was demonstrated that the synthesis of highly luminescent CdSe quantum dots at room temperature could be achieved using *F. oxysporum* (Kumar et al. 2007b). The fungus *F. oxysporum* was challenged with aqueous $CdCl_2$, $SeCl_4$ solution. The CdSe nanoparticle synthesis was shown after 96 h of reaction with a change in coloration to reddish brown. The UV-vis spectra of the reaction mixture displayed strong surface plasmon peak at 370 nm. The FTIR analysis revealed intense peaks at 100, 111, 220, 311, and 222 corresponding to Bragg reflections of CdSe particles. The TEM analysis of the CdSe particles indicated polydisperse particles in size range of 9–15 nm with spherical morphology. The X-ray photoelectron spectroscopy of the particles obviously showed presence of Cd, Se, C, O, N, and Na as prominent elements.

3.6 Characterization of Mycosynthesized Metal Nanoparticles

The advanced instrumentation techniques used for the characterization of synthesized metal nanoparticles have revealed a valuable insight into several morphological and structural features with respect to their size. UV-vis spectroscopy used to follow up the reaction process. Fourier transform infrared (FTIR) spectroscopy is a powerful tool for detecting types of chemical bonds in a molecule and analyzing the characteristic functional groups present in the synthesized nanoparticles. Information about particle size, crystal structure, and surface morphology is obtained using X-ray diffraction (XRD), transmission electron microscopy (TEM), and scanning electron microscopy (SEM). Atomic force microscopy (AFM) or scanning force microscopy (SFM) can be used to image and manipulate atoms and structures on a variety of surfaces. Energy-dispersive X-ray (EDX) spectroscopy is an analytical technique used for the elemental analysis or chemical characterization of nanoparticles. Dynamic light scattering (DLS) analysis was used to determine the size distribution profile of nanoparticles. The stability of the particles can be recognized by the reported techniques.

3.7 Mechanistic Basis for Myconanoparticle Synthesis

The study of biosynthesis of nanomaterials using the fungal system suggests a valuable role in material synthesis. Development of nanoscale approaches using biosynthesized nanomaterials and their use in a wide range of applications has recently fascinated researchers toward bionanotechnology (Rai et al. 2008; Mohanpuria et al. 2008; Prasad et al. 2016). Fungi can accumulate metal ions by physicochemical and biological mechanisms with extracellular binding by polymers and metabolites, binding to particular polypeptides, and metabolism-dependent accumulation (Holan and Volesky 1995). Fungi can reduce metal ions to nano-sized particles by two different mechanisms (Fig. 3.2).

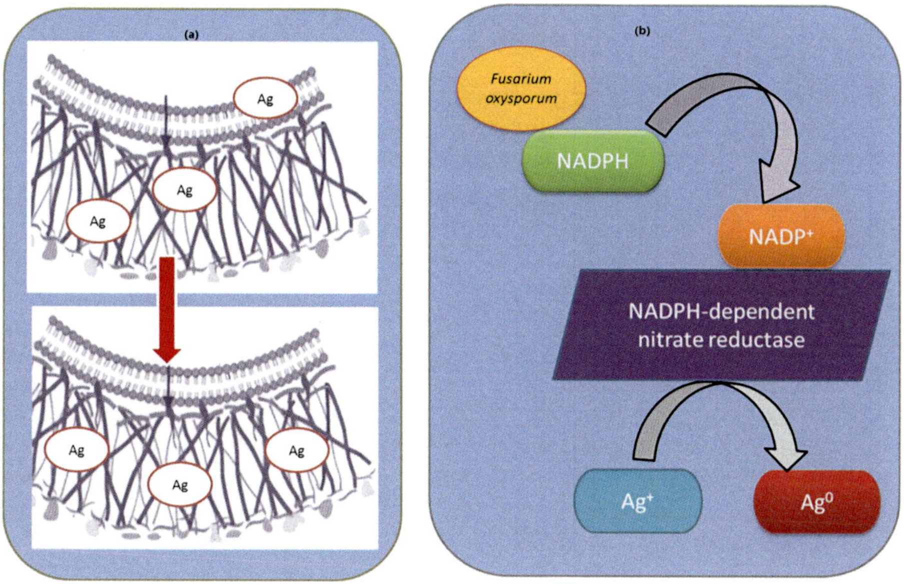

Fig. 3.2 Hypothetical mechanistic basis for myconanoparticle synthesis: (**a**) entrapment of silver ions in the fungal cell wall followed by reduction and formation of nanoparticles and (**b**) nitrate reductase assay

3.7.1 Mechanism I (The Cell Wall of Fungi)

The cell wall of fungi particularly cell wall sugars is likely to play a significant role in the metal ion reduction (Pebberdy 1990). The fungal cell wall is a dynamic structure, which modifies and changes at different stages in the fungal life cycle. It is consisted of a microfibrillar component found to the inner side of the wall and usually inserted in an amorphous matrix material. The main components of the fungal cell wall consist of β-linked glucans and chitin, while the matrix contains generally of polysaccharides that are commonly water soluble.

The fungal cell wall shows a very important role in the absorption of heavy metals. The intracellular synthesis of nanoparticles can be clarified using a stepwise mechanism. In the initial step of bioreduction, trapping of metal ions occurs at the fungal cell surface. This is possibly due to the electrostatic interaction of the positively charged groups in enzymes present on the mycelia cell wall. In the next step, the enzymes within the cell wall probably reduced the metal ions which lead to the aggregation of metal ions and synthesis of nanoparticles. The TEM studies of the fungus show the presence of nanoparticles on the cytoplasmic membrane in addition to cytoplasm. These display the probability of some particles to diffuse through the cell wall and are reduced by the cytoplasmic membrane enzymes and cytoplasm although some of the smaller nanoparticles diffuse through the fungal cell wall and get trapped in the cytoplasm (Sastry et al. 2003).

3.7.2 Mechanism II Nitrate Reductase Assay

The enzyme a-NADPH-dependent nitrate reductase was isolated from *Fusarium oxysporum* and used for the in vitro synthesis of silver nanoparticles. The extracellular synthesis method may possibly involve the NADPH-dependent nitrate reductase enzyme which is excreted by the fungi into the reaction medium. The process of reduction of metal ions to the nanolevel is accompanied simultaneous with conversion of NADPH to $NADP^+$. The UV-vis spectra of the control samples depicted absorption bands at 260–270 nm corresponding to proteins, a-NADPH and hydroxyquinoline. The spectra of reaction mixture displayed strong surface plasmon resonance at 413 nm which intensified with time while the absence of absorption band at 413 nm for the reaction mixture in the absence of enzyme evidently showed that the reduction of silver involves enzymatic reduction of nitrate to nitrite. It is hypothesized that the hydroxyquinoline shuttles the electrons generated during the enzymatic reaction, involving the conversion of nitrate to nitrite, to the Ag^+ ions, converting them to Ag^0. The absence of enzyme from the reaction medium leads to conspicuous disappearance of all the bands, thus validating the active enzymatic role in the whole process.

To discover the mechanism of extracellular synthesis of nanoparticles using fungi whether it is possibly due to reductase action or by electron shuttle quinones or both, the nitrate reductase assay test was conducted through the reaction of nitrite with 2,3-diaminophthalene. The emission spectrum confirmed two major peaks of fluorescence intensity at 405 and 490 nm relating to the emission maximum of nitrite and 2,3-diaminonapthotriazole (DAN), respectively. The intensity of these two bands was increased with the addition of a 0.1% KNO_3 solution, approving the presence of nitrate reductase. Thus, it was determined that the enzyme reductase is responsible for the reduction of Ag^+ ions and then formation of silver nanoparticles. The role of nitrate reductase in the synthesis of nanoparticles was also considered by (Kumar et al. 2007a). It was shown that the nitrate reductase test using commercially available nitrate reductase disks will explain the synthesis of silver nanoparticles. The color of the disk converted to reddish from white when tested with fungal filtrate signifying the presence of nitrate reductase. Thus, it can be determined that the enzyme a-NADH-dependent reductase is associated with the reduction of Ag^+ to Ag^0 in the instance of fungi (Ingle et al. 2008; Prasad et al. 2016).

3.8 Mycosynthesis Nanoparticles Strategies

The synthesis of metal nanoparticles is divided into two strategies: a bottom-up (self-assembly) and a top-down (Marchiol 2012). Bottom-up approach indicates the building of a structure atom-by-atom, molecule-by-molecule, or by self-organization. Top-down approach in which a suitable starting material is reduced in size by physical or chemical methods. Typical production techniques have been established to obtain particles on the nanoscale, e.g., cutting, grinding, and etching (Singh et al. 2011). The production of functional nanometer-sized objects and semiconductor QDs is a good model of bottom-up approach (Saravanan et al. 2008). One of the bottom-up advantages is the better probabilities to achieve nanostructures (nanotubes, nanorods, nanocubes, nanowires, nanosheets, etc.) with slight defects and more homogeneous chemical compositions (Thakkar et al. 2010). This is because the bottom-up approach is obtained mostly by the reduction of Gibbs free energy so that such synthesized nanostructures/nanomaterials are in a case closer to a thermodynamic equilibrium state (Behari 2010).

3.9 The Applications of Myconanoparticles in Agriculture

Nanotechnology has the potential to develop different sectors of the agriculture and allied parts that likely will produce myriads of nanostructured materials (Mandal et al. 2006; García et al. 2010; Gade et al. 2010; Goel 2015; Rai and Ingle 2012, 2015; Prasad 2014). Apart from its major application as antimicrobial agents for the plant protection, nanoparticles can assist as nanofungicides, nanopesticides, nanoinsecticides, and nanofertilizers (Prasad et al. 2014, 2017; Bhattacharyya et al. 2016). Some of the prominent applications in various sectors of agriculture have been described in following sections. Nanomaterials are also useful for the development of nanobiosensors which can be effectively used for sensing a wide variety of fertilizers, herbicide, pesticide, pathogens, moisture, and soil pH (Mousavi and Rezaei 2011; Prasad et al. 2014, 2017). These systems permit the farmer to recognize the best time for planting and harvesting to avoid of bad weather conditions.

3.9.1 Nanofungicides

Myconanotechnology represents a new trend in the development of nanofungicides. There are a large number of fungal species that are common plant pathogens and can be managed by nanomaterials (Ingle and Rai 2011, 2014; Singh et al. 2015; Bhattacharyya et al. 2016, Ismail et al. 2017). It was considered that the antifungal activity of double-capsulized nanosilver (1.5 nm) solution against rose powdery mildew is caused by *Sphaerotheca pannosa* var. *rosae*. A large area infected by rose powdery mildew was sprayed with nanosilver (1.5 nm) solution which was diluted up to 10 ppm. Two days after the spray, more than 95% of rose powdery mildew faded out and did not recur for a week (Kim et al. 2008a). The toxic effect of the silver nanoparticles of 5–24 nm on *Colletotrichum gloesporioides*, which causes anthracnose in a wide range of fruits, such as apple, avocado, mango, and papaya has also been studied. A significant delayed growth of *C. gloesporioides* in the presence of silver nanoparticles was reported. Consequently, nanosilver could be a substitute of fungicide to manage plant diseases (Aguilar-M'endez et al. 2010).

Mycelia free media of *Bipolaris nodulosa*, a phytopathogenic fungus of nodular group, can yield anisotropic silver nanoparticles which display efficient antimicrobial activity against representative organisms of public concern (Saha et al. 2010). These particles, when conjugated with proteins, may be catalyzing chemical reactions in aqueous solutions and be utilized as biolabeling agents. Spherically shaped monodispersed silver nanoparticles, produced from *Amylomyces rouxii* fungal strain KSU-09, indicated activity against *Shigella dysenteriae* type I, *Citrobacter* sp., *Escherichia coli*, *Pseudomonas aeruginosa*, *Staphylococcus aureus*, *Candida albicans*, and *Fusarium oxysporum*.

It was demonstrated that the different functional associated groups (C = O, C-N, N-H, O-H, N = O) make the synthesized zinc oxide nanoparticles as effective antimicrobial agent. Synthesized zinc oxide nanoparticles can be used as effective antifungal agent. These nanoparticles can be added to the food to reduce the food poisoning effect by the various *Aspergillus* sp., which is legally approved.

Particles from an endophytic fungus *Aspergillus clavatus* display spherical and hexagonal shapes with ease of down-steam processing and good antibacterial activity against *Candida albicans*, *Pseudomonas fluorescens*, and *Escherichia coli*. The average minimum inhibitory concentration is reported to be 5.83 μg ml^{-1} with minimum fungicidal concentrations of 9.7 μg ml^{-1} against *C. albicans* (Verma et al. 2010).

3.9.2 *Nanopesticides*

Nanopesticides represent a wide spectrum of chemical molecules that vary in terms of nature (organic/inorganic) and form (particles/micelles). Nanoformulation is the counterprocess of pesticide formulations and usually has similar aims: (a) increase solubility, (b) slow release of active compounds, and (c) increase persistence by protecting active compounds from rapid degradation. Nanopesticide has already been commercialized for quite a long time, since basic definition of nanopesticide will be inclusion of nanoparticles in pesticide to gain novel properties, which will already apply to many of common pesticides.

Metal nanoparticles have proved effectiveness against many plant pests, which will give them advantage to be included in new pesticide formulations (Barik et al. 2008; Goswami et al. 2010). Hollow silica nanoparticles (PHSNs) have shown efficiency in controlling the release of water-soluble pesticide and improve their delivery to targeted locations (Liu et al. 2006). Prolonged release of pesticide represents desired property that can be achieved by usage of PHSNs. Nanoemulsion was used in insecticide formulations and demonstrated higher effectiveness against insect pests (Wang et al. 2007).

Studies on the biocidal effect of nanosilica against insect pests verified their absorption into the cuticular lipids by physiosorption and therefore overcoming a variety of circular lipids that insects used to protect from water evaporation and cause lethal effect for insects (Barik et al. 2008). Hydrophobic nanosilica that ranges in size from 3 to 5 nm and carries surface charge may have strong biocidal effect against insect pests (Ulrichs et al. 2005). Rice weevil (*Sitophilus oryzae*) and grasserie disease (baculovirus *B. mori* nuclear polyhedrosis virus) were effectively controlled by the application of verity of nanoparticles like aluminum oxide, silver nanoparticles, titanium dioxide, and zinc oxide (Goswami et al. 2010).

The effect of nanoparticles on plant pathogenic fungi was particularly evaluated against sclerotia-forming fungi, and silver nanoparticles (4–8 nm) have shown higher antimicrobial effectiveness in comparison to bulk silver that was majorly attributed to the higher surface area and higher fraction of surface atoms of nanoparticles (Min et al. 2009). Separation of layers of hyphal walls in addition to collapse of hypha was shown by microscopic examination when fungal hypha was treated by silver nanoparticles with size ranging 4–8 nm.

Research on application of nanoparticles for crop protection has significantly accumulated over the last few years and encourages more studies to consider the effect of nanopesticide (Kah and Hofmann 2014). As toxicity of nanoparticles is vital issue for their application as antimicrobial compounds, silver has proved greater advantage particularly for their nontoxic effect to human (Elchiguerra et al. 2005; Yeo et al. 2003). Silver nanoparticles have proved induced antimicrobial effect due to their larger surface area-to-volume ratio, which increases contact and penetrability (Kim et al. 2008a, b). As plant pests and humans compete for vegetation resources, humans should safeguard their main food source by developing durable pest control strategies. Nanopesticides represent novel and promising plant protection products that can be useful tool in increasing global food production (Bouwmeester et al. 2009).

Studies have shown effectiveness of silver nanoparticles against oak wilt causal agent (*Raffaelea* sp.) and highlighted the damage caused to fungal hypha by application of silver nanoparticles. Also, silver nanoparticles interfere with microbial biological functions like nutrient absorption, growth, and conidial germination (Woo et al. 2009). Other reports showed silver nanoparticle biocidal characters against *R. solani*, *S. sclerotiorum*, and *S. minor* (Min et al. 2009). Similarly, Jo et al. (2009) reported silver nanoparticles antimicrobial effect against *B. sorokiniana* and *M. grisea* and showed their effect on spore germination assessed as colony formation and effect on disease progress. All such scientific reports show the possible advantage and superiority of novel nanoparticles over traditional pesticides.

3.9.3 Nanofertilizers

Continuous application of chemical fertilizers in traditional agriculture has shown to cause serious consequences on soil fertility and water contamination. Reasonable and targeted application of fertilizers will protect soil and water resources from the negative prolonged effect of chemical fertilizers. Nanofertilizers could serve as novel and effective alternative to conventional fertilization techniques and products, as they minimize unnecessary applications and reduce undesired drawbacks of chemical fertilizers.

Nanoformulation of fertilizers to achieve slow release or induced release through biological and physical activation will improve properties of fertilizers and allow overcoming complications to agriculture and environment. Meanwhile, nanofertilizers will improve nutrition efficiency of plants and will avoid toxicity of excess of chemical fertilizers. Therefore, that will particularly help developing countries to develop sustainable agriculture programs (Naderi and Danesh-Shahraki 2013; Prasad et al. 2017).

New conception was suggested for slow and continuous release of fertilizer by using chitosan nanoparticle in chemical fertilizers formulation (Corradini et al. 2010). That new conception will help in overcoming chemical fertilizers leaching into the environment because of excess application. Chinnamuthu and Boopathi (2009) have reported natural nanoparticles like nanoclay and zeolites can be applied as nanofertilizers. Urea-fertilized zeolite chips were assessed as slow-release nitrogen fertilizer (Millán et al. 2008). Ammonium-zeolite particle-carrying charges gain the potential to increase phosphate solubility and hence phosphorus uptake. Other reports showed that hexadecyltrimethylammonium can be used as carrier surfactant to modify zeolite particles and control nitrate release (Li 2003). Furthermore, coating potash fertilizers with plaster of Paris or wax will slow down the release and minimize leaching of fertilizers in the soil (Subbarao et al. 2013).

Application of nanotechnology in field of crop biotechnology has improved many essential agriculture practices such as herbicide treatments. Nanoparticles can help herbicide to penetrate through plant cuticle and tissue and maintain regular levels of release for extended effectiveness. Such nanocapsules can perform as "magic bullet" to improve delivery of active substances (herbicides, fertilizers, gene, etc.) into precise partitions in plants (Pérez-de-Luque and Rubiales 2009). Mesoporous silica nanoparticles (3 nm) were assessed as delivering carriers of DNA and chemical molecules into plant cells. They achieve improved delivering by comprising the active substances, passing through the plant cell wall, and transporting the target molecules while avoiding toxicity of traditional delivery systems. That delivery system was initially assessed to transport DNA to tobacco and corn plants (Torney et al. 2007).

3.9.4 Nanoparticles in Electrochemical Sensors and Biosensors

Utilization of nanoparticles as electrochemical and biosensors has received great attention recently and showed promising potential in different sensing systems. Different forms of nanoparticles were evaluated such as oxide, metal, and semiconductor nanoparticles. The substantial characteristics of nanoparticles that provide them with advantage as biosensors are their particular features of immobilization of biomolecules and catalyzing electrochemical reactions. Also, they can improve the electron transfer between electrodes and can act as markers for biomolecules while retaining their chemical activity. Gold nanoparticles were commonly used to immobilize proteins (Liu et al. 2003). Gold nanoparticles were bonded to gold electrode modified by monolayer of cysteamine and were used to immobilize horse radish peroxidase (Xiao et al. 1999). DNA molecules were also proved to be able to immobilize by nanoparticles and form DNA sensors. In order to achieve DNA immobilization, DNA was manipulated by adding meticulous function groups that can bind strongly with nanoparticles (Cai et al. 2001).

Acknowledgments This research was supported by the International Foundation for Science, Stockholm, Sweden, through a grant to Dr. Youssef Khamis (E5757) and Dr. Hashim Ayat (F5853).

References

Agnihotri M, Joshi S, Kumar AR, Zinjarde S, Kulkarni S (2009) Biosynthesis of gold nanoparticles by the tropical marine yeast *Yarrowia lipolytica* NCIM 3589. Mater Lett 63:1231–1234

Aguilar-Méndez AM, Martín-Martinez ES, Ortega-Arroyo L, Cobián-Portillo G, Saonchez-Espıondola E (2010) Synthesis and characterization of silver nanoparticles: effect on phytopathogen *Colletotrichum gloesporioides*. J Nanopart Res 13:2525–2532

Ahmad A, Mukherjee P, Mandal D, Senapati S, Khan MI, Kumar R et al (2002) Enzyme mediated extracellular synthesis of CdS nanoparticles by the fungus, *Fusarium oxysporum*. J Am Chem Soc 124:12108–12109

Ahmad A, Senapati S, Khan MI, Kumar R, Sastry M (2005) Extra-/intracellular biosynthesis of gold nanoparticles by an alkalotolerant fungus Trichothecium sp. J Biomed Nanotechnol 1:47–53

Balaji DS, Basavaraja S, Deshpandeb R, Mahesh DB, Prabhakara BK, Venkataraman A (2009) Extracellular biosynthesis of functionalized silver nanoparticles by strains of *Cladosporium cladosporioides* fungus. Colloids Surf B: Biointerfaces 68:88–92

Banu A, Rathod V, Ranganath E (2011) Silver nanoparticle production by *Rhizopus stolonifera* and its antibacterial activity against extended spectrum b-lactamase producing (ESBL) strains of *Enterobacteriaceae*. Mater Res Bull 46:1417–1423

Barik TK, Sahu B, Swain V (2008) Nanosilica-from medicine to pest control. Parasitol Res 103:253–258

Baskar G, Chandhuru J, Fahad KS, Praveen AS (2013) Mycological synthesis, characterization and antifungal activity of zinc oxide nanoparticles. Asian J Pharm Technol 3:142–146

Behari J (2010) Principles of nanoscience: an overview. Indian J Exp Biotechnol 48:1008–1019

Bhainsa KC, D'Souza SF (2006) Extracellular biosynthesis of silver nanoparticles using the fungus *Aspergillus fumigatus*. Colloids Surf B: Biointerfaces 47:160–164

Bhambure R, Bule M, Shaligram N, Kamat M, Singhal R (2009) Extracellular biosynthesis of gold nanoparticles using *Aspergillus niger*-its characterization and stability. Chem Eng Technol 32:1036–1041

Bhattacharyya A, Duraisamy P, Govindarajan M, Buhroo AA, Prasad R (2016) Nano-biofungicides: emerging trend in insect pest control. In: Prasad R (ed) Advances and applications through fungal nanobiotechnology. Springer International Publishing, Switzerland, pp 307–319

Binupriya AR, Kumar SM, Vijayaraghavan K, Yuna SI (2010) Bioreduction of trivalent aurum to nano-crystalline gold particles by active and inactive cells and cell free extract of *Aspergillus oryzae* var. *viridis*. J Hazard Mater 177:539–545

Bouwmeester H, Dekkers S, Noordam MY, Hagens WI, Bulder AS, Heer C, Voorde SECG et al (2009) Review of health safety aspects of nanotechnologies in food production. Regul Toxicol Pharmacol 53:52–62
Cai H, Xu C, He P, Fang Y (2001) Colloid Au-enhanced DNA immobilization for the electrochemical detection of sequence-specific DNA. J Electroanal Chem 510:78–85
Chauhan A, Zubair S, Tufail S, Sherwani A, Sajid M, Raman SC et al (2011) Fungus-mediated biological synthesis of gold nanoparticles: potential in detection of liver cancer. Int J Nanomedicine 6:2305–2319
Chen JC, Lin ZH, Ma XX (2003) Evidence of the production of silver nanoparticles via pretreatment of *Phoma* sp 32883 with silver nitrate. Lett Appl Microbiol 37:105–108
Chinnamuthu CR, Boopathi PM (2009) Nanotechnology and agroecosystem. Madras Agric J 96:17–31
Corradini E, Moura MRD, Mattoso LHC (2010) A preliminary study of the incorporation of NPK fertilizer into chitosan nanoparticles. Express Polym Lett 4:509–515
Duran N, Marcato PD, Alves OL, de Souza GIH, Esposito E (2005) Mechanistic aspects of biosynthesis of silver nanoparticles by several *Fusarium oxysporum* strains. J Nanobiotechnol 3:1–7
Elchiguerra JL, Burt JL, Morones JR, Camacho- Bragado A, Gao X, Lara HH, Yacaman MJ (2005) Interaction of silver nanoparticles with HIV-1. J Nanobiotechnol 3:1–10
Faramarzi MA, Forootanfar H (2011) Biosynthesis and characterization of gold nanoparticles produced by laccase from *Paraconiothyrium variabile*. Colloids Surf B: Biointerfaces 87:23–27
Fayaz AM, Balaji K, Girilal M, Yadav R, Kalaichelvan PT, Venketesan R (2011) Silver nanoparticles biosynthesis by *Trichoderma viridi* strains. Nanomedicine 6:103–109
Gade A, Ingle A, Whiteley C, Rai M (2010) Mycogenic metal nanoparticles: progress and applications. Biotechnol Lett 32:593–600
García M, Forbe T, Gonzalez E (2010) Potential applications of nanotechnology in the agro-food sector. Ciênc Tecnol Aliment Campinas 30:573–581
Gericke M, Pinches A (2006) Microbial production of gold nanoparticles. Gold Bull 39:22–28
Goel A (2015) Agricultural applications of nanotechnology. J Biol Chem Res 32:260–266
Goswami A, Roy I, Sengupta S, Debnath N (2010) Novel applications of solid and liquid formulations of nanoparticles against insect pests and pathogens. Thin Solid Films 519:1252–1257
Holan ZR, Volesky B (1995) Accumulation of cadmium, lead and nickel by fungal and wood biosorbents. Appl Biochem Biotechnol 53:133–146
Ingle AP, Rai MK (2011) Genetic diversity among Indian phytopathogenic isolates of *Fusarium semitectum* Berkeley and Ravenel. Adv Biosci Biotechnol 2:142–148
Ingle A, Gade A, Pierrat S, Sonnichsen C, Rai MK (2008) Mycosynthesis of silver nanoparticles using the fungus *Fusarium acuminatum* and its activity against some human pathogenic bacteria. Curr Nanosci 4:141–144
Ingle A, Duran N, Rai M (2014) Bioactivity, mechanism of action and cytotoxicity of copper-based nanoparticles: a review. Appl Microbiol Biotechnol 98:1001–1009
Ismail M, Prasad R, Ibrahim AIM, Ahmed ISA (2017) Modern prospects of nanotechnology in plant pathology. In: Prasad R, Kumar M, Kumar V (eds) Nanotechnology. Springer Nature Singapore Pte Ltd, Singapore, pp 305–317
Jo YK, Kim BH, Jung G (2009) Antifungal activity of silver ions and nanoparticles on phytopathogenic fungi. Plant Dis 93:1037–1043
Kah M, Hofmann T (2014) Nanopesticide research: current trends and future priorities. Environ Int 63:224–235
Kathiresan K, Manivannan S, Nabeel MA, Dhivya B (2009) Studies on silver nanoparticles synthesized by a marine fungus, *Penicillium fellutanum* isolated from coastal mangrove sediment. Colloids Surf B: Biointerfaces 71:133–137
Kim H, Kang H, Chu G, Byun G (2008a) Antifungal effectiveness of nanosilver colloid against rose powdery mildew in greenhouses. Solid State Phenom 135:15–18
Kim KJ, Sung WS, Moon SK, Choi JS, Kim JG, Lee DG (2008b) Antifungal effect of silver nanoparticles on dermatophytes. J Microbiol Biotechnol 18:1482–1484

Krumov N, Oder S, Perner - Nochta IP, Angelov A, Posten C (2007) Accumulation of CdS nanoparticles by yeasts in a fed-batch bioprocess. J Biotechnol 132:481–486
Kumar SA, Abyaneh MK, Gosavi SW, Kulkarni SK, Pasricha R, Ahmad A, Khan MI (2007a) Nitrate reductase-mediated synthesis of silver nanoparticles from $AgNO_3$. Biotechnol Lett 29:439–445
Kumar AS, Ansary AA, Ahmad A, Khan MI (2007b) Extracellular biosynthesis of CdSe quantum dots by the fungus *Fusarium oxysporum*. J Biomed Nanotechnol 3:190–194
Li Z (2003) Use of surfactant-modified zeolite as fertilizer carriers to control nitrate release. Microporous Mesoporous Mater 61:181–188
Liu S, Leech D, Ju H (2003) Application of colloidal gold in protein immobilization, electron transfer, and biosensing. Anal Lett 36:1–19
Liu F, Wen LX, Li ZZ, Yu W, Sun HY, Chen JF (2006) Porous hollow silica nanoparticles as controlled delivery system for water soluble pesticide. Mater Res Bull 41:2268–2275
Longoria EC, Nestorb ARV, Borja MA (2011) Biosynthesis of silver, gold and bimetallic nanoparticles using the filamentous fungus *Neurospora crassa*. Colloids Surf B: Biointerfaces 83:42–48
Mala G, Rose C (2014) Facile production of ZnS quantum dot nanoparticles by Saccharomyces cerevisiae MTCC 2918. J Biotechnol 170:73–78
Mandal D, Bolander ME, Mukhopadhyay D, Sarkar G, Mukherjee P (2006) The use of microorganisms for the formation of metal nanoparticles and their application. Appl Environ Biotechnol 69:485–492
Marchiol L (2012) Synthesis of metal nanoparticles in living plants. Ital J Agron 7:e37
Millán G, Agosto F, Vázquez M (2008) Use of clinoptilolite as a carrier for nitrogen fertilizers in soils of the Pampean regions of Argentina. Cien Inv Agr 35:293–302
Min JS, Kim KS, Kim SW, Jung JH, Lamsal K, Kim SB et al (2009) Effects of colloidal silver nanoparticles on Sclerotium-forming phytopathogenic fungi. Plant Pathol J 25:376–380
Mishra AN, Bhadauria S, Gaur MS, Pasricha R (2010) Extracellular microbial synthesis of gold nanoparticles using fungus *Hormoconis resinae*. JOM 62:45–48
Mishra A, Tripathy SK, Wahab R, Jeong S-H, Hwang I, Yang Y et al (2011) Microbial synthesis of gold nanoparticles using the fungus *Penicillium brevicompactum* and their cytotoxic effects against mouse mayo blast cancer C_2C_{12} cells. Appl Microbiol Biotechnol 92:617–630
Mohanpuria P, Rana NK, Yadav SK (2008) Biosynthesis of nanoparticles: technological concepts and future applications. J Nanopart Res 7:9275–9280
Mousavi SR, Rezaei M (2011) Nanotechnology in agriculture and food production. J Appl Environ Biol Sci 1:414–419
Mukherjee P, Ahmad A, Mandal D, Senapati S, Sainkar SR, Khan MI et al (2001) Bioreduction of $AuCl^{-4}$ ions by the fungus, *Verticillium* sp. and surface trapping of the gold nanoparticle formed. Angew Chem Int Ed 40:3585–3588
Mukherjee P, Roy M, Mandal BP, Dey GK, Mukherjee PK, Ghatak J et al (2008) Green synthesis of highly stabilized nanocrystalline silver particles by a nonpathogenic and agriculturally important fungus *T. asperellum*. Nanotechnology 19:075103–075104
Naderi MR, Danesh-Shahraki A (2013) Nanofertilizers and their roles in sustainable agriculture. Int J Agric Crop Sci 5:2229–2232
Pebberdy JF (1990) Fungal cell walls- a review. In: Kuhn PJ, Trinci APJ, Jung MJ, Goosey MW, Coping LG (eds) Biochemistry of cell walls and membranes in fungi. Springer, Germany, pp 5–30
Pérez-de-Luque A, Rubiales D (2009) Nanotechnology for parasitic plant control. Pest Manag Sci 65:540–545
Philip D (2009) Biosynthesis of Au, Ag and Au-Ag nanoparticles using edible mushroom extract. Spectrochim Acta Part A Mol Biomol Spectrosc 73:374–381
Prasad R (2014) Synthesis of silver nanoparticles in photosynthetic plants. J Nanoparticle Article ID 963961, https://doi.org/10.1155/2014/963961
Prasad R, Kumar V, Prasad KS (2014) Nanotechnology in sustainable agriculture: present concerns and future aspects. Afr J Biotechnol 13(6):705–713

Prasad R, Pandey R, Barman I (2016) Engineering tailored nanoparticles with microbes: quo vadis. WIREs Nanomedicine Nanobiotechnol 8:316–330. https://doi.org/10.1002/wnan.1363

Prasad R, Bhattacharyya A, Nguyen QD (2017) Nanotechnology in sustainable agriculture: recent developments, challenges, and perspectives. Front Microbiol 8:1014. https://doi.org/10.3389/fmicb.2017.01014

Rai M, Ingle A (2012) Role of nanotechnology in agriculture with special reference to management of insect-pests. Appl Microbiol Biotechnol 94:287–293

Rai M, Yadav A, Gade A (2008) Current trends in phytosynthesis of metal nanoparticles. Crit Rev Biotechnol 28:277–284

Rai M, Ingle AP, Yadav A, Birla S, Santos CAD (2015) Strategic role of selected noble metal nanoparticles in medicine. Crit Rev Microbiol. https://doi.org/10.3109/1040841X.2015.1018131

Rajan A, Cherian E, Baskar G (2016) Biosynthesis of zinc oxide nanoparticles using Aspergillus fumigatus JCF and its antibacterial activity. Int J Mod Sci Technol 1:52–57

Ramanathan VR, Kalishwaralal K, Gopalram S, Gurunathan S (2009) Nanosilver- the burgeoning therapeutic molecule and its green synthesis. Biotechnol Adv 27:924–937

Roy W, Mukherjee T, Chakraborty S, Das TK (2013) Biosynthesis, characterization & antifungal activity of silver nanoparticles synthesized by the fungus *aspergillus foetidus* mtcc8876s. Dig J Nanomater Biostruct 8:197–205

Saha S, Sarkar J, Chattopadhyay D, Patra S, Chakraborty A, Acharya K (2010) Production of silver nanoparticles by a phytopathogenic fungus bipolaris nodulosa and its antimicrobial activity. Dig J Nanomater Biostruct 5:887–895

Sanghi R, Verma P (2009) A facile green extracellular biosynthesis of CdS nanoparticles by immobilized fungus. Chem Eng J 155:886–891

Saravanan P, Gopalan R, Chandrasekaran V (2008) Synthesis and characterization of nanomaterials. Def Sci J 58:504–516

Sarkar J, Ray S, Chattopadhyay D, Laskar A, Acharya K (2011) Mycogenesis of gold nanoparticles using a phytopathogens *Alternaria alternata*. Bioprocess Biosyst Eng. https://doi.org/10.1007/s00449- 011-0646-4

Sastry M, Ahmad A, Khan MI, Kumar R (2003) Biosynthesis of metal nanoparticles using fungi and actinomycetes. Curr Sci 85:162–170

Shankar SS, Ahmad A, Pasricha R, Sastry M (2003) Bioreduction of chloroaurate ions by Geranium leaves and its endophytic fungus yields gold nanoparticles of different shapes. J Mater Chem 13:1822–1826

Singh M, Manikandan S, Kumaraguru AK (2011) Nanoparticles: a new technology with wide applications. Res J Nanosci Nanotechnol 11:1–11

Singh R, Shedbalkar UU, Wadhwani SA, Chopade BA (2015) Bacteriagenic silver nanoparticles: synthesis, mechanism, and applications. Appl Microbiol Biotechnol 99:4579–4593

Subbarao CV, Kartheek G, Sirisha D (2013) Slow release of potash fertilizer through polymer coating. Int J Appl Sci Eng 11:25–30

Syed A, Saraswati S, Kundu GC, Ahmad A (2013) Biological synthesis of silver nanoparticles using the fungus Humicola sp. and evaluation of their cytoxicity using normal and cancer cell lines. Spectrochim Acta A Mol Biomol Spectrosc 114:144–147. https://doi.org/10.1016/j.saa.2013.05.030

Thakkar KN, Mhatre SS, Parikh RY (2010) Biological synthesis of metallic nanoparticles. Nanomed Nanotechnol Biol Med 6:257–262

Torney F, Trewyn BG, Lin VS, Wang K (2007) Mesoporous silica nanoparticles deliver DNA and chemicals into plants. Nat Nanotechnol 2:295–300

Ulrichs C, Mewis I, Goswami A (2005) Crop diversification aiming nutritional security in West Bengal: biotechnology of stinging capsules in nature's water-blooms. Ann Tech Issue State Agric Technol Serv Assoc:1–18

Vahabi K, Mansoori GA, Karimi S (2011) Biosynthesis of silver nanoparticles by fungus *Trichoderma Reesei*. Insci J 1:65–79

Vardhana J, Kathiravan G (2015) Biosynthesis of silver nanoparticles by endophytic fungi *Pestaloptiopsis pauciseta* isolated from the leaves of *Psidium guajava* Linn. Int J Pharm Sci Rev Res 31:29–31

Varshney R, Mishra AN, Bhadauria S, Gaur MS (2009) A novel microbial route to synthesize silver nanoparticles using fungus *Hormoconis resinae*. Dig J Nanomater Biostruct 4:349–355

Verma VC, Kharwar RN, Gange AC (2010) Biosynthesis of antimicrobial silver nanoparticles by the endophytic fungus *Aspergillus clavatus*. Nanomedicine 5:33–40

Verma VC, Singh SK, Solanki R, Prakash S (2011) Biofabrication of anisotropic gold nanotriangles using extract of endophytic *Aspergillus clavatus* as a dual functional reductant and stabilizer. Nanoscale Res Lett 6:16

Vigneshwaran N, Kathe AA, Varadarajan PV, Nachane RP, Balasubramanya RH (2006) Biomimetics of silver nanoparticles by white rot fungus *Phaenerochaete chrysosporium*. Colloids Surf B: Biointerfaces 53:55–59

Wang L, Li Z, Zhang G, Dong J, Eastoe J (2007) Oil-in-water nanoemulsions for pesticide formulations. J Colloid Interfaces Sci 314:230–235

Woo KS, Kim KS, Lamsal K, Kim YJ, Kim SB, Jung M et al (2009) An in vitro study of the antifungal effect of silver nanoparticles on oak wilt pathogen *Raffaelea* sp. J Microbiol Biotechnol 19:760–764

Xiao Y, Ju H, Chen H (1999) Hydrogen peroxide sensor based on horseradish peroxidase-labeled Au colloids immobilized on gold electrode surface by cysteamine monolayer. Anal Chim Acta 391:73–82

Xiaorong Z, Xiaoxiao H, Kemin W, Yonghong W, Huimin L, Weihong T (2009) Biosynthesis of size-controlled gold nanoparticles using fungus, *Penicillium* sp. J Nanosci Nanotechnol 9:5738–5744

Yeo SY, Lee HJ, Jeong SH (2003) Preparation of nanocomposite fibers for permanent antibacterial effect. J Mater Sci 38:2143–2147

Chapter 4
Myconanotechnology in Agriculture

Sairengpuii Hnamte, Busi Siddhardha, and V. Venkateswara Sarma

Abstract Nanotechnology is a fast-growing field of science that involves synthesis and development of various nanomaterials, production, manipulation and use of materials ranging in size from less than a micron to that of individual atoms. Formation of nanoparticles employing fungi and their application in medicine, agriculture and other areas is known as myconanotechnology. Fungal nanoparticles could be used in various fields including agriculture, industry and medicine. In the present chapter, the status of research carried out on fungal nanoparticles in the area of agriculture is consolidated and presented. Myconanotechnology has emerged as one of the key eco-friendly technologies, and its use in management of bacterial and fungal diseases, pest control, preserved foods and beverages is constantly being explored. Thus, myconanotechnology provides a greener alternative to chemically synthesized nanoparticles. Mycosynthesized nanoparticles found their vast application in pathogen detection and control, wound healing, food preservation, textile fabrics and many more. The present chapter provides an appraisal on the application of myconanotechnology in agriculture and looks into the future prospects.

4.1 Introduction

There is a rapid growth in the field of nanotechnology which involves the synthesis and production of various nanomaterials and nanostructures. The synthesis of nanoparticles with fungi is known as myconanotechnology. Fungi produce large amount of enzymes which can be used for the synthesis of nanoparticles. There are three methods in the synthesis of metal nanoparticles such as physical, chemical and biological methods. Among these, biological synthesis is favoured than chemical and physical synthesis due to their fast synthesis and ability to control shape and

S. Hnamte • B. Siddhardha
Department of Microbiology, School of Life Sciences, Pondicherry University, Kalapet, Puducherry 605014, India

V. Venkateswara Sarma (✉)
Department of Biotechnology, School of Life Sciences, Pondicherry University, Kalapet, Puducherry 605014, India
e-mail: sarmavv@yahoo.com

R. Prasad (ed.), *Fungal Nanotechnology*, Fungal Biology,
DOI 10.1007/978-3-319-68424-6_4

size. Therefore, they are of cost-effective and environment friendly in nature (Saxena et al. 2014). Different fungi have been investigated for the synthesis of nanoparticles such as gold, silver, selenium, platinum, zinc oxide, titanium nanoparticles, etc. by various researchers. Fungi have numerous advantages from other organisms in the synthesis of nanoparticles. They are easy to isolate and handle and are capable of secreting extracellular enzymes (Mandal et al. 2006; Mohanpuria et al. 2007; Alghuthaymi et al. 2015) and have the ability to withstand flow pressure than bacteria and plants (Narayanan and Sakthivel 2010). In addition, the process of synthesis has a greener approach as it is nontoxic and occurs at low cost. Fungi-based synthesis of nanoparticles has received much attention to researchers due to their extensive advantages in different fields. Fungal nanoparticles can be used in various fields like agriculture, engineering, pharmaceuticals, environment, textiles, medicine, food industry, etc. (Dasgupta et al. 2015; Nanda and Majeed 2014; Yadav et al. 2015; Prasad et al. 2016). Thus myconanotechnology provides a greener alternative to chemically synthesized nanoparticles. The present chapter provides the application of myconanotechnology in agriculture and its future perspectives.

4.2 Synthesis of Fungal Nanoparticles

Fungal nanoparticles can be synthesized both intracellularly and extracellularly (Fig. 4.1). There have been several reports on the intracellular and extracellular synthesis of nanoparticles using fungi. The work done by researchers in the synthesis of fungal nanoparticles is listed in (Table 4.1). Nanoparticles are fabricated inside the cell of the fungus in intracellular where the biomass of the fungus is reacted with a metal, whereas in extracellular synthesis, the filtrate of the fungus reacts with the solution of a metal (Yadav et al. 2015). Electrostatic interactions occur during intracellular synthesis where ions of the metal bind upon the fungal cell. The ions of the metal are reduced by the enzymes present in the cell wall, and then formation of nanoparticles occurs due to aggregation of the metal ions. During extracellular synthesis, the fungus, when exposed to the metal ions, leads to release of reductase enzymes and formation of nanoparticles, which are highly stable (Kashyap et al. 2013). A rapid extracellular and intracellular biosynthesis of gold nanoparticles using the fungus *Penicillium* sp. was reported by Du et al. (2011). Intracellular synthesis of gold nanoparticles was obtained when $AuCl_4^-$ ions reacted with the cell filtrate of the fungus in 1 min, whereas extracellular synthesis occurred when solution of $AuCl_4^-$ incubated with fungal biomass for 8 h. There are two different methods for the preparation of extracellular biosynthesis, i.e. rapid synthesis and slow synthesis, whereas intracellular biosynthesis is a time-limiting factor which depends on in vivo synthesis of cells (Moghaddam 2010). Due to an additional step required to obtain the purified nanoparticles, the extracellular synthesis method is more favourable than intracellular method (Kuber et al. 2006).

The extracellular and intracellular syntheses of fungal nanoparticles reported by various researchers are described below.

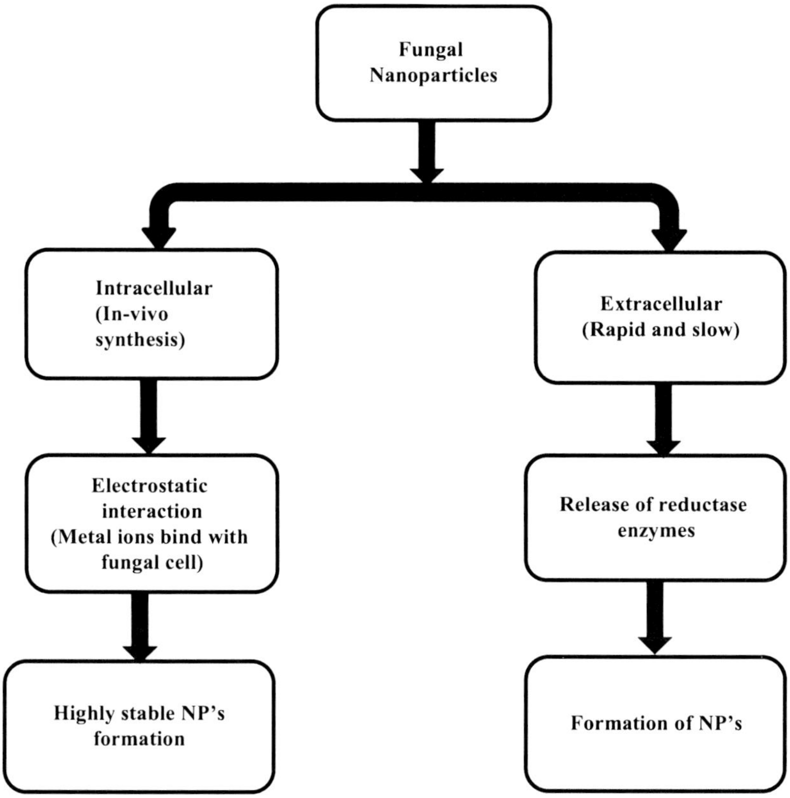

Fig. 4.1 Mycosynthesis of nanoparticles

Fatima et al. (2016) reported the mycosynthesis of silver nanoparticles using *Aspergillus flavus*. The synthesized nanoparticle was found to be spherical in shape with 50 nm in size which showed antimicrobial effect against pathogenic fungi and bacteria. Further, it is also reported to be a microbicidal agent in the field of agriculture. In another study conducted by Ingle et al. (2009), *Fusarium solani* was reported to be a new biological agent in the extracellular synthesis of silver nanoparticles. Fourier transform infrared spectroscopy (FTIR) revealed the silver nanoparticle to be highly stable due to the presence of capping agent. Extracellular synthesis of gold nanoparticles using *Fusarium oxysporum* f. sp. *cubense* JTI and its antimicrobial activity against *Pseudomonas* sp. was reported by Thakker et al. (2013). The mycelium reacted with auric chloride and changed the yellow colour to purple colour within 60 min indicating the formation of gold nanoparticles with particle size of 22 nm. The extracellular synthesis of gold nanoparticles using *Helminthosporium tetramera* was studied by Shelar and Chavan (2014). The synthesized gold nanoparticle was found to be polydispersed spherical with size range of 8–50 nm. This study would be appropriate for establishing a process for large-scale manufacturing of scanty AuNPs. Zinc oxide nanoparticles against two

Table 4.1 List of mycosynthesis of nanoparticles

Fungal species	Nanoparticle	Synthesis	Size (nm)	Reference
Aspergillus flavus (KF934407)	Silver	Extracellular	50	Fatima et al. (2016)
Fusarium solani	Silver	Extracellular	5–35	Ingle et al. (2009)
Fusarium oxysporum f. sp. *cubense* JT1	Gold	Extracellular	22	Thakker et al. (2013)
Helminthosporium tetramera	Gold	Extracellular	8–50	Shelar and Chavan (2014)
Botrytis cinerea and *Penicillium expansum*	Zinc oxide	Extracellular	70	He et al. (2011)
Aspergillus terreus	Zinc oxide	Extracellular	54.8–82.6	Baskar et al. (2013)
Coriolus versicolor	Silver	Intracellular and extracellular	10	Sanghi and Verma (2009a, b)
Saccharomyces cerevisiae	CdS	Extracellular	2.5–5.5	Prasad and Jha (2010)
Coriolus versicolor	CdS	Extracellular	10	Sanghi and Verma (2009a, b)
Aspergillus flavus	Titanium dioxide	Extracellular	62–74	Rajakumara et al. (2012)
Alternaria alternata	Selenium	Extracellular	13–15	Sarkar et al. (2011)
Fusarium oxysporum	Platinum	Extracellular	5–30	Syed and Ahmad (2012)
Mucor hiemalis	Silver		5–15	Aziz et al. (2016)

pathogenic fungi, i.e. *Botrytis cinerea* and *Penicillium expansum*, and its antifungal activity were investigated by He et al. (2011). The results suggest that zinc oxide nanoparticles could be used in agriculture as a productive fungicide and applications in food safety. Baskar and his co-workers also studied a greener synthesis of zinc oxide nanoparticles against *Aspergillus terreus*. The synthesized crystalline zinc nanoparticles were characterized by UV absorption spectrum, X-ray diffraction spectrum and Fourier transform infrared spectroscopy and were found to be spherical in shape, and scanning electron microscope revealed the size range from 54.8 to 82.6 nm. The synthesized zinc oxide nanoparticle was found to be a potent antifungal agent against fungal species (Baskar et al. 2013).

An intracellular synthesis and extracellular synthesis of silver nanoparticles using the fungus *C. versicolor* was demonstrated by Sanghi and Verma (2009a) reported for the first time. The fungus when treated with silver nitrate solution agglomerated the silver nanoparticles on its surface in 72 h. The reaction was quick and could easily continue at room temperature even without stirring under alkaline conditions. The resulting AgNPs displayed controllable structural and optical properties depending on the experimental parameters such as pH and reaction temperatures. The average size, morphology and structure of particles were characterized by

TEM, AFM, XRD and UV-vis spectrophotometry. The FTIR study imparts that the particles were bound by the amino groups which was responsible for the nanoparticle stability. Another study of intracellular and extracellular synthesis of gold nanoparticles using an alkalotolerant fungus *Trichothecium* sp. was investigated by Ahmad et al. (2005). Gold ions when reacted under stationary conditions, with the fungal biomass, produced extracellular synthesis, while the biomass reaction with agitating conditions resulted in intracellular growth of the nanoparticles. The synthesized gold nanoparticles were found to be spherical and triangular in morphology. They demonstrated that altering the conditions of the reactions of the fungal biomass and gold ions resulted in the intracellular and extracellular synthesis, where under stationary conditions, the enzymes and proteins are released into the medium but are not released under shaking conditions.

A rapid and eco-friendly method of biosynthesis of CdS nanoparticle using *S. cerevisiae* was reported by Prasad and Jha (2010). Another extracellular synthesis of CdS nanoparticle was studied by Sanghi and Verma (2009b) using an immobilized fungus *Coriolus versicolor*. The first report on fungus-based biosynthesis of titanium dioxide nanoparticle was investigated by Rajakumara et al. (2012). The synthesized nanoparticle was found to have good antibacterial property. Scanning electron microscopy disclosed the particle to be oval and spherical shapes with size being in the range of 62–74 nm. Sarkar et al. (2011) reported for the first time in synthesizing selenium nanoparticles with *Alternaria alternata*. The synthesized nanoparticles were characterized by atomic force microscopy, dynamic light scattering and transmission electron microscopy and revealed the nanoparticle size to be in the range of 30–15 nm. An extracellular synthesis of platinum nanoparticles using *Fusarium oxysporum* was studied by Syed and Ahmad (2012). *Fusarium oxysporum* reacted with hexachloroplatinic acid resulted in the formation of selenium nanoparticle with size in the range of 5–30 nm which are highly stable. Because of their high stability, ability not to flocculate and having a good monodispersity, they may find applications in various fields.

4.3 Factors That Affect Nanoparticles Mycosynthesis

There are several factors that affect the synthesis of fungal nanoparticles such as temperature, biomass, concentration and time in exposure of substrate, pH and the presence of particular enzyme (Fig. 4.2). These are known to be the major factors that affect the shape and size of nanoparticles (Kashyap et al. 2013). Khan et al. (2016) studied the optimization of various parameters such as pH, quantity of fungal biomass, temperature and concentration of silver nitrate in the synthesis of silver nanoparticles from *Aspergillus niger*. They concluded that optimizing the above parameters will enhance the silver nanoparticles synthesis as well as its yield. Optimizing the cultural and physical conditions in the synthesis of silver nanoparticles from *Fusarium oxysporum* has also been studied by Birla et al. (2013).

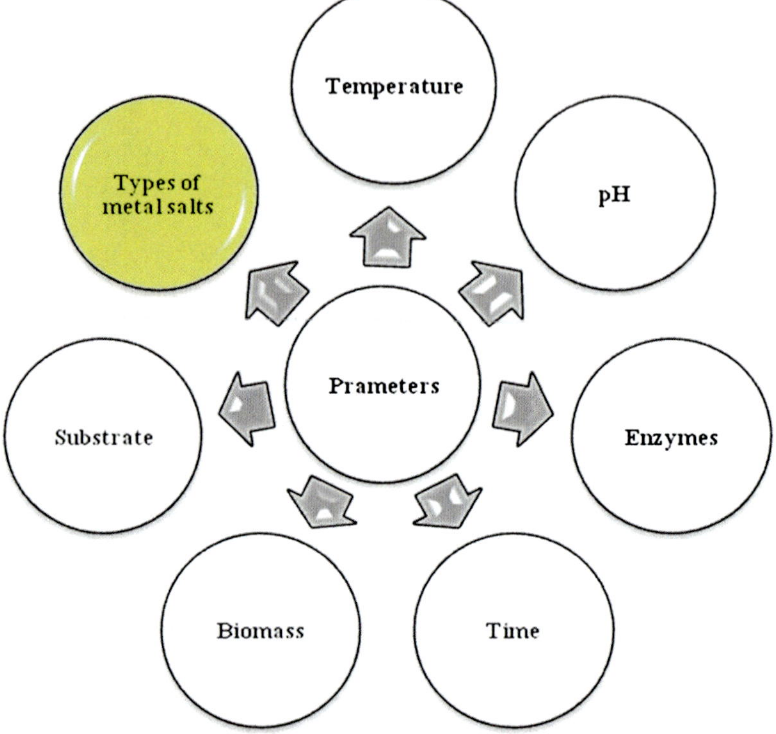

Fig. 4.2 Parameters effecting mycosynthesis of nanoparticles

One of the most important factors affecting the mycosynthesis of nanoparticles is pH. It greatly influences the nature and size of the nanoparticles synthesized (Gardea-Torresdey et al. 1999; Armendariz et al. 2004). The synthesis of gold nanoparticles with *Aspergillus terreus* and its antibacterial property against *Escherichia coli* was studied by Priyadarshini et al. (2014). The nanoparticle was found to be 10–19 nm at pH 10 based on UV-vis spectroscopy, XRD, TEM, EDX and FTIR results which revealed the characteristic property of the synthesized nanoparticle. Temperature plays an important role in regulating the activity of the fungus and the movement of the ions (Dhillon et al. 2012). The synthesis of nanoparticles with a greener approach requires temperatures less than 100 °C or ambient temperature. The effect of temperature in the synthesis of fungal based nanoparticles was studied by Fayaz et al. (2009). They found that increase in temperature of the reaction results in decrease of the nanoparticle size but increase in monodispersity. Incubation time is also another important factor affecting the synthesis of fungal nanoparticles. The time period in which the reaction medium incubates greatly enhances the type of nanoparticle synthesized and the quality (Darroudi et al. 2011). The incubation time might occur in different ways such as the particles may aggregate because they are stored for a longer period of time; therefore, the potential is affected (Baer 2011).

4.4 Strategies for Mycosynthesis of Nanoparticles

There are two widely used approaches or strategies for the synthesis of nanoparticles. It can be synthesized using the top-down approach and bottom-up approach. In top-down method, physical methods, like diffusion, thermal decomposition, irradiation, etc., are employed. Nanoparticles are fabricated by using biological approach like biological entities and chemical approach like chemical reduction, seeded growth method, electrochemical synthesis and polyol synthesis method in bottom-up approach (Tikariha et al. 2012). In top-down approach, etching and machining techniques are employed in top-down approach where the large materials are ruptured down slowly to nanosized materials (Lengke et al. 2011). The synthesis of fungal nanoparticles is a bottom-up approach. The main reaction of mycosynthesis of nanoparticles resembles bottom-up approach because it includes substrate oxidation or reduction resulting to colloidal structures. The possibilities to a greater degree for acquiring nanostructures such as nanotubes, nanorods, nanowires, nanocubes, nanosheets, etc. with better uniformity in chemical composition in bottom-up approach make more advantageous than other approaches. This is due to this approach, which is mainly steered by reduction of Gibbs free energy that brings those synthesized nanomaterials or nanostructures to a state closer to thermodynamic equilibrium state (Kashyap et al. 2013).

4.5 Applications of Mycosynthesis of Nanoparticles in Agriculture

Nanotechnology is a fast-developing industry which has an impact in the field of agriculture. Fungus-based synthesis of nanoparticles has attracted the researchers due to their enormous applications in various fields. Some of the applications in various fields of agriculture are illustrated below.

4.5.1 Nanofungicides

The common plant pathogens are fungi compared to viruses and bacteria. These plant pathogens such as species of *Aspergillus*, *Fusarium* and *Phytophthora* can be used as a nanomaterial for the synthesis of nanoparticles (Yadav et al. 2015). Kim et al. (2012) reported a silver nanoparticle of nanosize which can be used as an effective antifungal agent in the treatment of different plant pathogens. In vitro assay was performed on petri dish. They used 18 plant pathogenic fungi for treating the silver nanoparticles on malt extract agar, potato dextrose agar and cornmeal agar plates. The fungal inhibition was calculated to evaluate the antifungal activity of silver nanoparticles against the pathogens. The results revealed that silver nanoparticles possess antifungal properties against these plant pathogens at different levels. In vitro as well as field conditions trial was conducted by Lamsal et al. (2011). They demonstrated the effects of silver nanoparticles using *Colletotrichum* species and pepper anthracnose disease. Silver nanoparticle solution at different concentrations, viz. 10, 30, 50 and 100 ppm, was used. In vitro assay was performed. The maximum inhibition rate was at 100 ppm silver nanoparticles solution with a percentage of 93.50%, while the lowest inhibition was found to be 11.33% at 10 ppm. In the case of field trial analysis, an experiment was conducted before and after the pepper was infected. Two positive controls, i.e. commercial fungicide NSS-F and chemical fungicide Feneri, were used. The leaves of the plants were treated with silver nanoparticles 3–4 weeks before and after the outbreak of the disease. After the disease outbreak treatment, results were analysed 1 week after the final treatment. Before the disease outbreak treatment, results were obtained 4 weeks after final treatment. Each experiment was performed in triplicates. Moreover, after the disease outbreak treatments, disease incidence was higher compared to before the disease outbreak treatments with NSS-F 72.1% and Fenari 63.4%. The lowest disease incidence was noticed on plants treated with 50 ppm silver nanoparticles before the disease outbreak with 9.7%, whereas the highest disease incidence was observed on plants treated with NSS-F after the disease outbreak with 72.1%. The results showed that before the outbreak of the disease, silver nanoparticle treatment was applied which suppressed the pathogen attack. Park et al. (2006) also studied the efficiency of nanosized silica-silver in controlling plant pathogenic microbes. Silica-silver nanoparticles were prepared, and antifungal activity was performed against *Rhizoctonia solani*, *Botrytis cinerea*, *Magnaporthe grisea*, *Colletotrichum gloeosporioides* and *Pythium ultimum*. Antifungal effect was performed on powdery mildew in the field. The results suggest that since silver and silica are nontoxic and safe for human health, the cost is much lower than the commercial fungicide. This nanoformulation is highly useful for managing different fungal plant diseases (Bhattacharyya et al. 2016).

4.5.2 Nanopesticides

Plant diseases have reduced agriculture production. Various methods have been employed in combating the different diseases of plants such as natural or artificial methods. Excessive use of pesticides can cause environmental hazards. Therefore, scientists have been investigating for the replacement of chemical-based pesticides (Kim et al. 2012). Due to its durability, and high efficacy, nanopesticides represent the next-generation pesticides (Bhattacharyya et al. 2016). Nanopesticide can be prepared in two ways: organic ingredients which are polymers and inorganic, i.e. metal oxides (Yadav et al. 2015). Bramhanwade et al. (2016) studied a stable copper nanoparticle and their fungicidal activity against three crop pathogenic fungi. Cetyltrimethylammonium bromide and copper nitrate were used to synthesize stable copper nanoparticles at room temperature. The antifungal effect was determined against three crop pathogenic *Fusarium* spp., i.e. *F. oxysporum*, *F. culmorum* and *F. equiseti*, and found to exhibit significant antifungal activity. Therefore, copper nanoparticles can be used as antifungal agents.

4.5.3 Nanofertilizers

Nanofertilizers are those vital nutrients for the growth of the plants for improvement in their production and supporting their growth (Liu and Lal 2015; Prasad et al. 2014, 2017). Excessive consumption and continuous use of chemical fertilizers decrease the fertility of soil and in the crop production. Therefore, nanofertilizers can replace in regaining and protecting the fertility of soil. The use of nanofertilizers leads to an increase in nutrient efficiencies and reduces toxicity of the soil (Prasad et al. 2014, 2017). An eco-friendly and low-cost method was reported by Tarafdar et al. (2014). Zinc nanoparticle was employed as a nanofertilizer in pearl millet *Pennisetum americanum* L. for enhancing crop production. Zinc oxide solution reacted with the fungus *R. bataticola* for 62 h resulted in the extracellular synthesis of high monodispersed zinc nanoparticles with an average size of 18.5 nm as confirmed by transmission electron microscopy. In order to determine the effect of the synthesized zinc nanoparticle as a nanofertilizer, seeds of pearl millet were sown at 3 cm depth in the field. The field experiment was conducted with three treatments such as control, i.e. without any treatment, nanosize and normal size zinc oxide. After 2 weeks of germination, foliage was sprayed with the normal size zinc oxide and nanozinc. Results were observed after 4 weeks of spray in which significant improvement was observed in shoot length, root area and root length. In addition, the chlorophyll content, dry biomass plant, total soluble leaf protein, dehydrogenase, and enzyme activities of acid phosphatase were also estimated in 6-week-old plants. In another study conducted by Pandey et al. (2010), zinc oxide nanoparticle served as an efficient nanofertilizer for zinc, which is an essential nutrient for the growth of the plants. Singhal et al. (2017) 'reported that nano-embedded fungus' formed by impact of synergistic association of ZnO-nanorods and fungus *Piriformospora indica* DSM 11827, for growth of *Brassica oleracea* var. botrytis (Broccoli). Hydrothermal method was employed for the synthesis of zinc oxide nanoparticles and was characterized by powder X-ray diffraction, and field emission electron microscopy provided the size range with a diameter of 20–30 cm and is of spherical shape. Zinc oxide nanoparticle was used during the root growth and seed germination of *Cicer arietinum* which resulted in an increased level of Indole acetic acid in the roots, therefore bringing out an increase in the rate of plant's growth.

4.6 Conclusion and Future Perspectives

Mycosynthesis of nanoparticles has attracted a wide attention among researchers due to their desirable characteristics and nontoxicity. The chapter illustrates the various reports exhibited by the researchers on the extracellular and intracellular synthesis of fungal nanoparticles. The different parameters that affect nanoparticles mycosynthesis have been discussed. Moreover, the applications in the field of agriculture such as nanofungicide, nanopesticide and nanofertilizer are also exemplified. An extensive research of myconanotechnology in the field of agriculture will improve in the growth of the plants and crop protection. In-depth studies are essential on the mechanism of mycosynthesis of nanoparticles at the molecular and cellular level.

References

Ahmad A, Senapati S, Khan MI, Kumar R, Sastry M (2005) Extra-/Intracellular biosynthesis of gold nanoparticles by an Alkalotolerant fungus, *Trichothecium* sp. J Biomed Nanotechnol 1(7):47–53

Alghuthaymi MA, Almoammar H, Rai M, Said-Galiev E, Abd-Elsalame KA (2015) Myconanoparticles: synthesis and their role in phytopathogens management. Biotechnol Biotechnol Equip 29(2):221–236

Armendariz V, Herrera I, Peralta-Videa JR (2004) Size controlled gold nanoparticle formation by *Avena sativa* biomass: use of plants in nanobiotechnology. J Nanopart Res 6(4):377–382

Aziz N, Pandey R, Barman I, Prasad R (2016) Leveraging the attributes of *Mucor hiemalis*-derived silver nanoparticles for a synergistic broad-spectrum antimicrobial platform. Front Microbiol 7:1984. https://doi.org/10.3389/fmicb.2016.01984

Baer DR (2011) Surface characterization of nanoparticles: critical needs and significant challenges. J Surf Anal 17(3):163–169

Baskar G, Chandhuru J, Fahad KS, Praveen AS (2013) Mycological synthesis, characterization and antifungal activity of zinc oxide nanoparticles. Asian J Pharm 3(4):142–146

Bhattacharyya A, Duraisamy P, Govindarajan M, Buhroo AA, Prasad R (2016) Nano-biofungicides: emerging trend in insect pest control. In: Prasad R (ed) Advances and applications through fungal nanobiotechnology. Springer International Publishing, Cham, pp 307–319

Birla SS, Gaikwad SC, Gade AK, Rai MK (2013) Rapid Synthesis of Silver Nanoparticles from by Optimizing Physicocultural Conditions. The Scientific World Journal 2013:1–12

Bramhanwade K, Shende S, Bonde S, Gade A, Rai M (2016) Fungicidal activity of Cu nanoparticles against *Fusarium* causing crop diseases. Environ Chem Lett 14(2):229–235

Darroudi M, Ahmad MB, Zamiri R, Zak AK, Abdullah AH, Ibrahim NA (2011) Time-dependent effect in green synthesis of silver nanoparticles. Int J Nanomedicine 6(1):677–681

Dasgupta N, Ranjan S, Mundekkad D (2015) Nanotechnology in agro-food: from field to plate. Food Res Int 69:381–400

Dhillon GS, Brar SK, Kaur S, Verma M (2012) Green approach for nanoparticle biosynthesis by fungi: current trends and applications. Crit Rev Biotechnol 32:49–73

Du L, Xian L, Feng J-X (2011) Rapid extra-/intracellular biosynthesis of gold nanoparticles by the fungus Penicillium sp. J Nanopart Res 13(3):921–930

Fatima F, Verma SR, Pathak N, Bajpai P (2016) Extracellular mycosynthesis of silver nanoparticles and their microbicidal activity. J Glob Antimicrob Resist 7:88–92

Fayaz MA, Balaji K, Kalaichelvan PT, Venkatesan R (2009) Fungal based synthesis of silver nanoparticles-an effect of temperature on the size of particles. Colloids Surf B: Biointerfaces 74(1):123–126

Gardea-Torresdey JL, Tiemann KJ, Gamez G, Dokken K, Pingitore NE (1999) Recovery of gold (III) by alfalfa biomass and binding characterization using X-ray microfluorescence. Adv Environ 3(1):83–93

He L, Liu Y, Mustapha A, Lin M (2011) Antifungal activity of zinc oxide nanoparticles against *Botrytis cinerea* and *Penicillium expansum*. Microbiol Res 166:207–215

Ingle A, Rai M, Gade A, Bawaskar M (2009) *Fusarium solani*: a novel biological agent for the extracellular synthesis of silver nanoparticles. J Nanopart Res 11:2079–2085

Kashyap PL, Kumar S, Srivastava AK, Sharma AK (2013) Myconanotechnology in agriculture: a perspective. World J Microbiol Biotechnol 29(2):191–207

Khan NT, Jameel N, Rehman SUA (2016) Optimizing Physioculture conditions for the synthesis of silver nanoparticles from *Aspergillus niger*. J Nanomed Nanotechnol 7:5

Kim SW, Jung JH, Lamsal K, Kim YS, Min JS, Lee YS (2012) Antifungal effects of silver nanoparticles (AgNPs) against various plant pathogenic fungi. Mycobiology 40(1):53–58

Kuber C, Bhainsa SF, Souza D (2006) Extracellular biosynthesis of silver nanoparticles using the fungus *Aspergillus fumigatus*. Colloids Surf B: Biointerfaces 47:160–164

Lamsal K, Kim SW, Jung JH, Kim YS, Kim KS, Lee YS (2011) Application of silver nanoparticles for the control of *Colletotrichum* species *in Vitro* and pepper anthracnose disease in field. Mycobiology 39(3):194–199

Lengke MF, Sanpawanitchakit C, Southam G (2011) Biosynthesis of gold nanoparticles: a review. In: Rai MK, Duran N (eds) Metal nanoparticles in microbiology. Springer, New York, pp 37–74

Liu R, Lal R (2015) Potentials of engineered nanoparticles as fertilizers for increasing agronomic productions. Sci Total Environ 514:131–139

Mandal D, Bolander ME, Mukhopadhyay D, Sarkar G, Mukherjee P (2006) The use of microorganisms for the formation of metal nanoparticles and their application. Appl Microbiol Biotechnol 69:485–492

Moghaddam KM (2010) An introduction to microbial metal nanoparticle preparation method. J Young Investig 19:1–7

Mohanpuria P, Rana NK, Yadav SK (2007) Biosynthesis of nanoparticles, technological concepts and future applications. J Nanopart Res 7:9275–9280

Nanda A, Majeed S (2014) Enhanced antibacterial efficacy of biosynthesized AgNPs from *Penicillium glabrum* (MTCC1985) pooled with different drugs. Int J Pharm Tech Res 6:217–223

Narayanan KB, Sakthivel N (2010) Biological synthesis of metal nanoparticles by microbes. Adv Colloid Interface Sci 156:1–13

Pandey AC, Sanjay SS, Yadav RS (2010) Application of ZnO nanoparticles in influencing the growth rate of *Cicer arietinum*. J Exp Nanosci 5(6):488–497

Park HT, Kim SH, Kim HJ, Choi SH (2006) A new composition of nanosized silica-silver for control of various plant diseases. Plant Pathol J 22(3):295–302

Prasad K, Jha AK (2010) Biosynthesis of CdS nanoparticles: an improved green and rapid procedure. J Colloid Interface Sci 342(1):68–72

Prasad R, Bhattacharyya A, Nguyen QD (2017) Nanotechnology in sustainable agriculture: recent developments, challenges, and perspectives. Front Microbiol 8:1014. https://doi.org/10.3389/fmicb.2017.01014

Prasad R, Kumar V, Prasad KS (2014) Nanotechnology in sustainable agriculture: present concerns and future aspects. Afr J Biotechnol 13(6):705–713

Prasad R, Pandey R, Barman I (2016) Engineering tailored nanoparticles with microbes: quo vadis. WIREs Nanomedicine Nanobiotechnol 8:316–330. https://doi.org/10.1002/wnan.1363

Priyadarshini E, Pradhan N, Sukla LB, Panda PK (2014) Controlled synthesis of gold nanoparticles using *Aspergillus terreus* IF0 and its antibacterial potential against Gram negative pathogenic bacteria. J Nanotechnol 2014:1–9. http://dx.doi.org/10.1155/2014/653198.

Rajakumara G, Rahumana AA, Roopanb SM, Khannac VG, Elangoa G, Kamaraja C, Zahira AA, Velayuthama K (2012) Fungus-mediated biosynthesis and characterization of TiO_2 nanoparticles and their activity against pathogenic bacteria. Spectrochim Acta A Mol Biomol Spectrosc 91:23–29

Sanghi R, Verma P (2009a) Biomimetic synthesis and characterisation of protein capped silver nanoparticles. Bioresour Technol 100:501–504

Sanghi R, Verma PA (2009b) A facile green extracellular biosynthesis of CdS nanoparticles by immobilized fungus. Chem Eng J 155:886–891

Sarkar J, Dey P, Saha S, Acharya K (2011) Mycosynthesis of selenium nanoparticles. IET Micro Nano Lett 6(8):599–602

Saxena J, Sharma MM, Gupta S, Singh A (2014) Emerging role of fungi in nanoparticle synthesis and their applications. World J Pharm Pharm Sci 3(9):1586–1613

Shelar GB, Chavan AM (2014) Extracellular biological synthesis, characterization and stability of gold nanoparticles using the fungus *Helminthosporium tetramera*. Int J Pure App Biosci 2(3):281–285

Singhal U, Khanuja M, Prasad R, Varma A (2017) Impact of synergistic association of ZnO-nanorods and symbiotic fungus *Piriformospora indica* DSM 11827 on *Brassica oleracea* var. botrytis (Broccoli). Front Microbiol 8:1909. doi:10.3389/fmicb.2017.01909

Syed A, Ahmad A (2012) Extracellular biosynthesis of platinum nanoparticles using the fungus *F. oxysporum*. Colloids Surf B: Biointerfaces 97:27–31

Tarafdar JC, Raliya R, Mahawar H, Rathore I (2014) Development of zinc nanofertilizer to enhance crop production in pearl millet (*Pennisetum americanum*). Agric Res 3(3):257–262

Thakker JN, Dalwadi P, Dhandhukia PC (2013) Biosynthesis of gold nanoparticles using *Fusarium oxysporum* f. sp. *cubense* JT1, a plant pathogenic fungus. ISRN Biotechnol 2013:5

Tikariha S, Singh SS, Banerjee S, Vidyarthi AS (2012) Biosynthesis of gold nanoparticles, scope and application: a review. Int J Pharm Sci Res 3(6):1603–1615

Yadav A, Kon K, Kratosova G, Duran N, Ingle AP, Rai M (2015) Fungi as an efficient mycosystem for the synthesis of metal nanoparticles: progress and key aspects of research. Biotechnol Lett 37:2099–2120

Chapter 5
Fungus-Mediated Bioleaching of Metallic Nanoparticles from Agro-industrial By-Products

Madan L. Verma

Abstract Nanotechnology is a multidisciplinary area that involves the synthesis of advanced material at the nanoscale level. Nowadays applications of the nanomaterial are widely accepted and being observed in all the areas ranging from health to environment sectors. Hence there is a surge of nanotechnology for production of nanomaterial at the large scale without compromising the eco-hazardous concerns of the chemical synthesis routes of nanoparticles. The chapter describes the advances and applications of the fungi in bioleaching of waste materials ranging from fly ash to agro-industrial by-products for the production of metallic nanomaterial with reference to silica nanoparticle.

5.1 Introduction

Nanotechnology involves the manipulation of the matter with at least one dimension in the range of 1–100 nm that has various applications in different sectors ranging from electronics, optics, sensing, catalysis, drug delivery, bioseparation, etc. (Verma 2017; Verma et al. 2012, 2013a, b, c, d, 2016; Abraham et al. 2014; Puri et al. 2013; Prasad 2014; Prasad et al. 2014, 2016, 2017a, b). The application of nanomaterials is vast; however, to tap the maximum potential, efficient methods of nanofabrication are required (Paull et al. 2003). Generally, nanomaterial is synthesized via three routes: physical, chemical, and biological (Fig. 5.1). Among the different routes of nanofabrication, the physical route employs the top-down approach that involves the synthesis of nanomaterial by size reduction of bulk material in very

M.L. Verma (✉)
Center for Chemistry and Biotechnology, Deakin University,
Melbourne, Victoria 3216, Australia
e-mail: madanverma@gmail.com

R. Prasad (ed.), *Fungal Nanotechnology*, Fungal Biology,
DOI 10.1007/978-3-319-68424-6_5

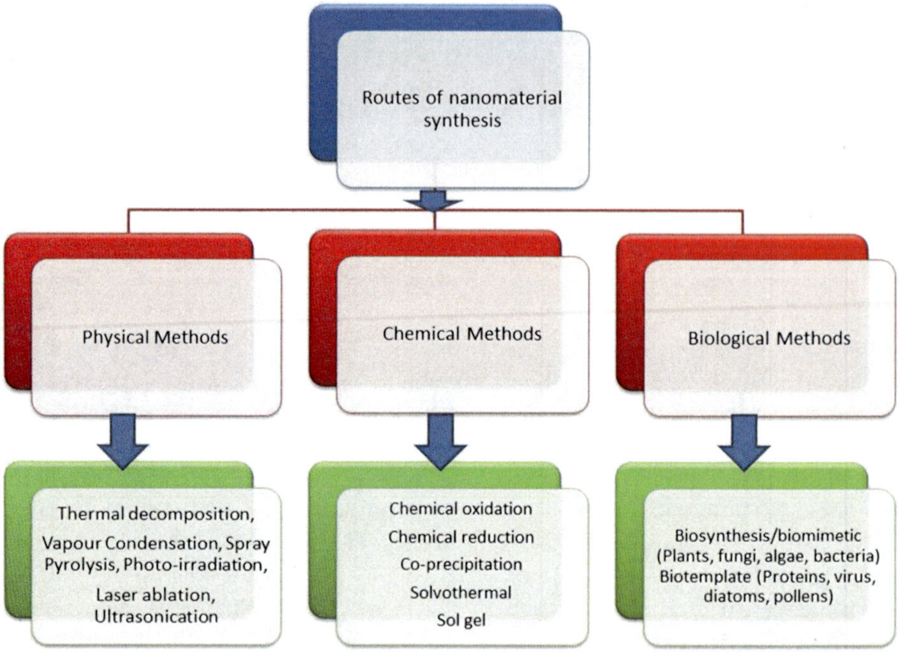

Fig. 5.1 Overview of methods employed for nanoparticle synthesis

specific physical engineering tools such as thermal decomposition, vapor condensation, spray pyrolysis, and photo-irradiation (Sakamoto et al. 2009; Burda et al. 2005; Swihart 2003; Sun et al. 2000), while the chemical method employs bottom-up approach for the synthesis of the nanomaterials. The wet chemical method was superior than physical method as size and shape are excellent manipulated via chemical routes. The chemical method or solution-based method involves oxidation/reduction of metal ions that resulted in assembly of atomic/molecular nanocrystals with the aid of capping agents, solvents, and template (Verma et al. 2016; Bansal et al. 2012). However, the negative concern to the environment is associated with the synthetic chemical methods due to the toxic chemicals and extreme reaction conditions of temperature, pH, and pressure (Verma 2017; Bansal et al. 2012; Thakkar et al. 2010). To overcome disadvantage of physicochemical methods, toxic chemical-free methods, typically known as "green" environmentally benign methods, have become more promising.

Massive generation of wastes from the industries and household activities causes a serious concern. Waste treatment methods such as pyrometallurgical and hydrometallurgical have been employed to overcome the menace caused to the environment (Park and Fray 2009; Lee et al. 2007). However, these methods can pose environment risks due to toxic reagent and excessive by-products. The synthetic route is comparatively not cost-effective in addition to eco-hazardous output. Thus, waste management through eco-friendly methods is most urgently needed. In this direction, bioleaching has provided a novel solution to extract wealth from waste using microorganisms.

Bioleaching is considered a green technology employed for extracting value-added metals from diverse waste materials (Pathak et al. 2017; Priya and Hait 2017; Wu and Ting 2006). Bioleaching of waste material rich in metal elements can be done by a variety of microorganisms. Although the different microbes used for the synthesis of nanoparticles, yet fungi are efficient candidates for biogenic fabrication of metal nanoparticles both intra- and extracellular (Adil et al. 2015; Shelar and Chavan 2014; Roy et al. 2013; Dhillon et al. 2012; Li et al. 2012; Pavani et al. 2012; Vahabi et al. 2011; Gade et al. 2010; Thakkar et al. 2010; Verma et al. 2010; Rai et al. 2009; Mohammadian et al. 2007; Sastry et al. 2003; Mulligan and Kamali 2003; Aziz et al. 2016). Fungi show the promising candidates among all the living organisms for syntheses of nanoparticles due to the unique properties such as high wall binding capacity, ease of culturing, and biomass separation (Vágó et al. 2016; Siddiqi and Husen 2016; Yadav et al. 2015; Vala 2015; Kitching et al. 2015; Maliszewska et al. 2014; Singh et al. 2014; Pantidos and Horsfall 2014; Durán et al. 2011; Jain et al. 2011; Durán et al. 2005; Mukherjee et al. 2001; Prasad et al. 2016).

Bioleaching is extraction of the metals by interaction of microorganisms with waste material enriched with metals elements (Khan et al. 2014). Solubilization of metals and biotransformation of solid waste yield economic recovery of value-added elements (Brombacher et al. 1997). It involves three steps to complete the bioleaching process: (a) transformation of inorganic/organic acids (proton), (b) oxido-reduction reaction, and (c) extraction of complex agents (Khan et al. 2014). Microorganisms endow with high metal tolerance that has been employed to synthesize metallic nanomaterials from the waste materials (Fig. 5.2; Bajaj et al. 2012; Dhanjal and Cameotra 2010; Bhattacharya and Gupta 2005).

In this article, we have discussed the role of fungi as a bioleaching carrier for the conversion of waste materials to the value-added silica nanoparticles (Table 5.1). Silica is an important inorganic material that has a plethora of applications in different fields such as enzyme technology, biocatalyst support, molecular sieves, resins, and biomedical carrier (Verma et al. 2012; Mousavi and Rezaei 2011; Barik et al. 2008; Hubert et al. 2000). In contrary to harsh environmental conditions of chemical synthesis of silica, various biological and biomimetic routes were employed for synthesis of crystalline as well as amorphous silica (Patwardhan 2011; Farrell et al. 2006; Sumper and Kroger 2004; Perry and Keeling-Tucker 2000, 2003; Pohnert 2002; Morse 1999; Mann and Ozin 1996; Mann 1995). Three cases of waste materials including rice husk, zircon sand, and fly ash have been discussed with reference to a special fungus *Fusarium oxysporum*, a plant mesophilic fungus. The same fungus can be easily grown on MGYP (malt extract, glucose, yeast extract, and peptone) media at ambient reaction conditions.

Fig. 5.2 Advantages of fungal mediated bioleaching of waste materials

Table 5.1 List of fungus employed for leaching nanoparticles from different natural/synthetic materials

Fungus name	Bioleaching materials	Product	References
Fusarium oxysporum	Sand	Silica nanoparticles	Bansal et al. (2005)
F. oxysporum	Rice husk	Nanocrystalline silica	Bansal et al. (2006)
F. oxysporum	Fly ash	Silica nanoparticles	Khan et al. (2014)
Humicola lanuginosa	Glass/modified glass	Silica nanoparticles	Kulkarni et al. (2008)
F. oxysporum	Zirconia sand	Silica nanoparticles	Bansal et al. (2007)

5.2 Fungus-Mediated Bioleaching of Fly Ash for Producing Extracellular Silica Nanoparticles

Fly ash is the product of coal burning or other materials. The waste material is produced in massive amount on daily basis in thermal power plant and deposited in the landfills/dump site that causes leaching out to the underground water body (Mizutani et al. 1996). It causes a number of environment hazards in particular to air pollution and soil pollution.

Fly ash is categorized into two types: high-rank fly ash (rich in silicon and aluminum) and low-rank fly ash (rich in calcium, magnesium, iron). However low-rank fly ash is not employed as additive to cement. High-grade fly ash has application in the cement industry (Sklivaniti et al. 2017; Behfarnia and Rostami 2017; Palla et al. 2017; Hemalatha and Ramaswamy 2017). Fly ash is composed of silica-rich inorganic minerals (Khan et al. 2014) such as quartz (SiO_2), mullite ($Al_6Si_2O_{13}$), sillimanite (Al_2SiO_5), and glassy compounds (silica glass and other oxides). Such waste material harbors a rich source of metals such as Al and Si and excellent material for synthesis of aluminosilicate compounds that have enhanced the speed of the reaction by acting as catalysts such as zeolite. Thus it may be considered as artificial ore (Bosshard et al. 1996).

Commercial utilization of the fly ash by extracting value-added elements (silica and aluminum) has provided technological advantage by utilization of waste material through chemical treatments such as organic acid, inorganic alkali, ultrasonication, or in combination (Gao et al. 2017). Such process are high efficient to separate elements by providing high purity (99%) and high yield (51%) of SiO_2. Gao et al. (2017) demonstrated application of extracted SiO_2 toward adsorption of cationic dyes (methylene blue and malachite green).

Khan et al. (2014) demonstrated the methodology for production of water-soluble silica nanoparticles from fly ash waste materials using mesophilic fungus *Fusarium oxysporum* at ambient reaction conditions. Silica nanoparticle was functionalized with bioleached fungal protein that exhibited crystalline and fluorescent properties. At short period of 1 day incubation, fungus leached out quasi-spherical silica nanoparticles from fly ash. Analytical techniques such as spectroscopy (UV-Vis, Fourier transform infrared), microscopy (transmission electron), X-ray (X-ray diffraction, energy-dispersive analysis of X-rays), and photoluminescence supported the bioleaching of unique silica nanoparticles.

Khan et al. (2014) employed isolated fungus *Fusarium oxysporum* from plant material for bioleaching the fly ash waste material. Bioprocessing for optimal fungal biomass was done using MGYP (malt, glucose, yeast, peptone) media. Subculturing was done using fresh slants from active growing stock culture. Three-day-old culture was employed for fermentation study. Fungus was grown in MYGP media with continuous high shaking at ambient temperature for 3 days. After completion of fermentation, fungal mycelial mass was separated by centrifugation and washed with excessive amount of distilled water three times. The fungal biomass and waste material were taken aseptically at the ratio of 6:1 for silica nanoparticles production and kept in the dark under optimized reaction conditions (pH 6.5, 25 ° C, 200 rpm).

Sophisticated instruments were employed for the characterization of the bioleached nanosilica. UV-Vis spectroscopy was employed for characterization of fungus-mediated fly ash bioleaching reaction at the resolution of 1 nm. The fly ash was unreacted and did not show any absorption peak in the entire range of 240–800 nm. However, fungus-mediated bioleaching of fly ash showed two peaks: prominent absorption (275 nm) and absorption edge (350 nm). Absorption peak of wavelength 275 nm was assigned due to aromatic amino acids such as tryptophan, tyrosine, and phenylalanine (Eftink and Ghiron 1981). Aromatic amino acids were presented in the protein secreted by fungus during bioleaching reaction. Such proteins/enzymes were secreted by fungus during stress conditions of element-rich waste materials of fly ash. The selected fungus reported for high affinity for silica elements (Bansal et al. 2007).

Bansal et al. (2006) showed the formation of initial bioleached product in the form of enzyme-silicic acid complex due to interaction of fungal secreted protein/enzyme with silicic acid component of fly ash. Initial bioleached complex was hydrolyzed by the fungal hydrolytic enzyme. The isoelectric point (pI:2) of silica is of very low that envisaged the mechanism of bioleaching extraction that involves cationic nature of fungal hydrolytic proteins.

Khan et al. (2014) performed the fly ash treatment at acidic reaction conditions (pH: 4) without fungus to explore the possibility of production of silica nanoparticles. The reaction product/filtrate was characterized by analytical techniques. There was no peak reported in UV-Vis and XRD. Photoluminescence study was done using fungal treated fly ash as well as untreated fly ash powder. The untreated fly ash powder did not show any emission bands at selected excited bands. However, fungal treated fly ash produced photoluminescence phenomenon at the interface of silica nanocrystal in SiO_2 (Pavesi et al. 2000). This unique property of bioleached silica nanoparticle can be applied in different fields such as diagnostics, biosensors, drug delivery (Khan et al. 2014).

The size and shape of the bioleached silica nanoparticle were done by transmission electron microscopy (TEM). TEM and HR-TEM study was done at 200 kV and 300 kV. In both the measurements, bioleached nanoparticles were prepared by drop coating the particles suspended in aqueous media on carbon-coated copper grids (Khan et al. 2014). Nanoparticles were polydispersed and quasi-spherical in shape and water soluble due to capping of biomolecules secreted by fungal. The particle size distribution histogram of nanoparticles was in the range of 20–26 nm with an average diameter of 22 nm. X-ray diffraction pattern of bioleached nanoparticle was aligned with crystalline polymorph of silica, and even the protein capping of the nanoparticles did not change the crystallinity as revealed from Bragg reflections. The XRD pattern and SAED spots were indexed with reference to the crystal structure from the PCPDF chart: silica (PCPDF card no. 040–1498).

Fourier transform infrared spectroscopy (FTIR) study of bioleached nanoparticles was done using KBR pellet. Fly ash powder showed different vibrational peaks of individual components. The characteristic peak corresponding to individual components were reported as follows: Si-O-Al stretching vibrations correspond to 600 wave number; 1098 wave number assigned to Si-O-Si asymmetric corresponding vibrations; and 1608 wave number corresponds to H-O-H bending vibrations (Frances et al. 2008; Innocenzi et al. 2003; Palomo et al. 1999). There was no amide region in the fly ash control sample. However, fungal treated fly ash showed all peaks of controlled fly ash in addition to two amide peaks of proteins (1641 and 1540 wave number corresponds to amide I and amide II peak) and thus confirmed silica nanoparticles leached out by fungus in the extracellular solutions.

The energy-dispersive analysis of X-rays for fly ash powder showed the presence of different components such as Si, Ca, Mg, Fe, Al, etc. Fungal mediated bioleaching of fly ash showed two prominent peaks corresponding to Si (1.74 KeV) and O (0.51 KeV), respectively. Higher amount of silica in bioleached sample confirmed higher purity of silica nanoparticle (72%). The results of XRD and FTIR studies confirmed the bioleaching of silica nanoparticles.

Khan et al. (2014) demonstrated the role of mesophilic fungus in the bioleaching of waste material that produced extracellular surface-functionalized and water-soluble metallic nanoparticles at ambient reaction conditions.

5.3 Fungus-Mediated Bioleaching of Sand for Producing Extracellular Silica Nanoparticles

Bansal et al. (2007) demonstrated bioleaching of silica present in zircon by using mesophilic fungus at ambient temperature. This approach has twofold benefits: firstly it has bioleached silicon impurities in the form of silica nanoparticle; secondly it has concentrated the zirconia in zircon sand. Bioleaching process by fungus was rapid and completed in 1 day. Bansal et al. (2007) developed the methodology for bioleaching of silica nanoparticles. The fungal biomass (20 g wet weight) was mixed aseptically with 10 g zircon sand in the ratio of 2:1 that suspended in distilled water under constant fast shaking. The fungal sand reaction was incubated for 1 day at 27 °C.

For the extracellular bioleaching of silica from zircon sand by *F. oxysporum*, the bioleached nanoproduct was pooled from the aqueous component by removing the fungal mycelia through filtration. The aqueous product of bioleached sample was made powder by low-pressure evaporation. The nanopowder was treated further by high-temperature calcination. The bioleached nanoproduct after the treatment of *Fusarium oxysporum*-zircon sand reaction mix at the pH of 3.5 for 1 day was analyzed by TEM. The nanoparticles depicted an overall quasi-spherical morphology and nanoparticle size range from 2 to 10 nm, with an average size of 5.5 nm (Bansal et al. 2007). The crystalline nature of nanoparticles was confirmed by SAED analysis, and the diffraction pattern could be indexed on the basis of the cristobalite polymorph of silica. The XRD and SAED patterns were indexed with reference to the crystal structures from the PCPDF charts: silica (PCPDF card nos. 03-0272,

32-0993, 45-0112, and 45-0131). FTIR analysis of nanoparticles from the fungus-zircon sand reaction medium showed two absorption amide I and II bands that confirmed the role of fungal secreted proteins in synthesis of quasi-spherical silica particles during reaction with zircon sand.

The XRD spectrum of nanosilica showed well-defined Bragg reflection characteristics of cristobalite polymorph of crystalline silica. In addition to well-defined Bragg reflections, the FTIR signatures corresponding to silica-entrapped proteins showed entrapped proteins in the silica particles did not interfere with crystallinity. Zircon sand was incubated at an acidic pH to explore the possibility of silica leaching out due to the acidic nature of the reaction medium, and then the filtrate was found absence of characteristic Si-O-Si vibrational modes of silica as well as signatures (Si-OH vibrational modes) from silicic acid (Innocenzi et al. 2003; Silverstein 1967).

In this sand-derived nanoparticle, authors demonstrated the use of fungus *Fusarium oxysporum* for selective bioleaching of silica present in zircon sand. The crystalline silica of the size ca. 5 nm was functionalized by stabilizing fungal proteins that secreted into solution by the fungus. Additionally, bioleaching of silica from zircon sand also enhanced significantly the zirconium component in zircon sand at ambient reaction conditions.

5.4 Fungus-Mediated Bioleaching of Agro-industrial By-Products for Production of Silica Nanoparticles

Rice husk is the by-product of rice (*Oryza sativa*) which is the principle cereal crops in India, China, Sri Lanka, Africa, and Latin America that fulfill the nutritional needs of a large mass of population (~70%) of these countries (Saha 2016). At world level, a rice crop is grown on 161 million hectares, with an annual production of about 679 million tons. Asia contributes a major, i.e., ~90% portion of rice production (FAOSTAT 2012; FAO 1995). In comparison to the production and consumption of rice in global scenario, India produced more than 100 million metric ton in the last 5 years, in particular to 2011–2012, and recent trends of rice production keep on increasing due to advance in technologies.

However, major cereal crop has produced post-harvesting by-products, i.e., rice husk. Among agro-industrial waste, rice husk is the primarily by-products from the rice harvesting as well as rice mill industry as it constitutes ~22% of its production. Rice husk contains high contents of silicon as SiO_2 in the range of 9–12% (Martin 1938; Ding et al. 2005; Aisuncion et al. 2005). The nature of silica in rice husk is of hydrated amorphous type. Bansal et al. (2006) demonstrated biotransformation of amorphous silica present in rice husk to its corresponding crystalline form at ambient reaction conditions of fermentation. The same researchers developed methodology for bioleaching of amorphous silica and simultaneously converted to value-added nanoparticles in the form crystalline morphology by using a mesophilic fungus. The selected fungus *Fusarium oxysporum* was used to extracellular leaching in the aqueous environment at ambient temperature. The biotransformation process mediated by fungus was quick to get the value-added product in the form crystalline silica from cheap silica present in the rice husk. The researchers claimed first type of the study as compared to previous studies (Kroger et al. 2002; Cha et al. 2002). Authors demonstrated the selected plant-pathogenic fungus as the excellent route for biogenic synthesis of metallic nanoparticles via bioleaching of precursor salts (Mukherjee et al. 2002).

Bansal et al. (2006) developed a methodology for the biotransformation of amorphous silica present in rice husk. Authors used 10 g rice husk and 20 g wet biomass of fungus *Fusarium oxysporum* and in the ratio of 1:2 and kept the suspension on continuous shaking condition at ambient temperature for 1 day. Then both the materials were separated by filtration process. The filtrate was treated with a biphasic solvent system (phenol-chloroform) to remove the extracellular proteins. The protein-free nanoparticle suspension was converted into powder form by low-temperature evaporation process. The nanopowders were treated for before and after calcination. Authors also checked the extracellular fungal proteins for the capacity to biosynthesize silica nanoparticles.

The morphology of silica present in the rice husk was also confirmed by EDX analysis. Chemically treated rice husk with hydrofluoric acid selectively released silica nanoparticles. EDX is a semiquantitative technique; thus, quantitative approach to measure silica bioleaching process was followed in the form of X-ray photoemission spectroscopy (XPS) analysis. Fungus-mediated reaction with rice husk resulted in the leaching out of 96% of silica from rice husk in aqueous form (Bansal et al. 2006).

The time-dependent FTIR analysis of nanoparticles from the fungus-rice husk reaction medium using KBr pellets was carried out to understand the leaching mechanism of silica after the regular interval of 4 h each till 24 h of reaction rice husk in the form of silica nanoparticles (Bansal et al. 2006). FTIR spectra of the filtrate containing silica nanoparticles synthesized by exposing rice husk to the fungus *Fusarium oxysporum* for 24 h, from silica nanoparticles calcined at 400 °C for 2 h, and from the filtrate obtained by exposing rice husk to water of pH 4.5 for 15 days. XRD patterns of silica particles synthesized by the exposure of rice husk to the fungus *Fusarium oxysporum* before and after calcination at 400 °C for 2 h supported the findings as in FTIR studies.

Bansal et al. (2006) demonstrated *Fusarium oxysporum* as an efficient fungus employed for biotransformation of amorphous silica present in rice husk to highly crystalline silica nanoparticles. The nanosilica of the size range 2–6 nm was functionalized by stabilizing fungal proteins. Calcination of the nanosilica particles leads to loss of occluded protein and apparently leads to a porous structure often of cubic morphology. The amorphous silica particles can also be bioleached from rice husk under in vitro conditions using cationic extracellular proteins; however, these proteins alone do not lead to biotransformation of amorphous silica into silica nanocrystallites. *Fusarium oxysporum* have demonstrated the biotransformation of amorphous silica present in rice husk, cheap source of inorganic nanosilica, into highly crystalline silica nanoparticles (Bansal et al. 2006).

Kulkarni et al. (2008) demonstrated leaching of silicates from borosilicate glass by using *Humicola* sp. Fungal mediated bioleaching of glass resulted the release of sub-10 nm monodispersed silicate nanoparticles. However, the presence of B-O and B-O-Si bonds disapproved the synthesis of pristine silica and confirmed the borosilicate type nanoparticles. Additionally, modified glass confirmed the silicate materials leaching by selected fungus.

5.5 Conclusion

Nanotechnology holds the current status of emerging cutting-edge technology that have plethora of applications. Although physicochemical routes of nanomaterial synthesis are the attractive routes, synthetic chemical routes have disadvantages of being eco-hazardous and extreme reaction conditions. Such limitations can be overcome with ease by employing eco-friendly routes. This has been achieved by using living organisms such as fungi, bacteria, actinomycetes, lichens, and viruses. This unique behavior of fungi has emerged as a key player for the nanofabrication via green chemistry route.

Through nanobiotechnology intervention, the conversion of waste materials to the value-added material has been feasible at the bench scale. However, there is urgency to upgrade the bench scale technology to the industrial scale. The ambient temperature synthesis of nanomaterials using fungi starting from potential cheap agro-industrial waste materials is an exciting possibility that an energy-conserving and economically viable green approach toward the large-scale synthesis of metallic nanomaterial. In the conclusion, an extension of such nonhazardous approach to technologically challenged large quantities waste materials such as magnetite, chromite, etc. also warrants further investigation.

References

Abraham RE, Verma ML, Barrow CJ, Puri M (2014) Suitability of ferrite nanoparticles immobilised cellulases in enhancing enzymatic saccharification of pretreated hemp biomass. Biotechnol Biofuels 7:90

Adil SF, Assal ME, Khan M (2015) Biogenic synthesis of metallic nanoparticles and prospects toward green chemistry. Dalton Trans 44:9709–9717

Asuncion MZ, Hasegawa I, Kampf JW, Laine RM (2005) The selective dissolution of rice hull ash to form $[OSiO_{1.5}]_8[R_4N]_8$ (R=Me,CH_2CH_2OH) octasilicates. Basic nanobuilding blocks and possible models of intermediates formed during biosilicification processes. J Mater Chem 15:2114–2121

Aziz N, Pandey R, Barman I, Prasad R (2016) Leveraging the attributes of *Mucor hiemalis*-derived silver nanoparticles for a synergistic broad-spectrum antimicrobial platform. Front Microbiol 7:1984. https://doi.org/10.3389/fmicb.2016.01984

Bajaj M, Schmidt S, Winter J (2012) Formation of Se (0) nanoparticles by *Duganella* sp. and *Agrobacterium* sp. isolated from Se-laden soil of North-East Punjab, India. Microb Cell Factories 11(0):64

Bansal V, Sanyal A, Rautaray D, Ahmad A, Sastry M (2005) Bioleaching of sand by the fungus *Fusarium oxysporum* as a means of producing extracellular silica nanoparticles. Adv Mater 17:889–892

Bansal V, Ahmad A, Sastry M (2006) Fungus-mediated biotransformation of amorphous silica in rice husk to nanocrystalline silica. J Am Chem Soc 128:14059–14066

Bansal V, Syed A, Bhargava SK, Ahmad A, Sastry M (2007) Zirconia enrichment in zircon sand by selective fungus mediated bioleaching of silica. Langmuir 23:4993–4998

Bansal V, Bharde A, Ramanathan R, Bhargava SK (2012) Inorganic materials using unusual micro-organisms. Adv Colloid Interf Sci 179–182:150–168

Barik TK, Sahu B, Swain V (2008) Nano-silica from medicine to pest control. Parasitol Res 103:253–258

Behfarnia K, Rostami M (2017) Effects of micro and nanoparticles of SiO_2 on the permeability of alkali activated slag concrete. Constr Build Mater 131:205–213

Bhattacharya D, Gupta RK (2005) Nanotechnology and potential of microorganisms. Crit Rev Biotechnol 25:199–204

Bosshard PP, Bachofen R, Brandl H (1996) Metal leaching of fly ash from municipal waste incineration by *Aspergillus niger*. Environ Sci Technol 30:3066–3070

Brombacher C, Bachofen R, Brandl H (1997) Biohydrometallurgical processing of solids: a patent review. Appl Microbiol Biotechnol 48:577–587

Burda C, Chen X, Narayanan R, El-Sayed MA (2005) Chemistry and properties of nanocrystals of different shapes. Chem Rev 105:1025–1102

Cha JN, Stucky GD, Morse DE, Deming TJ (2002) Biomimetic synthesis of ordered silica structures mediated by block polypeptides. Nature 403:289–292

Dhanjal S, Cameotra SS (2010) Aerobic biogenesis of selenium nanospheres by *Bacillus cereus* isolated from coalmine soil. Microb Cell Factories 9:52–52

Dhillon GS, Brar SK, Kaur S, Verma M (2012) Green approach for nanoparticles biosynthesis by fungi: current trends and applications. Crit Rev Biotechnol 32:49–73

Ding TP, Ma GR, Shui MX, Wan DF, Li RH (2005) Silicon isotope study on rice plants from the Zhejiang province, *China*. Chem Geol 218:41–50

Durán N, Marcato PD, Alves OL, DeSouza G, Esposito E (2005) Mechanistic aspects of biosynthesis of silver nanoparticles by several *Fusarium oxysporum* strains. J Nanobiotechnol 3:1–8

Durán N, Marcato PD, Durán M, Yadav A, Gade A, Rai M (2011) Mechanistic aspects in the biogenic synthesis of extracellular metal nanoparticles by peptide, bacteria, fungi, and plants. Appl Microbiol Biotechnol 90:1609–1624

Eftink MR, Ghiron CA (1981) Fluorescence quenching studies with proteins. Anal Biochem 114:199–227

FAO World Rice Information (1995) issue No. 1. FAO, Rome

FAOSTAT Database FAO (2012) Rome, www.faostat.fao.org

Farrell NO, Houlton A, Horrocks BR (2006) Silicon nanoparticles: applications in cell biology and medicine. Int J Nanomedicine 1(4):451–472

Frances IH, Denisse VA, Meghan EG (2008) High temperature aerogels in the Al_2O_3- SiO_2 system. American chemical society 236th national meeting, Philadelphia

Gade A, Ingle A, Whiteley C, Rai M (2010) Mycogenic metal nanoparticles: progress and applications. Biotechnol Lett 32(5):593–600

Gao M, MaQ LQ, Chang J, Ma H (2017) A novel approach to extract SiO_2 from fly ash and its considerable adsorption properties. Mater Des 116:666–675

Hemalatha T, Ramaswamy A (2017) A review on fly ash characteristics–towards promoting high volume utilization in developing sustainable concrete. J Clean Prod 147:546–559

Hubert DHW, Jung M, Fredrik PM, Bomans PHH, Meuldijk J, German AL (2000) Vesicle-directed growth of silica. Adv Mater 12:1286–1290

Innocenzi P, Falcaro P, Grosso D, Babonneau F (2003) Order-disorder transitions and evolution of silica structure in self-assembled mesostructured silica films studied through FTIR spectroscopy. J Phys Chem B 107:4711–4717

Jain N, Bhargava A, Majumdar S, Tarafdar JC, Panwar J (2011) Extracellular biosynthesis and characterization of silver nanoparticles using *Aspergillus flavus* NJP08: a mechanism perspective. Nanoscale 3:635–641

Khan SA, Uddin I, Moeez S, Ahmad A (2014) Bioleaching of waste material such as fly-ash as a means of producing extracellular protein capped fluorescent and water soluble silica nanoparticles. PLoS One 9(9):e107597

Kitching M, Ramani M, Marsili E (2015) Fungal biosynthesis of gold nanoparticles: mechanism and scale up. Microb Biotechnol 8:904–917

Kroger N, Lorenz S, Brunner E, Sumper M (2002) Self-assembly of highly phosphorylated silaffins and their function in biosilica morphogenesis. Science 298:584–586

Kulkarni S, Syed A, Singh S, Gaikwad A, Patil K, Vijayamohanan K, Ahmad A, Ogale S (2008) Silicate nanoparticles by bioleaching of glass and modification of the glass surface. J Non-Cryst Solids 354:3433–3437

Lee JC, Song HT, Yoo JM (2007) Present status of the recycling of waste electrical and electronic equipment in Korea. Resour Conserv Recycl 50(4):380–397

Li G, He D, Qian Y et al (2012) Fungus mediated green synthesis of silver nanoparticles using *Aspergillus terreus*. Int J Mol Sci 13:466–476

Maliszewska I, Juraszek A, Bielska K (2014) Green synthesis and characterization of silver nanoparticles using ascomycota fungi *Penicillium nalgiovense* AJ12. J Clust Sci 25:989–1004

Mann S (1995) Biomineralization and biomimetic materials chemistry. J Mater Chem 5:935–946

Mann S, Ozin GA (1996) Synthesis of inorganic materials with complex form. Nature 382:313–318

Martin JL (1938) The desilicification of rice hulls and a study of the products obtained. M.S. Thesis, Louisiana State University, Eunice

Mizutani S, Yoshida T, Sakai S, Tatatsuki H (1996) Release of metals from MSW 1 fly ash and availability in alkali condition. Waste Manag 16:53–544

Mohammadian A, Shaojaosadati SA, Rezee MH (2007) *Fusarium oxysporum* mediates photogeneration of silver nanoparticles. Sci Iran 14:323–326

Morse DE (1999) Silicon biotechnology: harnessing biological silica production to construct new materials. Trends Biotechnol 17:230–232

Mousavi SR, Rezaei M (2011) Nanotechnology in agriculture and food production. J Appl Environ Biol Sci 1:414–419

Mukherjee P, Ahmad A, Mandal D, Senapati S, Sainkar SR, Khan MI, Parischa R, Ajaykumar PV, Alam M, Kumar R, Sastry M (2001) Fungus mediated synthesis of silver nanoparticles and their immobilization in the mycelial matrix: a novel biological approach to nanoparticle synthesis. Nano Lett 1:515–519

Mukherjee P, Senapati S, Mandal D, Ahmad A, Khan MI, Kumar R, Sastry M (2002) Extracellular synthesis of gold nanoparticles by the fungus *Fusarium oxysporum*. Chembiochem 3(5):461–463

Mulligan CN, Kamali M (2003) Bioleaching of copper and other metals from low-grade oxidized mining ores by *Aspergillus niger*. J Chem Technol Biotechnol 78(5):497–503

Palla R, Karade SR, Mishra G, Sharma U, Singh LP (2017) High strength sustainable concrete using silica nanoparticles. Constr Build Mater 138:285–295

Palomo A, Grutzeck MW, Blanco MT (1999) Alkali-activated fly ashes a cement for the future. Cem Concr Res 29:1323–1329

Pantidos N, Horsfall LE (2014) Biological synthesis of metallic nanoparticles by bacteria, fungi and plants. J Nanomedicine Nanotechnol 5:5

Park YJ, Fray DJ (2009) Recovery of high purity precious metals from printed circuit boards. J Hazard Mater 164(2):1152–1158

Pathak A, Morrison L, Healy MG (2017) Catalytic potential of selected metal ions for bioleaching, and potential techo-economic and environmental issues: a critical review. Bioresour Technol 229:211–221

Patwardhan SV (2011) Biomimetic and bioinspired silica: recent developments and applications. Chem Commun 47:7567–7582

Paull R, Wolfe J, Hébert P, Sinkula M (2003) Investing in nanotechnology. Nat Biotechnol 21:1144–1147

Pavani KV, Kumar NS, Sangameswaran BB (2012) Synthesis of lead nanoparticles by *Aspergillus* species. Pol J Microbiol 61:61–63

Pavesi L, Negro LD, Nazzoleni C, Franzo G, Priolo P (2000) Optical gain in silicon nanocrystals. Nature 408:440–444

Perry CC, Keeling-Tucker T (2000) Biosilicification: the role of the organic matrix in structure control. J Biol Inorg Chem 5(5):537–550

Perry CC, Keeling-Tucker T (2003) Model studies of colloidal silica precipitation using biosilica extracts from *Equisetum telmateia*. Colloid Polym Sci 281(7):652–664

Pohnert G (2002) Biomineralization in diatoms mediated through peptide- and polyamine-assisted condensation of silica. Angew Chem Int Ed Eng 41(17):3167–3169

Prasad R (2014) Synthesis of silver nanoparticles in photosynthetic plants. J Nanopart 11:963955–963961

Prasad R, Kumar V, Prasad KS (2014) Nanotechnology in sustainable agriculture: present concerns and future aspects. Afr J Biotechnol 13:705–713

Prasad R, Pandey R, Barman I (2016) Engineering tailored nanoparticles with microbes: quo vadis. WIREs Nanomedicine Nanobiotechnol 8:316–330

Prasad R, Pandey R, Varma A, Barman I (2017a) Polymer based nanoparticles for drug delivery systems and cancer therapeutics. In: Kharkwal H, Janaswamy S (eds) Natural polymers for drug delivery. CAB International, UK, pp 53–70

Prasad R, Bhattacharyya A, Nguyen QD (2017b) Nanotechnology in sustainable agriculture: recent developments, challenges, and perspectives. Front Microbiol 8:1014. https://doi.org/10.3389/fmicb.2017.01014

Priya A, Hait S (2017). Feasibility of bioleaching of selected metals from electronic waste by *Acidiphilium acidophilum*. Waste Biomass Valor pp 1–7

Puri M, Barrow CJ, Verma ML (2013) Enzyme immobilization on nanomaterials for biofuel production. Trends Biotechnol 31:215–216

Rai M, Yadav A, Bridge P, Gade A (2009) Myconanotechnology: a new and emerging science. In: Rai MK, Bridge PD (eds) Applied mycology. CAB International, New York, pp 258–267

Roy S, Mukherjee T, Chakraborty S, Das TK (2013) Biosynthesis, characterization and antifungal activity of silver nanoparticles synthesized by the fungus *Aspergillus foetidus* MTCC8876. Dig J Nanomater Biostruct 8:197–205

Saha M (2016) An overview of rice diseases in India. Everyman's Sci 3:183–191

Sakamoto M, Fujistuka M, Majima T (2009) Light as a construction tool of metal nanoparticles: synthesis and mechanism. J Photochem Photobiol C: Photochem Rev 10(1):33–56

Sastry M, Ahmad A, Khan MI, Kumar R (2003) Biosynthesis of metal nanoparticles using fungi and actinomycetes. Curr Sci 85:162–170

Shelar GB, Chavan AM (2014) Fungus–mediated biosynthesis of silver nanoparticles and its antibacterial activity. Arch App Sci Res 6:111–114

Siddiqi KS, Husen A (2016) Fabrication of metal nanoparticles from fungi and metal salts: scope and application. Nanoscale Res Lett 11:98

Silverstein RM (1967) Spectrometric identification of organic compounds, 2nd edn. Wiley, New York, p 102

Singh D, Rathod V, Ninganagouda S, Hiremath J, Singh AK, Mathew J (2014) Optimization and characterization of silver nanoparticle by endophytic fungi *Penicillium* sp. isolated from *Curcuma longa* (turmeric) and application studies against MDR *E. coli* and *S. aureus*. Bioinorg Chem Appl 408021:8

Sklivaniti V, Tsakiridis PE, Katsiotis NS, Velissariou D, Pistofidis N, Papageorgiou D, Beazi M (2017) Valorisation of woody biomass bottom ash in Portland cement: a characterization and hydration study. J Environ Chem Eng 5(1):205–213

Sumper M, Kroger N (2004) Silica formation in diatoms: the function of long-chain polyamines and silaffins. J Mater Chem 14:2059–2065
Sun S, Murray CB, Weller D, Folks L, Moser A (2000) Monodisperse FePt nanoparticles and ferromagnetic FePt nanocrystal superlattices. Science 287:1989–1992
Swihart MT (2003) Vapor-phase synthesis of nanoparticles. Curr Opin Colloid Interface Sci 8(1):127–133
Thakkar KN, Mhatre SS, Parikh RY (2010) Biological synthesis of metallic nanoparticles. Nanomedicine 6:257–262
Vágó A, Szakacs G, Sáfrán G, Horvath R, Pécz B, Lagzi I (2016) One-step green synthesis of gold nanoparticles by mesophilic filamentous fungi. Chem Phys Lett 645:1–4
Vahabi K, Mansoori GA, Karimi S (2011) Biosynthesis of silver nanoparticles by the fungus *Trichoderma reesei*. Insci J 1:65–79
Vala AK (2015) Exploration on green synthesis of gold nanoparticles by a marine-derived fungus aspergillus sydowii. Environ Prog Sustain Energy 34:194–197
Verma ML (2017) Enzymatic nanobiosensors in the agricultural and food industry. In: Ranjan S., Dasgupta N., Lichtfouse E. (eds) Nanoscience in food and agriculture 4, Sustainable agriculture reviews series, vol 24. Springer International Publishing, Cham, pp 229–245
Verma VC, Kharwar RN, Gange AC (2010) Biosynthesis of antimicrobial silver nanoparticles by endophytic fungus *Aspergillus clavatus*. Biomedicine 5:33–40
Verma ML, Barrow CJ, Kennedy JF, Puri M (2012) Immobilization of β-galactosidase from *Kluyveromyces lactis* on functionalized silicon dioxide nanoparticles: characterization and lactose hydrolysis. Int J Biol Macromol 50:432–437
Verma ML, Barrow CJ, Puri M (2013a) Nanobiotechnology as a novel paradigm for enzyme immobilization and stabilisation with potential applications in biofuel production. Appl Microbiol Biotechnol 97:23–39
Verma ML, Chaudhary R, Tsuzuki T, Barrow CJ, Puri M (2013b) Immobilization of β-glucosidase on a magnetic nanoparticle improves thermostability: application in cellobiose hydrolysis. Bioresour Technol 135:2–6
Verma ML, Naebe M, Barrow CJ, Puri M (2013c) Enzyme immobilisation on amino-functionalised multi-walled carbon nanotubes: structural and biocatalytic characterisation. PLoS One 8(9):e73642
Verma ML, Rajkhowa R, Barrow CJ, Wang X, Puri M (2013d) Exploring novel ultrafine Eri silk bioscaffold for enzyme stabilisation in cellobiose hydrolysis. Bioresour Technol 145:302–306
Verma ML, Puri M, Barrow CJ (2016) Recent trends in nanomaterials immobilised enzymes for biofuel production. Crit Rev Biotechnol 36(1):108–119
Wu HY, Ting YP (2006) Metal extraction from municipal solid waste (MSW) incinerator fly ash-chemical leaching and fungal bioleaching. Enzym Microb Technol 38(6):839–847
Yadav A, Kon K, Kratosova G, Duran N, Ingle AP, Rai M (2015) Fungi as an efficient mycosystem for the synthesis of metal nanoparticles: progress and key aspects of research. Biotechnol Lett 37(11):2099–2120

Chapter 6
Synthesis and Applications of Nanofungicides: A Next-Generation Fungicide

Sonu Kumari and Suphiya Khan

Abstract With the increasing population, the pressure of enhancing food production and management of fungal diseases of food crops and fruits in agriculture sector needs urgent concern. Nanofungicides due to their vast physiochemical and functionalization properties could be easily applied for plant disease management. This chapter covers the different types of nanofungicide synthesis with mechanism. The chapter also gives comprehensive idea about fungal mycotoxins and its harmful effects on agricultural sector. Apart from it, this chapter also highlights the effects of nanoparticles (NPs) on mycotoxins produced by fungi and its mechanism of action.

6.1 Introduction

Nanoparticles (NPs) are defined as particles of size in range of 1–100 nm. Professor Norio Taniguchi of Tokyo Science University in 1974 first coined the term nanotechnology (Taniguchi 1974). The NPs nanosize provides great opportunities in several areas. Nanotechnology provides knowledge to synthesis nanostructures, and their unique properties show new application areas like environmental, pharmacology and medicine (Rai et al. 2009; Gupta et al. 2012; Aziz et al. 2015). In recent years, utilization of fungi as a source for NPs synthesis has been observed (Li et al. 2008).

NPs synthesis through different non-toxic and environment-friendly methods has been newly emerged. In this respect, there has been increasing research for NPs synthesis through microorganisms (Bernhardt et al. 2010). The NPs synthesis follows three steps that are appropriate solvent medium selection, reducing agent and stabilizing agents for NPs stabilization (Raveendran et al. 2003). Generally, this reduction mechanism is carried out by fungi enzymes for NPs synthesis (Kalimuthu et al. 2008; Kalishwaralal et al. 2008). Importantly, fungi have also secreted fairly

S. Kumari • S. Khan (✉)
Department of Bioscience and Biotechnology, Banasthali Vidyapith,
Banasthali Tonk, Rajasthan 304022, India
e-mail: suphiyakhan@gmail.com

R. Prasad (ed.), *Fungal Nanotechnology*, Fungal Biology,
DOI 10.1007/978-3-319-68424-6_6

large amount of proteins and secondary metabolites extracellularly, and hence, the fungal biomass could reduce the metal ions more easily leading to the rapid formation of NPs (Prasad et al. 2016). Because of these advances, the myco-based extracellular synthesis method is often considered as a better resource for higher productivity of NPs. The myconanoparticles play advantageous role over chemical antifungal agents as these have low tendency for microbial resistance. Among several inorganic NPs, silver has been comprehensively used because of its advantages over NPs such as copper, gold, zinc oxide NPs, etc. (Kalishwaralal et al. 2008).

Mycotoxins are secondary metabolites produced by fungi, and Bennett and Bentley (1989) defined it as "metabolic intermediates or products, found as a differentiation product in restricted taxonomic groups, not essential to growth and life of the producing organism, and biosynthesized from one or more general metabolites by a wider variety of pathways than is available in general metabolism". Two groups were formed for toxigenic fungi infecting crops as "field" and "storage" fungi. Among all mycotoxins, aflatoxins, patulin, ochratoxin A, fumonisins and citrinin are considered the most disadvantageous types for crops. The confirmed mechanism for NPs action on fungi is not well understood. Many possible mechanisms were given by researchers, but the exact mechanism has not been reported.

This chapter provides insight into mycosynthesis of NPs and its mechanism. The different mycotoxins and its effects on crops were compiled along with NPs effects on the produced mycotoxins. The mechanism of action of NPs on fungi is also included.

6.2 Synthesis of Myconanoparticles

Nowadays, fungi are used as substrate for the production of different NPs. The fabrication of NPs using fungi has some practical advantages. Fungi have proved to be more beneficial for NPs synthesis compared to other microorganisms. Laboratory scale synthesis of fungi is done which provide the same biomass density as bacteria. Xu et al. (1999) and Taherzadeh et al. (2003) reported the biomass yield of 0.31 g g^{-1} and 0.55 g g^{-1} for *Escherichia coli* and *Rhizopus oryzae* culture when grown with glucose batch bioreactor. Several fungi species were reported for extracellular production of different NPs, such as *Penicillium*, *Fusarium oxysporum*, *Aspergillus* and *Verticillium* (Mukherjee et al. 2001; Gericke and Pinches 2006). Fungi can be used as excellent source of various extracellular enzymes which influence NPs synthesis (Saxena et al. 2014). There are several advantages of NPs synthesis from fungi over bacteria and plants.

6.2.1 *Protein Secretion*

Fungi secrete high concentration of extracellular enzymes that catalyze the heavy metal ions and form NPs. This results in faster production of NPs using fungi than chemical synthesis (Rai et al. 2009).

6.2.2 Isolation and Culture

Due to simple nutritional requirements, fungi are easy to isolate and subculture. There are various isolation methods for fungi such as plating, hyphal extraction and serial dilutions. Fungi are totipotent, and therefore hyphae or spores can be used to grow fungus and can be subcultured to obtain pure isolate (Rai et al. 2009).

6.2.3 Growth Control

The several enzymes secreted by fungi can be used to synthesize NPs of defined size and shape. Fungi are able to maintain under high agitation and flow pressure as compared to bacteria and plants (Saha et al. 2010).

6.2.4 Extracellular Synthesis of NPs

Fungi can produce NPs extracellularly which is suitable for easier downstream processing and handling of biomass. Extracellular synthesis of silver NPs using *Aspergillus* sp. has been reported (Gade et al. 2008).

Fungi can produce NPs both intracellularly and extracellularly through two different mechanisms (Saxena et al. 2014).

Silver NPs were synthesized extracellularly using *Aspergillus fumigates* (Bhainsa and D'Souza 2006). The NPs synthesis mechanism is shown in Fig. 6.1. Gopinath and Arumugam (2014) investigated the synthesis of gold NPs from *F. solani* culture filtrate. TEM analysis showed the size of gold NPs was in the diameter range between 20 and 50 nm. Some researchers reported the biosynthesis of silver NPs by using *Trichoderma reesei*. In the production of silver NPs by *T. reesei*, mycelium contacts to the silver nitrate solution. This stimulates the fungus to secrete enzymes and metabolites for its survival mechanism. Further, the reduction of silver ions was done by secreted enzymes and metabolites. Chan and Don (2012) showed that *Schizophyllum commune* and *Pycnoporus sanguineus* could be used for biological synthesis of AgNPs. They used directly mycelia or culture supernatant of these white rot fungi for testing their reduction effect. The mycelia or culture supernatant produced AgNPs with different sizes (Chan and Don 2012). Cuevas et al. (2015) described the use of the extract of *Stereum hirsutum* for preparing copper and copper oxide NPs. In their study, three copper salts ($CuCl_2$, $CuSO_4$ and Cu $(NO_3)_2$) were used, and the effect of various pH on reduction activity of this extract was investigated. $CuCl_2$ (5 mM) gave the highest NPs formation at alkaline conditions. The extracellular protein in this extract may be responsible for NPs formation and stabilization. Sanghi and Verma (2009) described a continuous and extracellular formation of cadmium sulphide NPs (CdS) by immobilized *Coriolus versicolor* in a column reactor. The protein was reported as a capping agent. TEM analysis showed

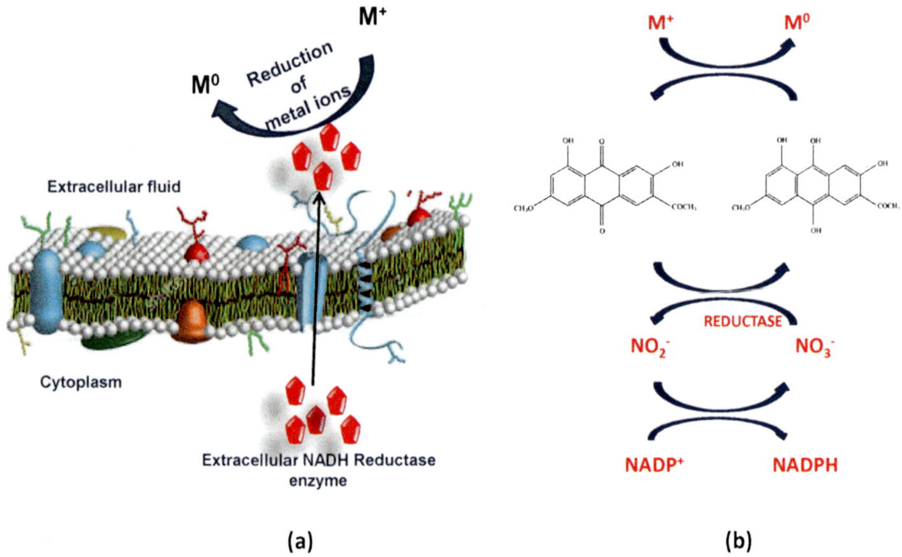

Fig. 6.1 Mechanism of extracellular NPs formation by fungi. (**a**) Extracellular NADH reductase enzyme action, (**b**) electron shuttle quinines action

the embedded NPs in the fungal matrix were well-dispersed spherical NPs with uniform size (about 5–9 nm) (Sanghi and Verma 2009). Bansal et al. (2005) showed that *F. oxysporum* secreted proteins competent of hydrolyzing zirconia ions at 30 °C. The research showed that metal halide precursors can be hydrolyzed by this fungus in acid medium (Bansal et al. 2005). To use as biosorbent agent, the group of SiO_2-NPs (N-Si) with *P. funiculosum* for the synthesis of N-Si-Pen was investigated. This biosynthesized NP was utilized to adsorb Pb(II). The maximum capacity value was 1266.7 μmol g^{-1} for N-Si-Pen combined particle. Sorption equilibrium was obtained in about 20 min.

Raliya and Tarafdar (2014) synthesized the magnesium, titanium and zinc NPs through the utilization of different fungus species, i.e., *A. niger*, *A. flavus*, *A. tubingensis*, *A. fumigates*, *A. terreus* and *A. oryzae* with various precursor salts of chlorides, nitrates, oxides and sulphates (Raliya and Tarafdar 2014; Klaus et al. 1999). The various factors of protocol were also optimized for increasing the production of NPs. The biosynthesis of silver NPs by the cell-free filtrate of *P. nalgiovense* was observed by the researcher (Maliszewska et al. 2014). The authors stated that proteins containing cysteine are responsible for reduction of metal ions for synthesis of NPs.

6.2.5 Intracellular NPs Synthesis

It was reported that NPs synthesized inside the organism could have smaller size as compared to extracellular NPs. Vigneshwaran et al. (2007) demonstrated the incubation for 72 h of *A. flavus* with silver nitrate for silver NPs synthesis. The NPs were obtained on its cell wall surface with average particle size of 8.92 nm (Vigneshwaran et al. 2007). In addition, *Verticillium* sp. produced silver NPs on exposure with aqueous silver nitrate solution (Mukheree et al. 2001). The *Verticillium* sp. was used for production of gold NPs, and NPs were detected on the surface and membrane of mycelium. Further, TEM analysis showed the hexagonal, spherical and triangular-shaped gold NPs on cell wall (Mukherjee et al. 2001). The gold NPs of less than 10 nm were synthesized using *V. luteoalbum* when incubated at pH 3.0 for 24 h (Gericke and Pinches 2006). Gold NPs were also synthesized intercellularly using *Trichothecium* sp. (Ahmad et al. 2005). The size of silver and gold NPs synthesized through fungi was in the range of 8.92–25 nm.

6.3 Mechanism of Mycosynthesis of NPs

The confirmed mechanism for NPs synthesis, for instance, of Ag NPs, in fungi is still not given, even though various researchers have worked to find possible mechanism for production of NPs (Duran et al. 2005; Meyer 2008). The mechanisms were reported for extracellular synthesis of NPs such as nitrate reductase action, electron shuttle quinones or both. The nitrate reductase method was observed due to reaction between nitrite and 2,3-diaminonaphthalene (Duran et al. 2005; Kumar et al. 2007). In many fungi species including *Penicillium*, the NPs synthesis was initiated by nitrate reductase, while some enzymes like extracellular shuttle quinine, alpha-NADPH-dependent reductases and nitrate-dependent reductases were responsible for silver NPs synthesis for *F. oxysporum*. The *A. flavus* was utilized for silver NPs production, which is possible due to "33 kDa" and cysteine protein which stabilizes the NPs by functioning as a capping agent (Jain et al. 2011). Most of the research paper favoured the nitrate reductase for NPs synthesis (Vigneshwaran et al. 2006; Perez-de-Luque et al. 2008; Deepa and Panda 2014). Fungal cell wall and cell wall sugars execute a major role for the absorption and reduction of metal ions (Sastry et al. 2003).

The proposed mechanisms by researchers for NPs intracellular synthesis consist of two steps (Kashyap et al. 2013):

(a) The binding of metal ions on the fungal cell wall surface via electrostatic interaction between negatively charged carboxylate groups in enzymes of the mycelia cell wall and positive charge of metal ions.
(b) Metals ions were reduced by enzymes, which lead to aggregation of NPs within the cell wall.

Extracellular NPs synthesis showed the interaction of metal ions and released enzymes (reductase), which leads to the formation of NPs in the solution (Kashyap et al. 2013). This method is advantageous than intracellular as its fungal cell lysis not required, NPs recovery and purification is easily achieved (Gade et al. 2008). The purification of NPs obtained through intracellular process is tedious task and requires time-consuming techniques. Kumar et al. (2007) showed the role of enzyme alpha-NADPH-dependent nitrate reductase in silver NPs synthesis.

6.4 Mycotoxins Produced by Fungi

Three genera of fungi were mainly reported for production of mycotoxins, namely, *Fusarium, Aspergillus* and *Penicillium* with dematiaceous fungal genera "*Alternaria, Helminthosporium, Drechslera, Phoma* and *Zygosporium*" (Ismaiel and Papenbrock 2014). The various fungi produced mycotoxins in plants during growing stage when environmental conditions are favourable. Toxigenic fungi in crops are grouped as:

(a) Field fungi: fungi which form toxins before crop harvest. It depends on factors like plant host and environmental interactions (insects).
(b) Storage fungi: these show problem after harvest in stored material. It depends on crop nutrients, moisture, temperature and insect's competition.

But the source of both field and storage fungi is field. Among the all known mycotoxins, aflatoxins, fumonisins, ochratoxin, citrinin and patulin are considered most harmful for food cereals. Aflatoxins are the major mycotoxins produced by genus *Aspergillus.* Almost 18 different aflatoxins were formed by *A. flavus* strains, and aflatoxin B1 is the extremely toxic and carcinogenic type (Ismaiel and Tharwat 2014; Richard et al. 2003). *Fusarium* sp. is responsible for the production of fumonisins (Thiel et al. 1991). Gelderblom et al. (1988) isolated the fumonisins B1 for the first time from *F. moniliforme* MRC 826. Ochratoxin was first discovered in 1965 in *A. ochraceus* (Van Der et al. 1965). Ochratoxin were mainly produced by *A. niger, A. carbonarius* and *P. verrucosum* fungal strains (Ciegler et al. 1977; Abarca et al. 1994; Horie 1995). Citrinin is another mycotoxin which was first obtained by Hetherington and Raistrick (1931) through *P. citrinum.* Several fungal species were identified for citrinin synthesis such as *P. verrucosum, M. ruber* and *A. terreus* (Ciegler et al. 1977). Chain et al. (1942) firstly isolated the patulin (PAT) mycotoxin from *P. claviforme* and named as calviformin, later renamed PAT due to production from *P. patulum.*

6.5 Phytotoxic Properties of Mycotoxins on Plant

Among all crop diseases, fungi are responsible for more than 70%, and fungi mycotoxins cause maximum crop loss in species such as cotton, rice, groundnut and wheat (Agrios 2005; Dhekney et al. 2007). Fungi inhibits the root parameters of plant more than shoot or mass causing various diseases such as browning, necrosis, lesions and wilt (McLean 1995; McLean et al. 1995). Several crop varieties were affected by mycotoxins produced by fungi such as cowpea, mung, sesame and gram. The carotenoid and chlorophyll synthesis and seed germination were also highly affected in some plants (Crisan 1973; Sinha and Kumari 1990; Sinha and Sinha 1993; Adekunle and Bassir 1997; Samuel and Valentine 2014). Inhibition of root and leaf development and chlorophyll synthesis was observed in some crop varieties due to aflatoxin (Reiss 1978). The lethal dose of aflatoxin was reported for barley, sorghum and wheat as 0.83 mg L^{-1}, 2.75 mg L^{-1} and 1.74 mg L^{-1}, respectively (Hasan 1999). Studies conducted showed the presence of necrosis and wilting when treated with concentration of 1000 µg·mL^{-1} of fumonisins for soybeans (Abbas and Boyette 1992). Fumonisins severally affect the maize and tomato seedlings compared to aflatoxin mycotoxin (Lamprecht et al. 1994). The cell death in plants was obtained by the ochratoxin produced by fungi (Wang et al. 2011). The exposure of *A. thaliana* with ochratoxin caused growth inhibition and necrotic lesions (Peng et al. 2010). When *A. thaliana* plants leaves were treated with ochratoxin solutions, the macroscopic lesions were seen within 2 days (Wang et al. 2011). Ochratoxin showed inhibitory effect on *Zea mays* embryos with 5 µg·mL^{-1} concentration (McLean et al. 1995). Citrinin showed phytotoxic effects in several trials conducted by researchers (White and Truelove 1972; Damodaran et al. 1975; Betina 1989; Maćıas et al. 2000). Bean, sorghum and cotton plants demonstrated wilting effects when treated with citrinin (Damodaran et al. 1975). The seed number and flower number of crops were affected by patulin mycotoxin presence (Ellis and McCalla 1973; Ismaiel et al. 2014).

6.6 Effect of Nanoparticles on Fungi Mycotoxins

6.6.1 Aflatoxin

Aspergillus flavus is responsible for the synthesis of maximum number of aflatoxins. The chlorophyll synthesis in crops is mostly inhibited by it and responsible for low crop production. Cereals are mostly exposed to *A. flavus* attack and micromycete produced by them, so requirement for antifungal agents is developed (Al-othman et al. 2014). Nabawy et al. (2014) reported the antifungal potential of ZnO and Fe_2O_3 NPs against isolated aflatoxigenic and non-aflatoxigenic *A. flavus* that were recovered from animal and poultry feeds associated with animal diseases using well and disc diffusion tests. The diameter zones of inhibition of non-aflatoxigenic strains were larger than in aflatoxigenic strains. The growth of all tested strains was not affected below the 25 µg ml^{-1} NPs concentration treatment. The well diffusion test proved better in studying of antifungal potential of NPs than disc diffusion. It is interesting to report here that the zone of *A. flavus* growth inhibition appeared at lower concentrations (50 µg ml^{-1}) of ZnO and Fe_2O_3 NPs, whereas, similar effects in traditional antifungals required relatively higher concentration (100–200 µg ml^{-1}). Also, it was reported that the antifungal effects of clove oil against *A. flavus* showed comparatively lower antifungal effects than NPs in the study. Significant correlation between growth of *A. flavus* and aflatoxins production was clearly observed. The decreasing levels of aflatoxin B1 were reported with inhibition of fungal colonies till complete inhibition of both (Nabawy et al. 2014). Several studies evaluated the antimicrobial activity of NPs of metal oxide particularly ZnO powder against fungi in culture media. Regarding the results of antifungal activities of

ZnO, moulds as *A. flavus*, *A. niger* and *A. ochraceus* required higher concentration of ZnO NPs to inhibit their growth. The diameters of zones of inhibition of ZnO NPs against *A. flavus* and *A. ochraceus* were 7 and 15 mm at the concentration of 300 μg ml^{-1}, whereas *A. niger* required relatively lower concentration (200 μg/ml) (6 mm) of NPs to inhibit its growth (Hassan et al. 2014). The antifungal activities of ZnO NPs against *P. expansum*, *Aspergillus* spp., *Rhizopus* spp. and yeast were investigated by other researchers (Violeta et al. 2011). The minimum inhibition concentration (MIC) of ZnO against *Aspergillus* spp. and *C. albicans* was reported to be 1.013–296 μg ml^{-1} and for SDS and fluconazole was 0.001–0.56 and 0.062–128 μg ml^{-1}, respectively. Moreover, it was added that the use of lower concentrations of ZnO NPs was the most effective antifungal and antibacterial. Furthermore, different studies conducted in different laboratories showed that the antimicrobial activity is influenced, not only by NPs concentration but also by the size of the ZnO particles (Violeta et al. 2011).

Silver NPs were also used in place of fungicides for reducing the growth and aflatoxin production by fungi. *A. terreus* was utilized for the synthesis of silver NPs, and experiment was conducted to analyse the effect of these on aflatoxin production by *A. flavus*. Inhibition in fungi growth and mycotoxin production was observed best at 150 ppm of NPs treatment. The fungi inhibition was seen through deformation of fungal hypae and decrement in spore numbers (Al-othman et al. 2014).

6.6.2 Citrinin

Citrinin is mainly produced in crops after harvest and presents mainly in storage products such as fruits, grains, spices, beans, etc. (Da Lozzo et al. 2002; Bragulat et al. 2008). Magro et al. (2016) conducted a study for deletion of citrinin from Monascus suspensions. For this experiment, “surface active maghemite nanoparticles” (SAMNs) were synthesized, and 0.1 g L^{-1} of Monascus suspension was treated with 1 g L^{-1} of SAMNs. The 1 g L^{-1} concentration of SAMNs removed 70% of citrinin and in second round with same concentration of NPs removed citrinin up to 0.2 mg L^{-1}. This showed the binding of citrinin on SAMNs and formation of mycotoxin complex with iron (III). The formation of mycotoxin-SAMNs complex mainly depends on the presence of iron-chelating group on citrinin like keto-enol group (Magro et al. 2016).

6.6.3 Fumonisin

Fusarium verticillioides, *F. nygamai* and *F. proliferatum* are responsible for the production of fumonisins (Thiel et al. 1991; Rheeder et al. 2002). The decreased fungi growth and mycotoxin formation was achieved by zinc NPs. Mycotoxins synthesis level was lowered with increase in NPs concentration. The fumonisin B1 synthesizing fungus was inhibited by 10 μg ml^{-1} zinc NPs treatment. SEM analysis revealed the rupture of fungi cell wall and reduction in the fumonisin production (Hassan et al. 2013).

6.6.4 Ochratoxin

Mouhamed et al. (2015) evaluated the antifungal potential of ZnO and Fe_2O_3 nanoparticles in comparison with some commercial antifungal feed additives (probiotic, propionic acid and clove oil) in inhibiting the growth of *Aspergillus ochraceus* and *Aspergillus niger* strains that were isolated from animal and poultry feeds using well and disc diffusion tests. The reported diameters of inhibition zones induced by metal NPs for non-ochratoxigenic strains were larger than that of ochratoxigenic strains, and the zone diameters increased when the NPs concentration increment. The 20 μg ml^{-1} NPs concentration did not affect the growth of all *A. ochraceus* and *A. niger* strains, whereas the zones of inhibition produced by the metal NPs required lower concentration (25 μg ml^{-1} and more) than that produced by the commercial antifungal feed additives (50 μg ml^{-1} and more). The ochratoxin A production by ochratoxigenic strains in liquid medium or on yellow corn was significantly diminished in parallel with the decline parameters in colony count of the treated ochratoxigenic strains. The field application of the used NPs and other drugs on commercial animal feed evidenced the availability to use ZnO and Fe_2O_3 NPs only as antifungal, but their antimycotoxin effect was limited to their use as feed additives during manufacture and before exposure of feeds to fungal contamination. Further studies were required for investigating the synergistic effects of combined antioxidant metal NPs and other commercial antimycotoxins to obtain dual synergistic actions in order to decrease the amount of used chemicals in the feed manufacture and to study the availability of its use in vivo (Mouhamed et al. 2015).

6.6.5 Patulin

Penicillium expansum and *F. oxysporum* fungi *are* considered as the major source of a patulin and commonly found in rotting apples. Yehia and Ahmed (2013) conducted an experiment on *F. oxysporum* and *P. expansum* for analysing the toxic efficiency of NPs. *P. expansum* showed high inhibitory effect against ZnO NPs treatment than *F. oxysporum.* The mechanism of NPs action was explained as fungi growth inhibition was due to fungal hypha structure deformation. It was observed that patulin production was decreased by both fungi with the enhancement of NPs concentration. Two postharvest fungi, *Botrytis cinerea* and *P. expansum,* were treated with zinc oxide NPs with concentration of 3, 6 and 12 mmol L^{-1}. *P. expansum* growth was more inhibited by NPs activity (Yehia and Ahmed 2013).

6.7 Fungicidal Mechanisms of NPs

Das et al. (2009) conducted a study to assess the gold NPs effects on *S. cerevisiae* and *Candida albicans*. The study showed the mechanism of action of gold NPs on fungi. SEM analysis confirmed the rupturing of fungi cell wall due to NPs interaction and action (Das et al. 2009). The copper-based NPs interaction resulted in formation of reactive oxygen species (ROS) and caused DNA disruption (Chen et al. 2006; Oberdürster 2000; Heinlaan et al. 2008). Shah et al. (2010) reported the reduction in lignocellulose-degrading enzymes. In addition, the interaction with NPs also caused mitochondria and protein damage (Shah et al. 2010). The lethal effects of NPs are described in Fig. 6.2.

6.8 Advantages

NPs synthesis through fungi has proved advantageous compared to other organisms (Fig. 6.3). The fungi are favourable due to trouble-free isolation and capability of extracellular enzyme secretion (Singh et al. 2014; Prasad et al. 2016). Also, the process proved environmental friendly and less time-consuming for metal ion reduction by the secreted proteins of fungi (Rai et al. 2009).

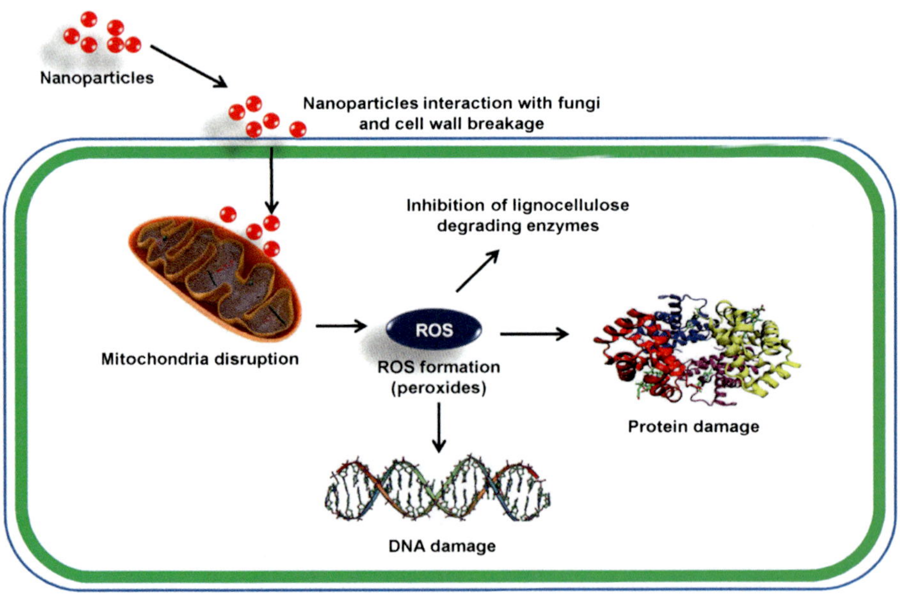

Fig. 6.2 Mechanism of antifungal action of NPs

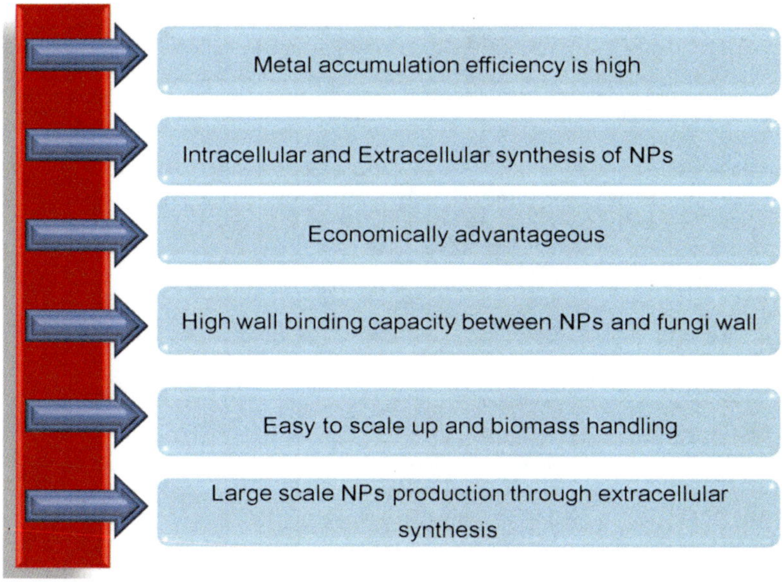

Fig. 6.3 Advantages of fungi for NPs synthesis

NPs are considered remarkable antifungal agents over chemical agents because of their low tendency to induce microbial resistance. The multiple modes of fungal inhibition by NPs also proved advantageous factor for using nanofungicides. The mentioned studies suggest that NPs can be used as an effective fungicide in agricultural and food safety applications (He et al. 2011; Bhattacharyya et al. 2016; Aziz et al. 2016; Ismail et al. 2017).

6.9 Conclusion

Nanotechnology is increasing in the agricultural sector for several applications. There is an increasing interest in NPs production through fungi due to its advantages over other sources. Several NPs were synthesized through fungi, but improvement in methods is required for controlled synthesis of NPs of required shape, composition and size. Mycotoxins were produced by several fungi, and effects on crops can be seen on basis of before harvest (field fungi) and after harvest (storage fungi). Aflatoxins, fumonisins, ochratoxin, citrinin and patulin were considered most harmful for crops produced by mainly *Aspergillus*, *Penicillium* and *Fusarium* species. The major disadvantage reported for agrochemicals used for various phytopathogenic fungi is the resistant developed pathogens. Therefore, NPs were explored as fungicides by several researchers. The studies suggested that the NPs are more effective fungicides as compared to agrochemicals used. The antimycotoxin effects of NPs were limited to NPs dose provided. The antifungal property of NPs can be helpful in agricultural sector and food storage industries. It can be concluded that NPs are "new-generation fungicides" and can be synthesized by advantageous mycosynthesis process.

Acknowledgement Authors gratefully acknowledge Prof. Aditya Shastri for providing us research facilities and the Bioinformatics Centre, Banasthali University, Rajasthan (India), for extensive use of computational facilities and also thankful to MHRD Department of Higher Education, Government of India, under the scheme of Establishment of Centre of Excellence for Training and Research in FAST (FAST 5-5/2014 TS VII) for their generous financial support.

References

Abarca ML, Bragulat MR, Castella G, Cabanes FJ (1994) Ochratoxin A production by strains of *Aspergillus niger* var. niger. Appl Environ Microbiol 60(7):2650–2652

Abbas HK, Boyette CD (1992) Phytotoxicity of fumonisin B1 on weed and crop species. Weed Technol 1:548–552

Adekunle AA, Bassir O (1997) The effects of aflatoxin B1 and G1 on the germination and leaf color of cowpea (*Vigna sinensis*). Mycopathol Mycol Appl 1997:1–2

Agrios GN (2005) Plant pathology. Elsevier Academic Press, San Diego

Ahmad A, Senapati S, Khan MI, Kumar R, Sastry M (2005) Extra–/intracellular biosynthesis of gold nanoparticles by an alkalotolerant fungus, *Trichothecium* sp. J Biomed Nanotechnol 1(1):47–53

Al-Othman MR, ARM AE, Mahmoud MA, Fifan SA, El-Shikh MM (2014) Application of silver nanoparticles as antifungal and antiaflatoxin B1 produced by *Aspergillus flavus*. Dig J Nanomater Bios 1(9):151–157

Aziz N, Faraz M, Pandey R, Shakir M, Fatma T, Varma A, Barman I, Prasad R (2015) Facile algae-derived route to biogenic silver nanoparticles: synthesis, antibacterial, and photocatalytic properties. Langmuir 31(42):11605–11612

Aziz N, Pandey R, Barman I, Prasad R (2016) Leveraging the attributes of *Mucor hiemalis*-derived silver nanoparticles for a synergistic broad-spectrum antimicrobial platform. Front Microbiol 7:1984. doi:10.3389/fmicb.2016.01984

Bansal V, Sanyal A, Rautaray D, Ahmad A, Sastry M (2005) Bioleaching of sand by the fungus *Fusarium oxysporum* as a means of producing extracellular silica nanoparticles. Adv Mater 17(7):889–892

Bennett JW, Bentley R (1989) What's in a name?-microbial secondary metabolism. Adv Appl Microbiol 34:1–28

Bernhardt ES, Colman BP, Hochella MF, Cardinale BJ, Nisbet RM, Richardson CJ, Yin L (2010) An ecological perspective on nanomaterial impacts in the environment. J Environ Qual 39(6):1954–1965

Betina V (1989) Mycotoxins. Chemical, biological and environmental aspects. Elsevier, Amsterdam

Bhainsa KC, D'Souza SF (2006) Extracellular biosynthesis of silver nanoparticles using the fungus Aspergillus fumigates. Colloids Surf B Biointerfaces 47:160–164

Bhattacharyya A, Duraisamy P, Govindarajan M, Buhroo AA, Prasad R (2016) Nano-biofungicides: emerging trend in insect pest control. In: Prasad R (ed) Advances and applications through fungal nanobiotechnology. Springer International Publishing, Cham, pp 307–319

Bragulat MR, Martínez E, Castellá G, Cabañes FJ (2008) Ochratoxin A and citrinin producing species of the genus *Penicillium* from feedstuffs. Int J Food Microbiol 126(1):43–48

Chain E, Florey HW, Jennings MA (1942) An antibacterial substance produced by *Penicillium claviforme*. Br J Exp Pathol 23(4):202
Chan Y, Don MM (2012) Characterization of Ag nanoparticles produced by white-rot fungi and its in vitro antimicrobial activities. Int Arab J Antimicrob Agents 2(3):1–8
Chen Z, Meng H, Xing G, Chen C, Zhao Y, Jia G, Wang T, Yuan H, Ye C, Zhao F, Chai Z (2006) Acute toxicological effects of copper nanoparticles in vivo. Toxicol Lett 163(2):109–120
Ciegler A, Vesonder RF, Jackson LK (1977) Production and biological activity of patulin and citrinin from *Penicillium expansum*. Appl Environ Microbiol 33(4):1004–1006
Crisan EV (1973) Effects of aflatoxin on seeding growth and ultrastructure in plants. Appl Microbiol 26(6):991–1000
Cuevas R, Durán N, Diez MC, Tortella GR, Rubilar O (2015) Extracellular biosynthesis of copper and copper oxide nanoparticles by *Stereum hirsutum*, a native white-rot fungus from chilean forests. J Nanomater 16(1):57
Da Lozzo EJ, Mangrich AS, Rocha ME, de Oliveira MB, Carnieri EG (2002) Effects of citrinin on iron-redox cycle. Cell Biochem Funct 20(1):19–29
Damodaran C, Kathirvel-Pandian S, Seeni S, Selvam R, Ganesan MG, Shanmugasundaram S (1975) Citrinin, a phytotoxin? Cell Mol Life Sci 31(12):1415–1417
Das SK, Das AR, Guha AK (2009) Gold nanoparticles: microbial synthesis and application in water. Hyg Manag 25(14):8192–8199
Deepa K, Panda T (2014) Synthesis of gold nanoparticles from different cellular fractions of *Fusarium oxysporum*. J Nanosci Nanotechnol 14(5):3455–3463
Dhekney S, Li A, Anaman M, Dutt M, Tattersall J, Gray D (2007) Genetic transformation of embryogenic cultures and recovery of transgenic plants in Vitis vinifera, Vitis rotundifolia and Vitis hybrids. Acta Hort 738:743–748
Durán N, Marcato PD, Alves OL, De Souza GI, Esposito E (2005) Mechanistic aspects of biosynthesis of silver nanoparticles by several *Fusarium oxysporum* strains. J Nanobiotechnol 3(1):8
Ellis JR, McCalla TM (1973) Effects of patulin and method of application on growth stages of wheat. Appl Microbiol 25(4):562–566
Gade AK, Bonde P, Ingle AP, Marcato PD, Duran N, Rai MK (2008) Exploitation of *Aspergillus niger* for synthesis of silver nanoparticles. J Biobased Mater Bio 2(3):243–247
Gelderblom WC, Jaskiewicz K, Marasas WF, Thiel PG, Horak RM, Vleggaar R, Kriek NP (1988) Fumonisins – novel mycotoxins with cancer-promoting activity produced by *Fusarium moniliforme*. Appl Environ Microbiol 54(7):1806–1811
Gericke M, Pinches A (2006) Biological synthesis of metal nanoparticles. Hydrometallurgy 83(1):132–140
Gopinath K, Arumugam A (2014) Extracellular mycosynthesis of gold nanoparticles using *Fusarium solani*. Appl Nanosci 4(6):657–662
Gupta S, Sharma K, Sharma R (2012) Myconanotechnology and application of nanoparticles in biology. Recent Res Sci Technol 4(8):36–38
Hasan HA (1999) Phytotoxicity of pathogenic fungi and their mycotoxins to cereal seedling viability. Mycopathologia 148(3):149–155
Hassan AA, Howayda ME, Mahmoud HH (2013) Effect of zinc oxide nanoparticles on the growth of mycotoxigenic mould. SCPT 1(4):66–74
Hassan AA, Oraby NA, Mohamed AA, Mahmoud HH (2014) The possibility of using zinc oxide nanoparticles in controlling some fungal and bacterial strains isolated from buffaloes. Egypt J Appl Sci 29(3):58–83
He L, Liu Y, Mustapha A, Lin M (2011) Antifungal activity of zinc oxide nanoparticles against *Botrytis cinerea* and *Penicillium expansum*. Microbiol Res 166(3):207–152
Heinlaan M, Ivask A, Blinova I, Dubourguier HC, Kahru A (2008) Toxicity of nanosized and bulk ZnO, CuO and TiO_2 to bacteria *Vibrio fischeri* and crustaceans *Daphnia magna* and *Thamnocephalus platyurus*. Chemosphere 71(7):1308–1316
Hetherington AC, Raistrick H (1931) Studies in the Biochemistry of Micro-organisms. On the production and chemical constitution of a new yellow colouring matter, citrinin, produced from glucose by Penicillium citrinum. Trans Roy Soc (Lond) B 220:269–295

Horie Y (1995) Productivity of ochratoxin a of *Aspergillus carbonarius* in *Aspergillus* section Nigri. Nippon Kingakukai Kaiho 36:73–76

Ismaiel AA, Papenbrock J (2014) The effects of patulin from *Penicillium vulpinum* on seedling growth, root tip ultrastructure and glutathione content of maize. Eur J Plant Pathol 139(3):497–509

Ismaiel AA, Papenbrock J (2014) The effects of patulin from Penicilllium vulpinum on seedling growth, root tip ultrastructure and glutathione content of maize. Eur J Plant Pathol 139:497–509

Ismail M, Prasad R, Ibrahim AIM, Ahmed ISA (2017) Modern prospects of nanotechnology in plant pathology. In: Prasad R, Kumar M, Kumar V (eds) Nanotechnology. Springer Nature Singapore Pte Ltd, Singapore, pp 305–317

Jain N, Bhargava A, Majumdar S, Tarafdar JC, Panwar J (2011) Extracellular biosynthesis and characterization of silver nanoparticles using *Aspergillus flavus* NJP08: a mechanism perspective. Nanoscale 3(2):635–641

Kalimuthu K, Babu RS, Venkataraman D, Bilal M, Gurunathan S (2008) Biosynthesis of silver nanocrystals by *Bacillus licheniformis*. Colloids Surf B Biointerfaces 65(1):150–153

Kalishwaralal K, Deepak V, Ramkumarpandian S, Nellaiah H, Sangiliyandi G (2008) Extracellular biosynthesis of silver nanoparticles by the culture supernatant of *Bacillus licheniformis*. Mater Lett 62(29):4411–4413

Kashyap PL, Kumar S, Srivastava AK, Sharma AK (2013) Myconanotechnology in agriculture: a perspective. World J Microbiol Biotechnol 29(2):191–207

Klaus T, Joerger R, Olsson E, Granqvist CG (1999) Silver-based crystalline nanoparticles, microbially fabricated. Proc Natl Acad Sci 96(24):13611–13614

Kumar SA, Abyaneh MK, Gosavi SW, Kulkarni SK, Pasricha R, Ahmad A, Khan MI (2007) Nitrate reductase-mediated synthesis of silver nanoparticles from $AgNO_3$. Biotechnol Lett 29(3):439–445

Lamprecht SC, Marasas WF, Alberts JF, Cawood ME, Gelderblom WC, Shephard GS, Thiel PG, Calitz FJ (1994) Phytotoxicity of fumonisins and TA-toxin to corn and tomato. Phytopathology 84(4):383–391

Li Q, Mahendra S, Lyon DY, Brunet L, Liga MV, Li D, Alvarez PJ (2008) Antimicrobial nanomaterials for water disinfection and microbial control: potential applications and implications. Water Res 42(18):4591–4602

Macías M, Ulloa M, Gamboa A, Mata R (2000) Phytotoxic compounds from the new coprophilous fungus *Guanomyces polythrix*. J Nat Prod 63(6):757–761

Magro M, Moritz DE, Bonaiuto E, Baratella D, Terzo M, Jakubec P, Malina O, Čépe K, de Aragao GM, Zboril R, Vianello F (2016) Citrinin mycotoxin recognition and removal by naked magnetic nanoparticles. Food Chem 203:505–512

Maliszewska I, Juraszek A, Bielska K (2014) Green synthesis and characterization of silver nanoparticles using ascomycota fungi *Penicillium nalgiovense* AJ12. J Clust Sci 25(4):989–1004

McLean M (1995) The phytotoxicity of selected mycotoxins on mature, germinating *Zea mays* embryos. Mycopathologia 132(3):173–183

McLean M, Watt MP, Berjak P, Dutton MF (1995) Aflatoxin B1-its effects on an in vitro plant system. Food Addit Contam 12(3):435–443

Meyer V (2008) Genetic engineering of filamentous fungi-progress, obstacles and future trends. Biotechnol Adv 26(2):177–185

Mouhamed AE, Hassan AA, Manal AH, El Hariri M, Refai M (2015) Effect of metal nanoparticles on the growth of Ochratoxigenic moulds and Ochratoxin A production isolated from food and feed. Int J Res Stud Biosci 3(9):1–14

Mukherjee P, Ahmad A, Mandal D, Senapati S, Sainkar SR, Khan MI, Parishcha R, Ajaykumar PV, Alam M, Kumar R, Sastry M (2001) Fungus-mediated synthesis of silver nanoparticles and their immobilization in the mycelial matrix: a novel biological approach to nanoparticle synthesis. Nano Lett 1(10):515–519

Nabawy GA, Hassan AA, Sayed El-Ahl RH, Refai MK (2014) Effect of metal nanoparticles in comparison with commercial antifungal feed additives on the growth of *Aspergillus flavus* and aflatoxin b1 production. J Glob Biosci 3:954–971

Oberdürster G (2000) Toxicology of ultrafine particles: in vivo studies. Philos Trans R Soc Lond A: Math Phys Eng Sci 358(1775):2719–2740

Peng XL, Xu WT, Wang Y, Huang KL, Liang ZH, Zhao WW, Luo YB (2010) Mycotoxin ochratoxin A-induced cell death and changes in oxidative metabolism of *Arabidopsis thaliana*. Plant Cell Rep 29(2):153–161

Perez-de-Luque A, Rubiales D, Marquina CI, Ibarra MR, De la Fuente JM (2008) Nanoparticles in agriculture, development of smart delivery systems for plant research. Nano-Spain, Braga

Prasad R, Pandey R, Barman I (2016) Engineering tailored nanoparticles with microbes: quo vadis? Wiley Interdiscip Rev Nanomed Nanobiotechnol 8(2):316–330

Rai M, Yadav A, Bridge P, Gade A, Rai MK, Bridge PD (2009) Myconanotechnology: a new and emerging science. Appl Mycol 14:258–2567

Raliya R, Tarafdar JC (2014) Biosynthesis and characterization of zinc, magnesium and titanium nanoparticles: an eco-friendly approach. Int Nano Lett 4(1):1. –0

Raveendran P, Fu J, Wallen SL (2003) Completely green synthesis and stabilization of metal nanoparticles. J Am Chem Soc 125(46):13940–13941

Reiss J (1978) Effects of mycotoxins on higher plants, algae, fungi and bacteria. Mycotoxic Fungi Mycotoxins Mycotoxicoses 3:118–144

Rheeder JP, Marasas WF, Vismer HF (2002) Production of fumonisin analogs by *Fusarium* species. Appl Environ Microbiol 68(5):2101–2105

Richard JL, Payne GA, Desjardin AE, Maragos C, Norred WP, Pestka JJ (2003) Mycotoxins, risks in plant, animal and human systems. In: CAST Task Force Report 139; Council for Agricultural Science and Technology: Ames, 101–103. 32

Saha S, Sarkar J, Chattopadhyay D, Patra S, Chakraborty A, Acharya K (2010) Production of silver nanoparticles by a phytopathogenic fungus Bipolaris nodulosa and its antimicrobial activity. Dig J Nanomater Biostruct 5(4):887–895

Samuel AT, Valentine IT (2014) Effect of total aflatoxin on the growth characteristics and chlorophyll level of sesame (*Sesamum indicum* L.) NY Sci J 7:8–13

Sanghi R, Verma P (2009) A facile green extracellular biosynthesis of CdS nanoparticles by immobilized fungus. Chem Eng J 155(3):886–891

Sastry M, Ahmad A, Khan MI, Kumar R (2003) Biosynthesis of metal nanoparticles using fungi and actinomycete. Curr Sci 85(2):162–170

Saxena J, Sharma MM, Gupta S, Singh A (2014) Emerging role of fungi in nanoparticle synthesis and their applications. World J Pharm Sci 3:1586–1613

Shah V, Dobiášová P, Baldrian P, Nerud F, Kumar A, Seal S (2010) Influence of iron and copper nanoparticle powder on the production of lignocellulose degrading enzymes in the fungus *Trametes versicolor*. J Hazard Mater 178(1):1141–1145

Singh D, Rathod V, Ninganagouda S, Hiremath J, Singh AK, Mathew J (2014) Optimization and characterization of silver nanoparticle by endophytic fungi *Penicillium* sp. isolated from *Curcuma longa* (turmeric) and application studies against MDR *E. coli* and *S. aureus*. Bioinorg Chem Appl 2014:1–8

Sinha KK, Kumari P (1990) Some physiological abnormalities induced by aflatoxin B1 in mung seeds (*Vigna radiata* variety *Pusa Baishakhi*). Mycopathologia 110(2):77–79

Sinha KK, Sinha AK (1993) Effect of aflatoxin B1 on germination index and seedling growth in wheat varieties. Mycopathologia 123(3):165–169

Taherzadeh MJ, Fox M, Hjorth H, Edebo L (2003) Production of mycelium biomass and ethanol from paper pulp sulfite liquor by *Rhizopus oryzae*. Bioresour Technol 88(3):167–177

Taniguchi N (1974) On the basic concept of nanotechnology. In: Proceedings of the international conference on production engineering, Tokyo, Part II, Japan Society of Precision Engineering. pp 18–23

Thiel PG, Marasas WF, Sydenham EW, Shephard GS, Gelderblom WC, Nieuwenhuis JJ (1991) Survey of fumonisin production by *Fusarium* species. Appl Environ Microbiol 57(4):1089–1093

Van der Merwe KJ, Steyn PS, Fourie L, Scott DB, Theron JJ (1965) Ochratoxin A, a toxic metabolite produced by *Aspergillus ochraceus* Wilh. Nature 205(4976):1112–1113

Vigneshwaran N, Ashtaputre NM, Varadarajan PV, Nachane RP, Paralikar KM, Balasubramanya RH (2006) Biological synthesis of silver nanoparticles using the fungus Aspergillus flavus. Mater Lett 61:1413–1418

Vigneshwaran N, Ashtaputre NM, Varadarajan PV, Nachane RP, Paralikar KM, Balasubramanya RH (2007) Biological synthesis of silver nanoparticles using the fungus *Aspergillus flavus*. Mater Lett 61(6):1413–1418

Violeta V, Catalin P, Constantin F, Monica A, Marius B (2011) Nanoparticles applications for improving the food safety and food processing. In: 7th international conference on materials science and engineering, Bramat, Brasov, vol 77, 24–26 Feb 2011

Wang Y, Peng X, Xu W, Luo Y, Zhao W, Hao J, Liang Z, Zhang Y, Huang K (2011) Transcript and protein profiling analysis of OTA-induced cell death reveals the regulation of the toxicity response process in *Arabidopsis thaliana*. J Exp Bot 29:err447

White AG, Truelove B (1972) The effects of aflatoxin B1, citrinin, and ochratoxin A on amino acid uptake and incorporation by cucumber. Can J Bot 50(12):2659–2664

Xu B, Jahic M, Blomsten G, Enfors SO (1999) Glucose overflow metabolism and mixed-acid fermentation in aerobic large-scale fed-batch processes with *Escherichia coli*. Appl Microbiol Biotechnol 51(5):564–571

Yehia RS, Ahmed OF (2013) In vitro study of the antifungal efficacy of zinc oxide nanoparticles against *Fusarium oxysporum* and *Penicillium expansum*. Afr J Microbiol Res 7(19):1917–1923

Chapter 7
Enzymes and Nanoparticles Produced by Microorganisms and Their Applications in Biotechnology

Emad Abada, Zarraq Al-Faifi, and Mohamed Osman

Abstract This chapter reports the most important microbial enzymes that have been studied extensively due to their production, low cost, purification, and characterization of their properties. Certain microorganisms have already been isolated from extreme sources under strict culture conditions. The aim of this process is to obtain isolated microbes that would have the capability to bio-synthesize special enzymes. Enzymes having special features and characteristics are needed by some bio-industries for their applications in preparation of necessary substrates and some renewable raw materials for production. The microbial enzymes serve as biocatalysts to make reactions in bioprocesses in an economical, safe, and eco-friendly way instead of utilizing chemical catalysts. Researchers benefited from the special features and characteristics of enzymes for their commercial interest and industrial applications, which involve thermotolerance, thermophilic nature, tolerance to a varied range of pH, stability of enzyme activity over a range of temperature and pH, and other harsh reaction conditions. It is assured that such enzymes have the utility in bio-applications such as food industry, leather and textile production, animal feed manufacturing, and in bioconversions.

E. Abada (✉)
Botany and Microbiology Department, Faculty of Science, Helwan University, Ain Helwan, Cairo, Egypt

Biology Department, Faculty of Science, Jazan University, Jazan, King Saudi Arabia (KSA)
e-mail: emadm_abada@yahoo.co.uk; eabada@jazanu.edu.sa

Z. Al-Faifi
Biology Department, Faculty of Science, Jazan University, Jazan, King Saudi Arabia (KSA)

M. Osman
Botany and Microbiology Department, Faculty of Science, Helwan University, Ain Helwan, Cairo, Egypt

R. Prasad (ed.), *Fungal Nanotechnology*, Fungal Biology,
DOI 10.1007/978-3-319-68424-6_7

Abbreviations

Abs	Antibodies
AgNPs	Silver nanoparticles
AuNPs	Gold nanoparticles
BacMPs	Bacterial magnetic particles
BMs	Bacterial magnetosomes
Brec	Bovine retinal endothelial cells
CdS-NPs	CdS nanoparticles
CSE	Cell-soluble extract
GTPase	Guanosine triphosphatase
HRP	Horseradish peroxidase
MRI	Magnetic resonance imaging
MTB	Magnetotactic bacteria
PHB	Polyhydroxybutyrate BRECs
TEM	Transmission electron microscope

7.1 Introduction

An enzyme is considered as a biological catalyst that has important function in biochemical reactions and metabolism of cell organism. The enzyme activity depends on its capability to bind the reactants (substrates) at specific place called active side, and the enzyme converts it into a product (Brahmachari et al. 2017). The active site of the enzyme is specific for a certain substrate without any interference of others. A group of enzymes requires nonprotein chemical compound or metallic ion called cofactor or helper molecules including metals such as Zn and Fe or organic compounds such as vitamins that support the biological reactions (Fig. 7.1). In nature, the enzymes have a certain role in synthesis or biodegradation of compounds during

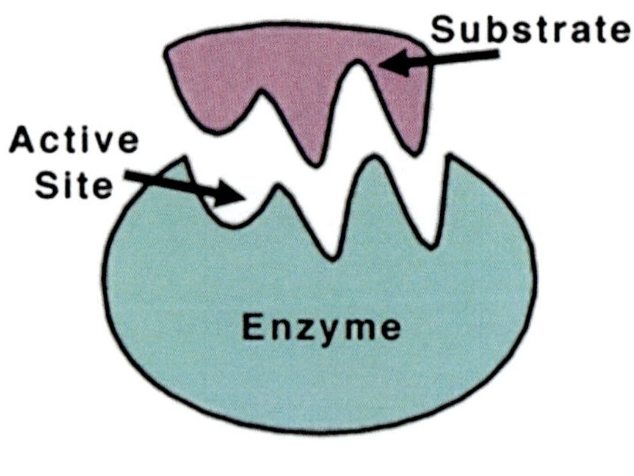

Fig. 7.1 The structure of enzyme

catabolic or anabolic process. It has been reported that the purification and production on a large scale of enzymes could be used in a versatile field of applications. The ancient uses of enzymes were focused in brewing and leather industries. Many enzymes have a special role and could be used as catalysts in many industrial applications. Mostly, microbial enzymes are excellent enzymes for industrial applications and commercial way (Table 7.1). In the twentieth century, many studies have been done about induction and characterization of microbial enzymes ranging from small- to large-scale production. The most common sources of microbial enzymes are fungi, yeast, and bacteria (Pandey et al. 1999). The microbial enzymes could be used as free or in immobilized form, which depend on the enzyme specificity. Due to rapid development of biotechnology, many microbial enzymes have been designed or engineered using molecular techniques. These enzymes have improved activity ready to use in many industrial applications (Chirumamilla et al. 2001). Consequently, many new products have been synthesized commercially in worldwide markets. Proteases are the most commercially applicable enzymes which are synthesized by bacteria such as *Bacillus*, *Pseudomonas*, and *Clostridium* and few fungi (Kumar and Takagi 1999). Moreover, xylanases have higher activity under a temperature range of 40–60 °C which were produced by many fungal genera *Trichoderma*, *Penicillium*, and *Aspergillus* (Ahmed et al. 2009). According to

Table 7.1 An overview of the most important microbial enzymes that have special industrial characteristics and applications

Enzyme	EC number	Sources	Extraction	Biotechnological use
Bacterial enzymes				
∞-Amylase	3.2.1.1	*Bacillus* sp.	Extracellular	Starch industry
β- Amylase	3.2.1.2	*Bacillus* sp.	Extracellular	Starch industry
Asparaginase	3.5.1.1	*E.coli*	Intracellular	Health industry
Glucose isomerase	5.3.1.5	*Bacillus* sp.	Intracellular	Fructose syrup
Penicillin amidase	3.5.1.11	*Bacillus* sp.	Extracellular	Pharmaceutical industry
Protease	3.4.21.14	*Bacillus* sp.	Extracellular	Detergent
Pullulanase	3.2.1.41	*Klebsiella*	Extracellular	Starch
Fungal enzymes				
α-Amylase	3.2.1.1	*Aspergillus*	Extracellular	Baking
Aminoacylase	3.5.1.14	*Aspergillus*	Intracellular	Pharmaceutical
Glucoamylase	3.2.1.3	*Aspergillus*	Extracellular	Starch
Catalase	1.11.1.6	*Aspergillus*	Intracellular	Food
Cellulase	3.2.1.4	*Trichoderma*	Extracellular	Waste
Dextranase	3.2.1.11	*Penicillium*	Extracellular	Food
Glucose oxidase	1.1.3.4	*Aspergillus*	Intracellular	Food
Lipase	3.1.1.3	*Rhizopus*	Extracellular	Food
Pectinase	3.2.1.15	*Aspergillus*	Extracellular	Drinks
Yeast enzymes				
Invertase	3.2.1.26	*Saccharomyces*	I/E	Confectionery

resistance of different temperature ranges, enzymes are classified as thermophilic, mesophilic, and psychrophilic, while enzymes resistant to different pH ranges are classified as acidophile, neutrophile, and alkalophile. Thermostable microbial enzymes that resist higher reaction temperatures decrease the probability of contamination by microorganisms through the production of enzyme at large scale in the industry (Banat et al. 1992; Wati et al. 1996; Wang et al. 2012). For example, the complete hydrolysis and saccharification of valuable polysaccharides of agricultural residues need more time to react that is frequently accompanied with huge risks of contamination by time. Thus, the search of active, effective mass transfer and stable enzymes at higher temperature during the reaction processes is recommended (Berka et al. 2011; Cai et al. 2011).

7.2 Recent Important Microbial Enzymes from Microorganisms

7.2.1 Protease

Proteases are hydrolytic enzymes which are considered as one of the most commercial viable group enzymes that have been widely discussed and studied (Mukherjee et al. 2008; Rahman et al. 2003; Chudasama et al. 2010; Genckal and Tari 2006; Gupta et al. 2002; Sony and Potty 2017; Vijayalakshmi et al. 2011). Proteases produced from microorganisms may be acidophile, neutral, and alkalophile. *Alkalophilic* protease is very active at high pH conditions, and their active site is a serine (Gupta et al. 2002). *Alkalophilic* proteases have particular importance since its stable in abnormal utmost reaction conditions (high temperature and pH) even in existence of inhibitory components (Gupta et al. 2002, 2008; Vijayalakshmi et al. 2011; Johnvesly and Naik 2001). *Bacillus* has already optimized the culture conditions for producing thermophilic and alkalophilic proteases (Vijayalakshmi et al. 2011). It has been reported that alkalophilic proteases have high stability by using detergents (Vijayalakshmi et al. 2011; Johnvesly and Naik 2001; Hadj-Ali et al. 2007). Moreover, the alkalophilic proteases produced from other microorganisms have many applications in bio-industries such as tannery, washing powders, leather processing, food industry, peptide synthesis, and pharmaceutical industries (Pandey et al. 1999; Kumar and Takagi 1999).

7.2.2 Keratinases

Keratin is fibrous and an insoluble protein which are the main ingredients in wool and feathers (Gopinath et al. 2015; Gushterova et al. 2005). The keratin containing materials is difficult to hydrolyze by widespread proteases. Keratinase is a serine proteolytic enzyme that is able to hydrolyze insoluble keratinous compounds.

Recently, the keratinase is increasingly used in varied industrial applications as leather production, detergent industry, animal feed manufacturing, medicine, and textile. Keratinases have been produced and purified from fungi and actinomycetes (Gushterova et al. 2005; Brandelli et al. 2010). It has been reported that the gene encoding the keratinase in different expression systems has been organized, cloned, and expressed (Gupta et al. 2013). The degradation of wool and feathers required high temperature that could be possibly done by utilizing thermophilic keratinase. The keratin solubility could be raised by the thermal stability of keratinase; it also could boost equilibrium expiration in endothermic reaction. The modification of keratinase thermal stability by tentative mutation or through computer design programs has been reported (Baihong et al. 2013). Keratinase of *Bacillus subtilis* mutant exhibited additive and synergistic impacts at 60 °C besides doubling the enzyme activity by 8.6-fold. In addition, the substitution of N122Y showed higher activity of 5.6-fold due to increase in its catalytic activity in comparison to the wild-type enzyme. It has been assured that an alkalophilic *Streptomyces albidoflavus* strain produces a certain type of protease extracellularly that is able to degrade keratin (Indhuja et al. 2012). As standardized, the biosynthesis of this enzyme was done by using immersed batch culture at pH 10.5, and the production of enzyme was enhanced by the use of chicken feathers as inducer substrate. At a percentage 0.8% of the concentration of keratin as a catalyst the keratinase activity could be boosted by sixfold. The highest keratinase activity was achieved at temperature between 60 and 70 °C and pH 9.0. These types of keratinases have considerable roles in industrial applications due to its tolerance to solvents and detergents (Indhuja et al. 2012). A recombinant alkaline keratinase of *Bacillus licheniformis* was expressed in *Bacillus subtilis* expression system which showed its ability to be used in the treatments of wool fiber (Liu et al. 2013).

7.2.3 Amylase

Amylase is an enzyme that belongs to the glycosidic hydrolases which catalyzes the hydrolysis of starch to simple sugars. It has been reported that the amylase enzyme has important applications in starch liquefaction; in production of corn syrup, fructose syrup, and glucose syrup; and finally in clarification of fruit juices (Nigam and Singh 1995; Pandey et al. 2000a, b; Khusro et al. 2017). In addition, amylases are used in textile fiber industry and as digestive tablets to enhance digestion of carbohydrates. Most of the amylase enzymes produced from microorganisms including bacteria and fungi. Based on the structural analysis, the molecular studies, and the kinetic of the reactions, the enzyme was proved to belong to the family α-amylases and grouped as cyclodextrin (Sivaramakrishnan et al. 2006; Kumar et al. 2000; Singh et al. 1995). It has been reported that a thermostable isoamylase of a *Bacillus* sp. has been characterized, cloned, and expressed. The produced amylase was novel since it showed optimum activity at a temperature rate of 30–70 °C and at pH range of 5.5–9.0 (Li et al. 2013). A

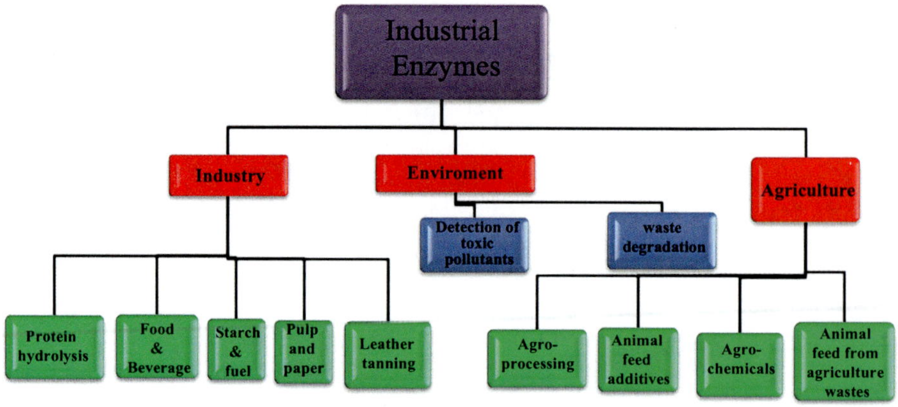

Fig. 7.2 Industrial applications of the microbial enzymes

molecular characterization study of α-amylase isolated from a thermophilic *Geobacillus* sp. has been reported. Diethylaminoethyl cellulose and Sephadex columns were used to purify the enzyme preparation. The results of this study showed that the enzyme is a novel because it has maximum activity at a pH 8.0 and at temperature of 90 °C with stability of 10 min (Gurumurthy and Neelagund 2012). Industrial applications of the microbial enzymes are shown in Fig. 7.2.

7.2.4 Xylanase

The agricultural residues are remnant of agriculture plants after harvesting such as rice, maize, wheat straw, husk, and cotton stalks. The main components of this residue are cellulose, hemicellulose, pectin, and lignin (Chakdar et al. 2016; Nigam and Pandey 2009). Xylan is a β-1,4-linked D-xylopyranosyl sugar which is considered as the main constituent of hemicellulose. The xylan of plant residues could be hydrolyzed by using a mixture of xylanase named β-D-xylosidase and endo-β-1,4-xylanase (Polizeli et al. 2005). It has been assured that xylanases have a great developed application in the field of biotechnology including the processes of clarifying fruit juice, producing fuels, and enhancing the digestion of rumen (Nigam and Pandey 2009). Xylanases have been extensively studied especially xylanases isolated from extremophilic microorganisms (Collins et al. 2005). One of the most important uses of xylanases is the treatment of wood pulp that is used in paper industry (Srinivasan and Rele 1995; Garg et al. 1998). It has been reported that the xylanase used in paper industry has maximum activity at higher temperature and higher pH; in addition, it has no activity against cellulose which is so important for the brightness of the paper. Xylanase enzyme of a thermophilic *Actinomyces thermophilus* has been produced at a temperature of 50 °C and pH of 8.5 and thermostable at 65 °C for 125 min (Kohli et al. 2001).

In addition, an alkalotolerant and thermotolerant xylanase of *Bacillus* sp. has been reported (Marques et al. 1998). Many expression systems including *Pichia*, *Escherichia coli*, and *Bacillus* sp. have been used to clone and express xylanases to be used in a commercial way (Jhamb and Sahoo 2012; Prade 1996; Luo et al. 2012). Also, acidic-stable xylanases of an acidophilic fungus *Bispora* have been reported (Luo et al. 2009). Mamo et al. (2009) have studied the catalytic activity of a xylanase at high pH. Moreover, three thermophilic xylanases of *Humicola* sp. named XynA, XynB, and XynC have been characterized for their possible use in brewing industry (Du et al. 2013). XynA has been found to be stable and adapt alkaline conditions and higher temperatures. Also, XynA possessed higher specificity and catalytic efficiency against various substrates. Another study has shown that XynA, XynB, and XynC have excellent performance in mashing, filtration, and viscosity reduction of substrate in the brewing industry if compared to the performance of the commercial xylanase of Novozyme (Du et al. 2013).

7.2.5 Laccase

Laccase is a copper-containing oxidase enzyme found in many microorganisms, plants, and fungi. Laccase is classified as lignin-modifying enzyme that plays an important role in the biodegradation of lignin. Laccase (ligninase) enzyme is used in the hydrolysis of lignocellulosic compounds of the agricultural residues (Chen et al. 2017). It has been reported that laccase is a synergistic and versatile enzyme; this means it could be used in industrial applications including control of pollution and bioremediation (Nigam et al. 1987a, b; Dahiya et al. 1998, 2001; Robinson et al. 2001a, b; Robinson and Nigam 2008). The cleaner of paper and pulp industry requires an important step of degradation and separation of lignin from agriculture residues, which is established by using laccases. It has been reported that the combination of laccase and xylanase could be used in enhancement of pulp industry (Dwivedi et al. 2010). In addition, isoforms of thermophilic laccase isolated from *Opuntia vulgaris* and *Cereus pterogonus* have been biophysically characterized by Gali and Kotteazeth (2012, 2013) and Kumar and Srikumar (2011, 2012). Several studies have been done to isolate and characterize novel laccases from *Steccherinum ochraceum* and *Polyporus versicolor* that can be used in biotechnological applications (Quaratino et al. 2007; Uthandi et al. 2010; Papinutti et al. 2008; Mishra and Kumar 2009).

7.2.6 Cellulase

Cellulase is an enzyme chiefly produced by bacteria, protozoans, and fungi which catalyze the decomposition of cellulose and other related polysaccharides. Cellulase breaks down the cellulose molecule into monosaccharides (simple sugars) such as beta-glucose, oligosaccharides, and shorter polysaccharides (Sani et al. 2015). Worldwide, cellulase enzyme is considered as the third most important enzyme for industrial uses. Researchers have focused on the production of cellulases for commercial production of glucose from the cellulosic plant residues (Pandey et al. 1999). One of the most important cellulases is those used in bioethanol production from cellulosic plant residues (Xue et al. 2017). Due to the amorphous and crystalline structure of cellulose, cellulose possesses three forms of enzymes that work synergistically. The first enzyme called endoglucanase acts on amorphous cellulose fibers, randomly hydrolyzes the glucose-polymer chain, and finally releases nonreducing and free-reducing ends. Then, the exoglucanase enzyme shall hydrolyze the free ends of the chain that produces cellobiose. Finally, the third form of cellulase is β-glucosidase that hydrolyzes the cellobiose to glucose as a final product of cellulose saccharification. Thermostability of cellulases is so important in the saccharification process of cellulose at higher temperatures, which is necessary to be maintained for the completion of the process. It has been reported that the thermal activity and stability of cellulases isolated from two *Basidiomycetes* cultures were studied. The results of this study showed that the treatment of enzyme preparation by heat causes an activation of endo- and/or exoglucanase activities and improves the enzyme stability during saccharification process. Many studies have been done for cellulase biosynthesis having high activity and high-yield preparation.

7.2.7 Diverse Enzymes

Other microbial enzymes not mentioned or described previously have a significant role in bio-industries. Lipase is an enzyme, which catalyzes the hydrolysis of lipids. Lipase has been extensively studied for their structures and applications in biotechnological industries (Pandey et al. 1999). In addition, pectinase has determined its role in the juice and fruit industries. In pharmaceutical industry, certain enzyme is specifically involved in the production of analytical assays and diagnostic kits. The most important enzymes and their applications in biotechnology are shown in Table 14.1.

7.3 Biological Synthesis of Nanoparticles by Microorganisms

Inorganic materials and biological molecules have been found in continuous interaction with each other since the beginning of life on the earth. Because of this interaction, life could incur on this planet with a well-organized ore of minerals. Recently, scientists have become more concerned in the interaction between biological and inorganic molecules (Reddy et al. 2017). It has been clarified that some microorganisms are able to produce inorganic nanoparticles (NPs)

Table 7.2 Metals nanoparticles produced by different microorganisms

Microorganisms	Nanoparticles name	Size (nm)	Shape	Place of production
Rhodococcus sp.	Au	5–15	Spherical	Intracellular
Shewanella oneidensis	Au	12 ± 5	Spherical	Intracellular
Escherichia coli	Au	20–30	Triangles	Extracellular
Bacillus cereus	Ag	4–5	Spherical	Intracellular
Corynebacterium glutamicum	Ag	5–50	Irregular	Extracellular
Aspergillus flavus	Ag	8.9 ± 1.61	Spherical	Extracellular
Fusarium oxysporum	Ag	5–50	Spherical	Extracellular
Shewanella algae	Pt	5	Not available	Intracellular
Enterobacter sp.	Hg	2–5	Spherical	Intracellular
Shewanella sp.	Se	181 ± 40	Spherical	Extracellular
Escherichia coli	CdTe	2.0–3.2	Spherical	Extracellular
Fusarium oxysporum	Au-Ag alloy	4–18	Spherical	Extracellular
Neurospora crassa	Au, Au/Ag	20–50	Spherical	Intracellular, extracellular

extracellularly or intracellularly. The production of many nanoparticles through microorganisms has already been discussed involving metallic nanoparticles such as silver, alloy, gold, and oxide nanoparticles consisting of nonmagnetic, magnetic, and sulfide (Table 7.2).

7.3.1 Metallic Nanoparticles

7.3.1.1 Gold Nanoparticles

During the Roman era, gold nanoparticles (AuNPs) were used as an ornamental and a decorative material to stain glass. In addition, AuNPs were used in curing many human diseases. AuNPs were synthesized 150 years ago by the researcher Michael Faraday, who was the first to observe that colloid gold solution has characteristic, which differs from piece of gold (Hayat 1989). Nanoparticle synthesis by microorganisms is an emerging field in bionanotechnology that received a considerable attention due to the growing demand for environment-friendly materials. It has been reported that the AuNPs formed extracellularly by *Fusarium oxysporum* as well as actinomycete *Thermomonospora* sp. (Mukherjee et al. 2002; Ahmad et al. 2003a, c). Furthermore, the preparation of AuNPs intracellularly by *Verticillium* sp. has already been discovered (Mukherjee et al. 2001a, b). Another study has been assured that the AuNPs of nanoscale dimensions may be precipitated inward of bacterial cell by incubation of the cells with Au^{3+} ions (Li et al. 2016; Southam and Beveridge 1996). Under extreme biological conditions such as alkaline and somewhat higher temperature, monodisperse AuNPs have been synthesized by using alkalotolerant *Rhodococcus* sp. (Ahmad et al. 2003b). Different shapes of AuNPs have been synthesized including cubic, octahedral, and spherical by cyanobacteria from Au_2Cl_6 and $Na_3Au(S_2O_3)_2$ complexes (Lengke et al. 2006a, b). The growth of nanoalloys and nanocrystals by using *Lactobacillus* has been shown (Nair and Pradeep 2002; Singaravelu et al. 2007; Suresh et al. 2011; Gericke and Pinches 2006, Du et al. 2007; Agnihotri et al. 2009; Husseiny et al. 2007; He et al. 2007; Konishi et al. 2007a, b).

7.3.1.2 Silver Nanoparticles

Silver nanoparticles (AgNPs) have size ranging between 1 and 100 nm. AgNPs have shown antibacterial activity not only against both Gram-negative and Gram-positive bacteria but also against highly multiresistant such as *Staphylococcus aureus* (Panácek et al. 2006). Many microorganisms are able to reduce the Ag^+ ions into spherical AgNPs (Mukherjee et al. 2001; Ahmad et al. 2003a, b, c; Fayaz et al. 2010). It has been reported that *Pseudomonas stutzeri* AG259 and *E. coli* were able to produce AgNPs within their periplasmic space when it sets on concentrated solution of $AgNO_3$ (Klaus et al. 1999; Orlov et al. 2016). In case of using *Fusarium oxysporum*, *Aspergillus flavus*, *Verticillium*, and *Mucor hiemalis*, AgNPs were prepared as a film or accumulated in their cell (Jain et al. 2011; Vigneshwaran et al. 2006, 2007; Bhainsa and D'Souza 2006; Senapati et al. 2004; Kalishwaralal et al. 2010; Castro-Longoria et al. 2011; Kalimuthu et al. 2008; Gurunathan et al. 2009; Sneha et al. 2010; Mohammed et al. 2009; Juibari et al. 2011; Babu and Gunasekaran 2009; Aziz et al. 2016) (Fig. 7.3).

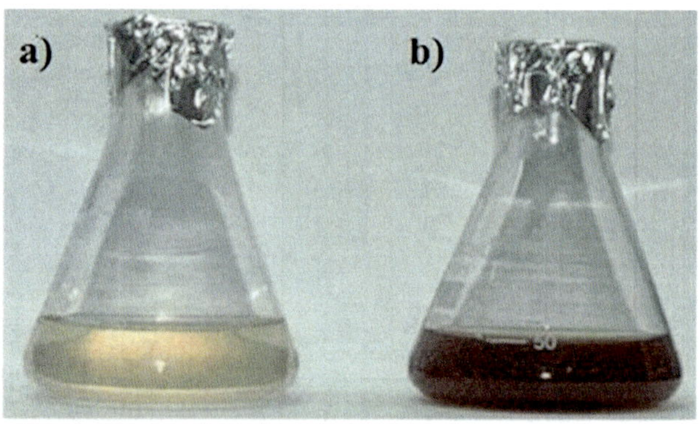

Fig. 7.3 Silver nanoparticles produced by microorganisms (**a**) before synthesis (control), (**b**) after synthesis (synthesized silver nanoparticles)

7.3.1.3 Alloy Nanoparticles

The result of preparation of two mixed nanoparticles is called alloy nanoparticles that are very interesting to researcher because of their applications in catalysis, electronics, as optical materials, and coatings (Senapati et al. 2005; Zheng et al. 2010a, b).It has reported that NADH plays an important role in biosynthesis and composition of Au-Ag alloy by *F. oxysporum* nanoparticles (Senapati et al. 2005). In addition, yeast cell is able to biosynthesize Au-Ag alloy nanoparticles (Zheng et al. 2010a, b). Characterization of Au-Ag alloy NPs by transmission and fluorescence electron microscope revealed that it is synthesized extracellularly and found in irregular polygonal NPs (Tripathi et al. 2015). Interestingly, it has been shown that the use of Au-Ag alloy NPs in vanillin carbon electrode is able to enhance the electrochemical response of the electrode by five times. It has been demonstrated that the Au-Ag alloy NP suspensions of *Fusarium semitectum* were stable for weeks (Sawle et al. 2008). Scanning electron microscope (SEM) photography shows the different shapes and sizes of nanoparticles produced by microorganisms (Fig. 7.4).

7.3.1.4 Other Metallic Nanoparticles

Generally, heavy metals are those metals with relatively high atomic number, densities, or atomic weights. Heavy metals are toxic to various microorganisms. Microorganisms able to resist the effect of heavy metals by chemical detoxification could be used as chemiosmotic cation, proton anti-transporters, and ATPase.

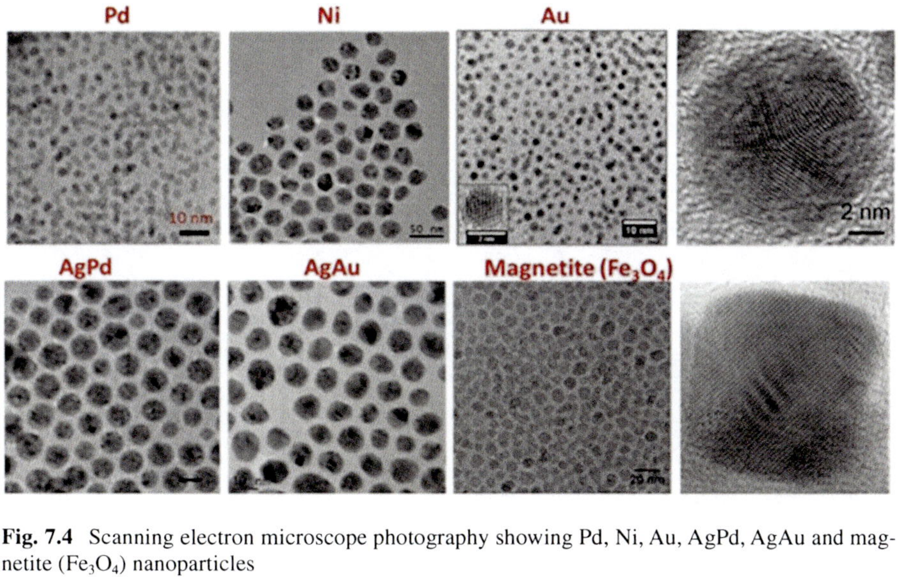

Fig. 7.4 Scanning electron microscope photography showing Pd, Ni, Au, AgPd, AgAu and magnetite (Fe_3O_4) nanoparticles

Moreover, aqueous $PtCl_6^{2-}$ ion was reduced into platinum nanoparticles of 5 nm size by *Shewanella* sp. when lactate was used as an electron donor (Konishi et al. 2007a, b). Also, HgNPs of size 2–5 nm were synthesized at pH 8 and low concentration of Hg by *Enterobacter* sp. (Sinha and Khare 2011). Interestingly, a hyperthermophilic and an anaerobic *Pyrobaculum islandicum* has the ability to reduce many heavy metals such as Tc(VII), U(VI), Cr(VI), Mn(IV), and Co(III) when hydrogen is used as an electron donor (Kashefiand and Lovley 2000). Moreover, Pd-NAPs could be synthesized by *Desulfovibrio desulfuricans*, a sulfate-reducing bacteria (Lloyd et al. 1998; Yong et al. 2002; De Windt et al. 2005; Lee et al. 2007; Bao et al. 2010).

7.3.1.5 Metal Oxide Nanoparticles

Metal oxide nanoparticle (MO-NP) synthesis represents a field of material chemistry that attracts considerable interest due to the potential technological applications of these compounds. The effects of these materials on fields including information technology, medicine, energy storage, sensing, and catalysis have paid the attention of much research in developing synthetic pathways to such nanostructures. There are two types of MO-NPs, nonmagnetic oxide and magnetic oxide nanoparticles (Table 7.3).

Table 7.3 Oxide and sulfide nanoparticles produced by different microorganisms

Microorganisms	Nanoparticles name	Size (nm)	Shape	Place of production
Rhodopseudomonas palustris	CdS	8	Cubic	Intracellular
Multicellular prokaryotes	FeS_4	Not available	Not available	Intracellular
Rhodobacter sphaeroides	CdS	8	Hexagonal	Intracellular
Rhodobacter sphaeroides	ZnS	10.5 ± 0.15	Spherical	Extracellular
Fusarium oxysporum	CdS	5–20	Spherical	Extracellular
Fusarium oxysporum	$SrCO_3$	10–50	Needle like	Intracellular
Yeasts	$Zn_3(PO_4)_2$	10–80 × 80–200	Rectangular	Extracellular
Shewanella oneidensis	Fe_3O_4	40–50	Rectangular, rhombic	Extracellular
Lactobacillus sp.	$BaTiO_3$	20–80	Tetragonal	Extracellular
Fusarium oxysporum	TiO_2	6–13	Spherical	Extracellular
Saccharomyces cerevisiae	Sb_2O_3	2–10	Spherical	Intracellular
Lactobacillus sp.	$BaTiO_3$	20–80	Tetragonal	Extracellular

7.3.1.6 Magnetic Nanoparticles

New materials of magnetic nanoparticles have been established due to their unique particular properties and micro-configuration such as high coercive force and superparamagnetic force and their promising applications in biomedicine fields and biological separation. Fe_2O_3 (maghemite) and Fe_3O_4 (magnetite) are considered as magnetic nanoparticles, and both of them are biocompatible. Both Fe_2O_3 and Fe_3O_4 were investigated to target stem cell sorting and cancer medication and treatment (magnetic hyperthermia) and to apply in DNA analysis, magnetic resonance imaging (MRI), and gene therapy (Fan et al. 2009). It has been reported that magnetotactic bacteria are able to synthesize magnetic particles intracellularly including iron sulfides and/or iron oxide (Bazylinski et al. 1994, 1995). Bacterial magnetite particles (BacMPs) that are arranged at an aligned row through the bacterium cell may act as biological compass needles, which help the bacterium cell to move across oxygen gradients in aquatic environments by the action of the Earth's magnetic field (Arakaki et al. 2008; Blakemore 1975; Shimoshige et al. 2017). Furthermore, BacMPs include a single magnetite or magnetic domain that has magnetic properties (Thornhill et al. 1995). Magnetotactic bacteria possess various morphological shapes such as spirilla; vibrios; ovoid, rod-shaped, cocci bacteria; and multicellular bacteria which possess unique properties (Thornhill et al. 1995; Spring and Schleifer 1995). It has been shown that magnetotactic cocci are microaerophilic and have high diversity at the water sediment surface. Three facultative anaerobic marine vibrio bacteria were named as vibrio strain MV-1, MV-2, and MV-4 that have been isolated from estuarine salt marshes. The three bacteria have been proven to belong to α-Proteobacteria and family *Rhodospirillaceae*. They were observed to synthesize a truncated hexoctahedron shape of BacMPs when increased as chemolithoautotrophic or chemoorganoheterotrophic. Otherwise, the family member of Magnetospirillaceae could be isolated by magnetic isolation techniques especially magnetotactic bacteria (Dinali et al. 2017). It has been reported that the uncultured magnetotactic bacteria were observed in numerous habitats that are able to grow at or below 30 °C (Lefevre et al. 2010a, b). The cultured magnetotactic bacteria were mesophilic and couldn't grow over 30 °C anymore. Some researchers have described the isolation of thermophilic magnetotactic bacteria (Marcano et al. 2017). A magnetotactic bacteria strain was isolated from hot spring with a temperature range of 32–63 °C and named as HSMV-1 (Lefevre et al. 2010a, b). The transmission electron microscope (TEM) images of unstained cells of HSMV-1 indicated a chain of bullet-shaped magnetosomes with a polar flagellum. The magnetosome crystal number was 12 ± 6/cell and 113 ± nm by 40 ± 5 nm dimension. The results of this study have elucidated that the magnetotactic bacteria are considered as moderately thermophilic, since the growth and deposition of magnetosome magnetite were around 63 °C (Lefevre et al. 2010a, b). In addition, the mesoporous structure of magnetic Fe_3O_4 was synthesized by yeast cells by using coprecipitation method (Zhou et al. 2009; Perez-Gonzalez et al. 2010; Zhu et al. 2010; Amemiya et al. 2007; Li et al. 2007; Bose et al. 2009).

7.3.1.7 Nonmagnetic Oxide Nanoparticles

Also, some researches discussed other oxide nanoparticles such as Sb_2O_3, TiO_2, $BaTiO_3$, ZrO_2, and SiO_2 nanoparticles (Jha et al. 2009a, b; Jha and Prasad 2010; Bansal et al. 2004, 2005, 2006; Narayanan and Sakthivel 2010). It has been shown that a low-cost green *Saccharomyces cerevisiae* is able to biosynthesize Sb_2O_3NPs at room temperature (Jha et al. 2009a, b). The analysis of Sb_2O_3NPs indicated that they have a spherical shape aggregate of a size ranging from 2 to 10 nm (Jha et al. 2009a, b). *Fusarium oxysporum* could produce TiO_2 and SiO_2 nanoparticles from a complex solution of TiF_6^{2-} and SiF_6^{2-}, respectively (Bansal et al. 2005). Moreover, they prepare quasi-spherical ZrO_2 and tetragonal $BaTiO_3$ nanoparticles from *F. oxysporum* with a size range of 3–11 nm and 4–5 nm, respectively (Bansal et al. 2004, 2006) (Table 14.3).

7.3.1.8 Sulfide Nanoparticles

Also, sulfide nanoparticles have been discussed because of their great technical and research usage and applications as cell labeling and fluorescent biomarkers on account of their optical or electronic properties (Yang et al. 2005). Microorganisms are able to synthesize CdS nanocrystal that is a typical type of sulfide nanoparticle. It has been found that *Clostridium thermoaceticum* is able to precipitate CdS in the medium and on their cell surfaces from $CdCl_2$ in the presence of cysteine hydrochloride in the growth medium using cysteine as a sole source of sulfide (Cunningham and Lundie 1993). Also, when *Klebsiella pneumoniae* Cd^{2+} ions are present in the growth medium, it form CdS NPs of size range between 20 and 200 nm on the cell surface. Interestingly, *Escherichia coli* was able to form wurtzite crystal intracellular CdS nanocrystals when incubated with Na_2SO_4 and $CdCl_2$ (Sweeney et al. 2004). In addition, the growth phase of *E. coli* affects nanocrystal formation to 20-fold when grown in the stability stage versus that grown in the late logarithmic stage. A research group has utilized *C. glabrata* and *S. pombe* to produce CdS nanoparticles intracellularly when used in cadmium solution (Dameron et al. 1989). It has been shown that *Desulfobacteraceae* and *Rhodobacter sphaeroides* are able to use ZnSNPs intracellularly in a range of 8 and 2–5 nm, respectively (Bai et al. 2006; Labrenz et al. 2000). In addition, PbSNPs were synthesized by *Rhodobacter sphaeroides* (Bai and Zhang 2009). Fungi, for example, eukaryotic microorganisms, are an excellent good choice for the preparation of sulfide nanoparticles (Ahmad et al. 2002). ZnS, CdS, MoS_2, and PbS have been found to be produced extracellularly by *F. oxysporum* when metal sulfate solution is used. Fe_3S_4 has been reported to be produced by magnetotactic uncultured bacteria (Bazylinski et al. 1995). Finally, magnetic FS nanoparticles have been reported to be produced by sulfate-reducing bacteria (Watson et al. 1999; Arakaki et al. 2010a, b; Bai et al. 2009; Sanghi and Verma 2009; Prasad and Jha 2010; Sweeney et al. 2004; Kowshik et al. 2002). Recently, fluorescent lead (IV) sulfide nanoparticles synthesized by *Idiomarina* sp. strain PR8–8 have been reported (Srivastava and Kowshik 2017).

7.3.1.9 Other Nanoparticles

Another study discussed that nanoparticles including $Zn_3(PO_4)_2$, $PbCO_3$, $SrCO_3$, CdSe, and $CdCO_3$ have been recorded to be prepared by microorganisms (Sanyal et al. 2005; Rautaray et.al. 2004; Pandian et al. 2009; Yan et al. 2009; Kumar et al. 2007). It has been reported that $SrCO_3$NPs were obtained by *Fusarium oxysporum* when it grow in a solution of Sr^{2+} ions (Rautaray et al. 2004). Also, $Zn_3(PO_4)_2$ NPs have been shown to be synthesized at a size ranging from 10 to 80 nm by yeast (Pandian et al. 2009; Yan et al. 2009). CdSeNPs have been shown that it could be synthesized by *F. oxysporum* at room temperature (Kumar et al. 2007).

7.3.2 Mechanisms of Nanoparticle Formation by Microorganisms

Nanoparticles are produced by microorganisms by different mechanisms. Firstly, metal ions are trapped inside or on the surface of microbial cells. In the presence of enzymes, the metal ions are reduced to nanoparticles. Generally, microorganisms influence the NP formation by two special ways: firstly, the composition of the solution becomes more supersaturated and, secondly, by impacting mineral formation through the microorganisms (Benzerara et al. 2010).

7.3.2.1 How Microorganisms Synthesize Nanoparticles

It has been proposed that the silver and gold nanoparticles are prepared by algal biomass or *Verticillium* sp. through trapping the silver and gold ions on the surface of the fungal cells. Then, an interaction occurs between the negatively charged cell wall from the carboxylate groups and the ions in the enzymes. At this point, the enzymes decreased the metal ions to form silver or gold nuclei that grow by further accumulation and reduction subsequently (Sneha et al. 2010). Moreover, another study postulated that the synthesis of AgNPs in *B. licheniformis* is catalyzed by nitrate reductase (Kalishwaralal et al. 2008). NADH-dependent nitrate reductase and NADH are considered as important factors that are involved in nanoparticle formation. It has been stated that *B. licheniformis* reduces Ag^+ to Ag^0, and then AgNPs are formed by NADH and NADH-dependent nitrate reductase (Husseiny et al. 2007; Prasad et al. 2016). For example, Cd^{2+}, Hg^{2+}, Co^{2+}, Ag^+, Cu^{2+}, Ni^{2+}, CrO_4^{2+}, Zn^{2+}, and Pb^{2+} have a toxic effect on microorganism's survival. Microorganisms are developing proteomic and genetic responses that enable cell to detoxify the effect of heavy metals by many mechanisms including efflux, reductive precipitation, or complexation (Reith et al. 2007; Nies 1999). As a result metallophilic microorganisms are able to thrive in the presence of higher concentrations of heavy metal ions (Mergeay et al. 2003). In addition, the biomineralization of BacMP by molecular mechanisms is a multistep process (Arakaki et al. 2008). Firstly, an invagination of the cytoplasmic membrane and the vesicle formed serve as the precursor of the BacMP membrane. However, the mechanism of envelope formation stays unclear. It has been assumed that BacMP vesical formation is probably mediated by a specific GTPase, which is similar to eukaryotes. Then, the formed vesicle will be assembled as a linear chain within a cytoskeletal filament. Secondly, the ferrous ions are collected inside the BacMP biomineralized vesicle through the transmembrane iron transporters. Finally, BacMP-bound proteins are able to enhance morphology and/or nucleation of magnetite crystal (Ghashghaei and Emtiazi 2016). The BacMP membrane together with other associated proteins may have a functional role in magnetite formation such as iron supersaturation, oxidation of iron, and maintenance of reductive conditions (Arakaki et al. 2008). It has been proposed that *Shewanella oneidensis* can form magnetites by utilizing active as well as passive mechanisms (Perez-Gonzalez et al. 2010). It has been suggested that CdSNPs are formed through cysteine bond and may involve the cleavage of S–H bond and formation of a new bond, which is (−S–Cd) bond of Cd-thiolate (Cd–S–CH_2COOH) complex on the nanoparticle surface (Sanghi and Verma 2009). It has been shown that cadmium-thiolate complexes reacted with H bound but did not react with $-NH_2$ groups of protein. As a result, the archived CdS NPs are bonded to $-NH_2$ groups by hydrogen by one oxygen atom of the carboxylic group (−COOH) (Lover et al. 1997; Tang et al. 2005).

7.3.2.2 Control of Size and Morphology of Nanoparticles

The optical and electronic properties of nanoparticles are mostly dependent on the shape and size. Most of the researchers are interested in managing the shape and size of nanoparticles. Microorganisms have the ability to control the shape and size of biological nanoparticles (Zhang et al. 2016). It has been shown that the synthesis of AuNPs of various sizes and morphologies may be obtained by *V. luteoalbum* (Gericke and Pinches 2006). The particle size and formation could be controlled by many parameters such as concentration of gold, temperature, and pH (Gurunathan et al. 2009). The scanning electron microscope pictures showed that great variations in size of particle range from 1 to 100 nm in diameter. Also, the results of the study suggested that the triangular- and hexagonal-shaped NPs are bigger than spherical NPs. On the synthesis of Pt nanoparticles, Riddin and coworkers found that in the absence of the spatial restrictions of the cell wall, the cell-soluble extract (CSE) was able to reduce Pt(IV) to form nanoparticles, which are stabilized in solution by bound proteins and exhibit both geometric and irregular morphologies (Riddin et al. 2010). An associated protein with bacterial magnetites called Mms6 was discovered in *Magnetospirillum magneticum* AMB-1 (Arakaki et al. 2010a, b). The group found the Mms6 protein involved in the formation of cuboctahedral magnetite crystals. In addition, the recombinant Mms6 bacterial protein was used to synthesize the magnetite crystals in solution under low temperatures in aqueous solution using recombinant Mms6 magnetotactic bacterial protein. Moreover, the aggregates of Mms6 have a sequence motif with high affinity to iron ions. The crystals exhibit similar sizes (20 nm) and morphologies (cuboctahedral) versus crystals formed in the absence of Mms6. This proposes that Mms6 has a huge impact in regulating the size as well as shape of nanoparticles during the synthesis process (Amemiya et al. 2007).

7.3.2.3 Applications of Nanoparticles

Nanomedicine is one of the most hopeful research fields especially in the treatment and diagnosis of diseases (Fadeel and Garcia-Bennett 2010). Usually, the dispersed types of nanoparticles are employed as biological fluorescent labels, gene delivery, detection of pathogen, engineering of tissues, destruction of tumor by heating, contrast enhancement of MRI, and pharmacokinetic studies (Chan and Nie 1998; Tian et al. 2008; Cui et al. 2007; Pantarotto et al. 2003; Edelstein et al. 2000; De la Isla et al. 2003; Ma et al. 2003; Shinkai et al. 1999; Weissleder et al. 1990; Parak et al. 2002; Prasad et al. 2016, 2017a, b). The usage of nanoparticles in the field of nanomedicine in various research papers and reviews has been reported (Emerich and Thanos 2006; Alanazi et al. 2010; Rodríguez and Villaverde 2010; Vaidyanathan et al. 2009; Mahmoudi et al. 2010; Dias et al. 2011; Shen et al. 2011; Piao et al. 2011; Chakravarthy et al. 2010; Xiang et al. 2007a, b; Prasad et al. 2017a, b). Here, we provide some examples to illustrate these applications.

7.3.2.4 Drug Delivery

Drug design accurately and safely to get the target sites in a timely manner to achieve maximum therapeutic effect is the key issue in the design and development of systems for the introduction of new medicines. In addition, it should be the nano-carriers transmitted through blood tissue barriers to reach the target cells. They must enter target cells via endocytotic transport and transcytotic techniques and mechanisms across cellular barriers (Fadeel and Garcia-Bennett 2010; Sun et al. 2014).

Because of their small size, nano-drug carriers are nanoparticles which exceed the blood-brain barrier and epithelial tight junctions of the skin that usually hinder the delivery of drugs to the target site. Secondly, as a result of the high space for the size of the nano-carriers, they show improved pharmacological and bio-distribution of therapeutic agents, thus reducing the toxicity of the accumulation in the target site (Vaidyanathan et al. 2009). Nano-carriers enhance the solubility of hydrophobic compounds and make them easy to be administered. Moreover, higher nano-carriers stabilize many therapeutic materials such as oligonucleotide and peptides (Emerich and Thanos 2006). It has been known that nanoparticles such as Fe_3O_4 (maghemite) and Fe_2O_3 (magnetite) are compatible and were investigated to target cancer disease, drug delivery, stem cell, MRI and gene therapy (Fan et al. 2009). The toxicity of magnetosome of *Magnetospirillum gryphiswaldense* has been applied to fibroblasts of mouse, and the results showed that the sterilized and purified magnetosomes were not toxic to mouse fibroblasts in vitro (Xiang et al. 2007a, b). The effect of magnetic bacterial nanoparticles on the immune response of mouse has been reported (Meng et al. 2010). In this study, the antigen of ovalbumin was used and was mixed with an adjuvant. BacMps were used for immunization of mouse BALB/C. The IgG of anti-ovalbumin, T lymphocyte proliferation, IFN-gamma, and the expression of ILs were estimated. The results of this study revealed that the BMPs have no influence on immune responses of mouse, while the magnetosomes could be used as a novel gene carriers or drug for the therapy of tumor. Moreover, the load of doxorubicin onto BMs was reported to study the ability of these particles to suppress the growth of tumor (Sun et al. 2007). The pharmacokinetics and biocompatibility of BMs were estimated in urine, serum and main organs after the injection of BMs into Sprague-Dawley (SD) rat (Sun et al. 2009). The results of this study revealed that the BMs were found only in the livers, while there was any evidence for the existence of BMs in the urine, after 72 h of intravenous administration (Sun et al. 2009). It has been shown that magnetotaxis was applied to change magnetotactic bacteria (MTB) that are embedded together with the flagella and magnetite nanoparticles to move inside the small blood capillaries (Felfoul et al. 2007). It has been reported that the utilization of MTB nanoparticles to deliver gene using Polyethyleneimine-associated (PEI-associated) with MTB for delivering the β-galactosidase reporter plasmid both in vivo and in vitro (Xie et al. 2009). The study revealed that the MTB-PEI nanoparticle systems are less toxic and more efficient in comparison with PEI alone. It has been shown that the gold and its derivative compounds were used since ancient time of world civilization as medicinal agents (Mukherjee et al. 2007; Mahdihassan 1988; Higby 1982; Patra et al. 2010; Giljohann et al. 2010). In addition, the biological and chemical applications of AuNPs were reported. In addition, Ag-NPs were used as a therapeutic drug because of its properties as antifungal, antibacterial, anti-inflammatory and antiviral. It has been shown that Ag-NPs of *Bacillus licheniformis* have a potential application as anti-angiogenic (Kalishwaralal et al. 2009). It has been expected that NPs aiming drug delivery would significantly decrease the dose of antitumor drugs with excellent specificity and less toxicity and improve efficiency. Nearly in the future, we expect the increase of applications in the field of diagnostic and therapeutic nanotechnology. It has been reported that the magnetic nanoparticles could be used for treatment of cancer hyperthermia. The treatment of cancer hyperthermia would be involving the administration of the nanoparticles inside the cancer tissue sites of the body. The heating of specific sites locally shall be empowered by application of a magnetic field (Chertok et al. 2008).

7.3.2.5 Antibacterial Agent

Due to increase and prevalence, the resistance of microorganisms against antiseptics and antibiotic has been confirmed recently. It has been reported that Ag-NPs were synthesized by *Trichoderma viride*. This study showed that Ag^{+} ions were reduced to AgNPs by using *T. viride* filtrate with a size ranging from 5 to 40 nm. The antimicrobial activity of AgNPs was tested together with many antibiotics against

Fig. 7.5 Different applications of biologically synthesized nanoparticles

Gram-negative and Gram-positive bacteria. The results revealed the increase antibacterial activity of kanamycin, ampicillin, chloramphenicol, and erythromycin when AgNPs were added against tested strains. The results concluded that the addition of AgNPs with antibiotics has improved antimicrobial activity that opens the gate to develop antimicrobial agents (Prasad 2014; Aziz et al. 2015, 2016; Prasad et al. 2016; Abbasifar et al. 2017). It has been reported that the production of AgNPs extracellularly by *Fusarium oxysporum* could be integrated into textile industries to minimize or prevent deterioration or contamination with different bacteria and fungi (Duran et al. 2007). Different applications of biologically synthesized nanoparticles are shown in Fig. 7.5.

7.3.2.6 Biosensor

Interestingly, nanoparticles have optic and electronic properties that could be used in applications of biosensor. It has been reported that SeNPs were formed by the *Bacillus subtilis* having a diameter ranging from 50 to 400 nm (Wang et al. 2010).

These SeNPs were spherical and monoclinic that become one-dimensional, trigonal, and anisotropic structure at room temperature. Moreover, SeNPs with high ratio surface-volume, biocompatibility, adhesive ability, and biocompatibility were utilizing as settled material to build and enhance horseradish peroxidase (HRP) biosensor. The modification of SeNP electrode will be probably promising for application in food, clinical, pharmaceutical, environmental, and industrial analyses, as well as detection of H_2O_2 in food. Another study has been shown that the biosensor of AuNP glucose oxidase (GOx) was improved. The AuNPs was shown an increase of GOx activity, and able to detect the glucose content in a glucose injection (Zheng et al. 2010a, b). Efforts have been done for developing novel nanomaterial-based biosensors for various targets, as well as the enormous potentials provided by these biosensors (Hou et al. 2016).

7.3.2.7 Reaction Rate Enhancement Agent

It has been reported that nanoparticles were used widely for improvement of various reactions as catalyst and/or reductants because of their special characteristics and huge surface areas (Hildebrand et al. 2008). In addition, magnetic nanoparticles have been reported to be improving the rate of various microbiological reactions. Not only the magnetic nanoparticles were used for their catalytic behavior but also for their ability to disband. A microbial cell of *Pseudomonas delafieldii* was coated with Fe_3O_4 magnetic nanoparticles to accomplish de-sulfurization of dibenzothiophene (Shan et al. 2005).

7.3.2.8 Magnetic Separation and Detection

The magnetic nanoparticles incorporated with biomolecules that are used as a label material in biological assays have been proposed. In addition, antibodies carried on BacMPs were improved for sensitive and rapid determination of tiny molecules such as hormones, toxic detergents and environmental pollutants by using chemiluminescence immunoassays (Matsunaga et al. 2003; Tanaka et al. 2004). Xenoestrogens, like alkyl phenol, bisphenol, and alkyl benzene, were detected by the use of monoclonal antibodies immobilized on BacMPs. The modification of magnetic nanoparticle surface is a very interesting research area with many applications. The surface of BacMP could be changed with amino saline to improve the DNA extraction by magnetic nanoparticle system. The magnetic nanoparticles are very promising biomolecules as an adsorbent solid phase that are suitable for DNA extraction because it can be easily handled by application of simple magnet.

7.4 Future Prospects

Recently, the development of nanoparticle production from microorganisms and their applications has been reported over the past years. Even so, intensive studies are needed to progress the control and efficiency synthesis of nanoparticle morphology and size. It has been known that the nanoparticle synthesis from microorganisms is a very slow process that may take several hours to few days in comparison to chemical and physical approaches. The reduction of time for nanoparticle synthesis would make the biosynthesis way much attractive. Both monodispersity and particle size are very important points in the estimation of nanoparticle synthesis.

Therefore, effective control of the particle size and monodispersity must be extensively investigated. It has been shown that nanoparticles synthesized from microorganisms might be hydrolyzed by time. As a result, there is a need for more studies to prolong and enhance the stability of nanoparticles (Xiang et al. 2007a, b; Hergt et al. 2005; Hergt and Dutz 2007). Changing parameters such as growth phase, microorganism strain, physical conditions, growth medium, source of nanoparticles, substrate concentrations, and nontarget ions might be sufficient to control monodispersity and particle size. Nowadays, researchers are studying microbial cells at proteomic and genomic level to understand the mechanism of nanoparticle synthesis on a molecular and cellular level. As an example identification and purification of the molecules that are able to reduce nanoparticles shorten the reaction time and increase efficiency of synthesis.

7.5 Conclusion

Nowadays, the research of nanomedicine becomes a growing field of research with huge possibilities for enhancing the diagnosis and treatment of human diseases. It is believed that the biosynthesis of nanoparticles by microbes is pure, clean, and nontoxic; it is necessary to implement eco-environmentally acceptable "green chemistry" proceedings. There are two ratings for the use of microorganisms including bacteria, yeast, fungi, and actinomycetes: intracellular and extracellular synthesis. This rating depends on the location where nanoparticles are composed. It's worth mentioned that controlling parameters such as pH, temperature, and substrate concentration could manipulate the rate of intracellular particle formation and therefore the size of the nanoparticles. Nowadays, research is conducted manipulating microorganisms at both the genomic and proteomic ranges. With the new progress and the continuing efforts in enhancing particle synthesis effectively and finding out their biomedical applications, the usage of these methods and approaches on a wide range and their commercial applications in medicine and health care will hopefully occur in the next years.

References

Abbasifar A, Ghani S, Irvani MA, Rafiee B, Kaji BV, Akbari A (2017) Antibacterial activity of silver nanoparticles synthesized by using extract of hedera helix. Zahedan J Res Med Sci. 10.17795/zjrms-5920

Agnihotri M, Joshi S, Kumar AR, Zinjarde S, Kulkarni S (2009) Biosynthesis of gold nanoparticles by the tropical marine yeast *Yarrowia lipolytica* NCIM 3589. Mater Lett 63(15):1231–1234

Ahmad A, Mukherjee P, Mandal D (2002) Enzyme mediated extracellular synthesis of CdS nanoparticles by the fungus, *Fusarium oxysporum*. J Am Chem Soc 124(41):12108–12109

Ahmad A, Mukherjee P, Senapati S (2003a) Extracellular biosynthesis of silver nanoparticles using the fungus *Fusarium oxysporum*. Colloids Surf B: Biointerfaces 28(4):313–318

Ahmad A, Senapati S, Khan MI (2003b) Intracellular synthesis of gold nanoparticles by a novel alkalotolerant actinomycete, *Rhodococcus species*. Nanotechnology 14(7):824–828

Ahmad A, Senapati S, Khan MI, Kumar R, Sastry M (2003c) Extracellular biosynthesis of monodisperse gold nanoparticles by a novel extremophilic actinomycete, *thermomonospora sp*. Langmuir 19(8):3550–3553

Ahmed S, Riaz S, Jamil A (2009) Molecular cloning of fungal xylanases: an overview. Appl Microbiol Biotechnol 84:19–35

Alanazi FK, Radwan AA, Alsarra IA (2010) Biopharmaceutical applications of nanogold. Saudi Pharm J 18(4):179–193

Amemiya Y, Arakaki A, Staniland SS, Tanaka T, Matsunaga T (2007) Controlled formation of magnetite crystal by partial oxidation of ferrous hydroxide in the presence of recombinant magnetotactic bacterial protein Mms6. Biomaterials 28(35):5381–5389

Arakaki A, Nakazawa H, Nemoto M, Mori T, Matsunaga T (2008) Formation of magnetite by bacteria and its application. J R Soc Interface 5(26):977–999

Arakaki A, Shibusawa M, Hosokawa M, Matsunaga T (2010a) Preparation of genomic DNA from a single species of uncultured magnetotactic bacterium by multiple-displacement amplification. Appl Environ Microbiol 76(5):1480–1485

Arakaki A, Masuda F, Amemiya Y, Tanaka T, Matsunaga T (2010b) Control of the morphology and size of magnetite particles with peptides mimicking the Mms6 protein from magnetotactic bacteria. J Colloid Interface Sci 343(1):65–70

Aziz N, Faraz M, Pandey R, Sakir M, Fatma T, Varma A, Barman I, Prasad R (2015) Facile algae-derived route to biogenic silver nanoparticles: synthesis, antibacterial and photocatalytic properties. Langmuir 31:11605–11612. https://doi.org/10.1021/acs.langmuir.5b03081

Aziz N, Pandey R, Barman I, Prasad R (2016) Leveraging the attributes of *Mucor hiemalis*-derived silver nanoparticles for a synergistic broad-spectrum antimicrobial platform. Front Microbiol 7:1984. https://doi.org/10.3389/fmicb.2016.01984

Babu MMG, Gunasekaran P (2009) Production and structural characterization of crystalline silver nanoparticles from *Bacillus cereus* isolate. Colloids Surf B: Biointerfaces 74(1):191–195

Bai HJ, Zhang ZM (2009) Microbial synthesis of semiconductor lead sulfide nanoparticles using immobilized Rhodobacter sphaeroides. Mater Lett 63(9–10):764–766

Bai HJ, Zhang ZM, Gong J (2006) Biological synthesis of semiconductor zinc sulfide nanoparticles by immobilized *Rhodobacter sphaeroides*. Biotechnol Lett 28(14):1135–1139

Bai HJ, Zhang ZM, Guo Y, Yang GE (2009) Biosynthesis of cadmium sulfide nanoparticles by photosynthetic bacteria *Rhodopseudomonas palustris*. Colloids Surf B: Biointerfaces 70(1):142–146

Baihong L, Juan Z, Zhen F, Lei G, Xiangru L, Guocheng D, Jian C (2013) Enhanced thermostability of keratinase by computational design and empirical mutation. J Ind Microbiol Biotechnol 40:697–704

Banat IM, Nigam P, Marchant R (1992) Isolation of a thermotolerant, fermentative yeasts growing at 52°C and producing ethanol at 45°C & 50°C. World J Microbiol Biotechnol 8:259–263

Bansal V, Rautaray D, Ahmad A, Sastry M (2004) Biosynthesis of zirconia nanoparticles using the fungus *Fusarium oxysporum*. J Mater Chem 22:3303–3305

Bansal V, Rautaray D, Bharde A (2005) Fungus-mediated biosynthesis of silica and titania particles. J Mater Chem 15(26):2583–2589

Bansal V, Poddar P, Ahmad A, Sastry M (2006) Room temperature biosynthesis of ferroelectric barium titanate nanoparticles. J Am Chem Soc 128(36):11958–11963

Bao H, Lu Z, Cui X (2010) Extracellular microbial synthesis of biocompatible CdTe quantum dots. Acta Biomater 6(9):3534–3541

Bazylinski DA, Garratt-Reed AJ, Frankel RB (1994) Electron microscopic studies of magnetosomes in magnetotactic bacteria. Microsc Res Tech 27(5):389–401

Bazylinski DA, Frankel RB, Heywood BR (1995) Controlled biomineralization of magnetite (Fe_3O_4) and greigite (Fe_3S_4) in a magnetotactic bacterium. Appl Environ Microbiol 61(9):3232–3239
Benzerara K, Miot J, Morin G, Ona-Nguema G, Skouri-Panet F, Ferard C (2010) Significance, mechanisms and ´ environmental implications of microbial biomineralization. Compt Rendus Geosci 343(2–3):160–167
Berka RM, Grigoriev IV, Otillar R, Salamov A, Grimwood J, Reid I, Ishmael N, John T, Darmond C, Moisan MC (2011) Comparative genomic analysis of the thermophilic biomass-degrading fungi *Myceliophthora thermophila* and *Thielavia terrestris*. Nat Biotechnol 29:922–927
Bhainsa KC, D'Souza SF (2006) Extracellular biosynthesis of silver nanoparticles using the fungus *Aspergillus fumigatus*. Colloids Surf B: Biointerfaces 47(2):160–164
Blakemore R (1975) Magnetotactic bacteria. Science 190(4212):377–379
Bose S, Hochella MF, Gorby YA et al (2009) Bioreduction of hematite nanoparticles by the dissimilatory iron reducing bacterium *Shewanella oneidensis* MR-1. Geochim Cosmochim Acta 73(4):962–976
Brahmachari G, Adrio JL, Demain AL (2017) Biotechnology of microbial enzymes: production, biocatalysis and industrial application. Elsevier Academic press, Amsterdam, pp 1–11
Brandelli A, Daroit DJ, Riffel A (2010) Biochemical features of microbial keratinases and their production and applications. Appl Microbiol Biotechnol 85:1735–1750
Cai H, Shi P, Bai Y, Huang H, Yuan T, Yang P, Luo H, Meng K, Yao B (2011) A novel thermoacidophilic family 10 xylanase from *Penicillium pinophilum* C1. Process Biochem 46:2341–2346
Castro-Longoria E, Vilchis-Nestor AR, Avalos Borja M (2011) Biosynthesis of silver, gold and bimetallic nanoparticles using the filamentous fungus *Neurospora crassa*. Colloids Surf B: Biointerfaces 83(1):42–48
Chakdar H, Kumar M, Pandiyan K, Singh A, Nanjappan K, Kashyap PL, Srivastava AK (2016) Bacterial xylanases: biology to biotechnology. 3 Biotech 6(2):150–165
Chakravarthy KV, Bonoiu AC, Davis WG (2010) Gold nanorod delivery of an ssRNA immune activator inhibits pandemic H1N1 influenza viral replication. Proc J Nanomater Nati Acad Sci USA 107(22):10172–10177
Chan WCW, Nie S (1998) Quantum dot bioconjugates for ultrasensitive nonisotopic detection. Science 281(5385):2016–2018
Chen T, Xu Y, Peng Z, Li A, Liu J (2017) Simultaneous enhancement of bioactivity and stability of laccase by Cu2+/PAA/PPEGA matrix for efficient biosensing and recyclable decontamination of pyrocatechol. Anal Chem 89(3):2065–2072
Chertok B, Moffat BA, David AE (2008) Iron oxide nanoparticles as a drug delivery vehicle for MRI monitored magnetic targeting of brain tumors. Biomaterials 29(4):487–496
Chirumamilla RR, Muralidhar R, Marchant R, Nigam P (2001) Improving the quality of industrially important enzymes by directed evolution. Mol Cell Biochem 224:159–168
Chudasama CJ, Jani SA, Jajda HM, Pate HN (2010) Optimization and production of alkaline protease from bacillus thuringiensis CC7. J Cell Tissue Res 10:2257–2262
Collins T, Gerday C, Feller G (2005) Xylanases, xylanase families and extremophilic xylanases. FEMS Microbiol Rev 29:3–23
Cui D, Tian F, Coyer SR (2007) Effects of antisense-myc-conjugated single-walled carbon nanotubes on HL-60 cells. J Nanosci Nanotechnol 7(4–5):1639–1646
Cunningham DP, Lundie LL (1993) Precipitation of cadmium by *Clostridium thermoaceticum*. Appl Environ Microbiol 59(1):7–14
Dahiya JS, Singh D, Nigam P (1998) Characterisation of laccase produced by *Coniothyrium minitans*. J Basic Microbiol 38:349–359
Dahiya J, Singh D, Nigam P (2001) Decolourisation of synthetic and spentwash-melanoidins using the white-rot fungus *Phanerochaete chrysosporium* JAG-40. Bioresour Technol 78:95–98
Dameron CT, Reese RN, Mehra RK (1989) Biosynthesis of cadmium sulphide quantum semiconductor crystallites. Nature 338(6216):596–597

De la Isla A, Brostow W, Bujard B (2003) Nanohybrid scratch resistant coatings for teeth and bone viscoelasticity manifested in tribology. Mater Res Innov 7(2):110–114

De Windt W, Aelterman P, Verstraete W (2005) Bioreductive deposition of palladium nanoparticles on *Shewanella oneidensis* with catalytic activity towards reductive dechlorination of polychlorinated biphenyls. Environ Microbiol 7(3):314–325

Dias AMGC, Hussain A, Marcos AS, Roque ACA (2011) A biotechnological perspective on the application of iron oxide magnetic colloids modified with polysaccharides. Biotechnol Adv 29(1):142–155

Dinali R, Ebrahiminezhad A, Manley-Harris M, Ghasemi Y, Berenjian A (2017) Iron oxide nanoparticles in modern microbiology and biotechnology. Crit Rev Microbiol 10:1–15

Du L, Jiang H, Liu X, Wang E (2007) Biosynthesis of gold nanoparticles assisted by *Escherichia coli* DH5α and its application on direct electrochemistry of hemoglobin. Electrochem Commun 9(5):1165–1170

Du Y, Shi P, Huang H, Zhang X, Luo H, Wang Y, Yao B (2013) Characterization of three novel thermophilic xylanases from *Humicola insolens* Y1 with application potentials in the brewing industry. Bioresour Technol 130:161–167

Duran N, Marcato PD, De Souza GIH, Alves OL, Esposito E (2007) Antibacterial effect of silver nanoparticles produced by fungal process on textile fabrics and their effluent treatment. J Biomed Nanotechnol 3(2):203–208

Dwivedi P, Vivikanand V, Pareek N, Sharma A, Singh RP (2010) Bleach enhancement of mixed wood pulp by xylanase-laccase concoction derived through co-culture strategy. Appl Biochem Biotechnol 160:255–268

Edelstein RL, Tamanaha CR, Sheehan PE (2000) The BARC biosensor applied to the detection of biological warfare agents. Biosens Bioelectron 14(10–11):805–813

Emerich DF, Thanos CG (2006) The pinpoint promise of nanoparticle-based drug delivery and molecular diagnosis. Biomol Eng 23(4):171–184

Fadeel B, Garcia-Bennett AE (2010) Better safe than sorry: understanding the toxicological properties of inorganic nanoparticles manufactured for biomedical application. Adv Drug Deliv Rev 62(3):362–374

Fan TX, Chow SK, Zhang D (2009) Biomorphic mineralization: from biology to materials. Prog Mater Sci 54(5):542–659

Fayaz MA, Balaji K, Girilal M, Yadav R, Kalaichelvan PT, Venketesan R (2010) Biogenic synthesis of silver nanoparticles and their synergistic effect with antibiotics: a study against gram-positive and gram-negative bacteria. Nanomedecine 6(1):e103–e109

Felfoul O, Mohammadi M, Martel S (2007) Magnetic resonance imaging of Fe3O4 nanoparticles embedded in living magnetotactic bacteria for potential use as carriers for in vivo applications. In: Proceedings of the 29th annual international conference of the IEEE Engineering in Medicine and Biology Society (EMBS '07), p 1463–1466

Gali NK, Kotteazeth S (2012) Isolation, purification and characterization of thermophilic laccase from xerophyte *Cereus pterogonus*. Chem Nat Compd 248:451–456

Gali NK, Kotteazeth S (2013) Biophysical characterization of thermophilic laccase from the xerophytes: *Cereus pterogonus* and *Opuntia vulgaris*. Cellulose 20:115–125

Garg AP, Roberts JC, McCarthy A (1998) Bleach boosting effect of cellulase free xylanase of *Streptomyces thermoviolaceus* and its comparison with two commercial enzyme preparations on birchwood Kraft pulp. Enzym Microb Biotechnol 22:594–598

Genckal H, Tari C (2006) Alkaline protease production from alkalophilic Bacillus sp. isolated from natural habitats. Enzym Microb Technol 39:703–710

Gericke M, Pinches A (2006) Biological synthesis of metal nanoparticles. Hydrometallurgy 83(1–4):132–140

Ghashghaei S, Emtiazi G (2016) The methods of nanoparticle synthesis using bacteria as biological nanofactories, their mechanisms and major applications. Current Biotechnol 1(1):3–17

Giljohann DA, Seferos DS, Daniel WL, Massich MD, Patel PC, Mirkin CA (2010) Gold nanoparticles for biology and medicine. Angew Chem Int Ed 49(19):3280–3294

Gopinath SCB, Anbu P, Lakshmipriya T (2015) Biotechnological aspects and perspective of microbial keratinase production. Biomed Res 2015:1–10

Gupta R, Beg QK, Lorenz P (2002) Bacterial alkaline proteases: molecular approaches and industrial applications. Appl Microbiol Biotechnol 59:15–32

Gupta A, Joseph B, Mani A, Thomas G (2008) Biosynthesis and properties of an extracellular thermostable serine alkaline protease from *Virgibacillus pantothenticus*. World J Microbiol Biotechnol 24:237–243

Gupta R, Sharma R, Beg QK (2013) Revisiting microbial keratinases: next generation proteases for sustainable biotechnology. Crit Rev Biotechnol 33:216–228

Gurumurthy DM, Neelagund SE (2012) Molecular characterization of industrially viable extreme thermostable novel alpha-amylase of *Geobacillus sp* Iso5 isolated from geothermal spring. J Pure Appl Microbiol 6:1759–1773

Gurunathan S, Kalishwaralal K, Vaidyanathan R (2009) Biosynthesis, purification and characterization of silver nanoparticles using *Escherichia coli*. Colloids Surf B: Biointerfaces 74(1):328–335

Gushterova A, Vasileva-Tonkova E, Dimova E, Nedkov P, Haertle T (2005) Keratinase production by newly isolated Antarctic actinomycete strains. World J Microbiol Biotechnol 21:831–834

Hadj-Ali NE, Rym A, Basma GF, Alya SK, Safia K, Moncef N (2007) Biochemical and molecular characterization of a detergent stable alkaline serine protease from a newly isolated *Bacillus licheniformis* NH1. Enzym Microb Technol 40:515–523

Hayat MA (1989) Colloidal gold: principles, methods, and applications. Academic, San Diego

He S, Guo Z, Zhang Y, Zhang S, Wang J, Gu N (2007) Biosynthesis of gold nanoparticles using the bacteria *Rhodopseudomonas capsulate*. Mater Lett 61(18):3984–3987

Hergt R, Dutz S (2007) Magnetic particle hyperthermia biophysical limitations of a visionary tumour therapy. J Magn Magn Mater 311(1):187–192

Hergt R, Hiergeist R, Zeisberger M (2005) Magnetic properties of bacterial magnetosomes as potential diagnostic and therapeutic tools. J Magn Magn Mater 293(1):80–86

Higby GJ (1982) Gold in medicine: a review of its use in the west before 1900. Gold Bull 15(4):130–140

Hildebrand H, Mackenzie K, Kopinke FD (2008) Novel nano-catalysts for waste water treatment. Global NEST J 10:47–53

Hou S, Zhang A, Su M (2016) Nanomaterials for biosensing applications. Nanomaterials 6:58. https://doi.org/10.3390/nano6040058

Husseiny MI, El-Aziz MA, Badr Y, Mahmoud MA (2007) Biosynthesis of gold nanoparticles using *Pseudomonas aeruginosa*. Spectrochim Acta 67(3–4):1003–1006

Indhuja S, Shiburaj S, Pradeep NS, Thankamani V, Abraham TK (2012) Extracellular keratinolytic proteases from an Alkalophilic *Streptomyces albidoflavus* TBG-S13A5: enhanced production and characterization. J Pure Appl Microbiol 6:1599–1607

Jain N, Bhargava A, Majumdar S, Tarafdar JC, Panwar J (2011) Extracellular biosynthesis and characterization of silver nanoparticles using *Aspergillus flavus* NJP08: a mechanism perspective. Nanoscale 3(2):635–641

Jha AK, Prasad K (2010) Ferroelectric $BaTiO_3$ nanoparticles: biosynthesis and characterization. Colloids Surf B: Biointerfaces 75(1):330–334

Jha AK, Prasad K, Prasad K (2009a) A green low-cost biosynthesis of Sb2O3 nanoparticles. Biochem Eng J 43(3):303–306

Jha AK, Prasad K, Kulkarni AR (2009b) Synthesis of TiO2 nanoparticles using microorganisms. Colloids Surf B Biointerfaces 71(2):226–229

Jhamb K, Sahoo DK (2012) Production of soluble recombinant proteins in Escherichia coli: effects of process conditions and chaperone co-expression on cell growth and production of xylanase. Bioresour Technol 123:135–143

Johnvesly B, Naik GK (2001) Studies on production of thermostable alkaline protease from thermophilic and alkaliphilic Bacillus sp. JB-99 in a chemical defined medium. Process Biochem 37:139–144

Juibari MM, Abbasalizadeh S, Jouzani GS, Noruzi M (2011) Intensified biosynthesis of silver nanoparticles using a native extremophilic *Ureibacillus thermosphaericus* strain. Mater Lett 65(6):1014–1017

Kalimuthu K, Suresh Babu R, Venkataraman D, Bilal M, Gurunathan S (2008) Biosynthesis of silver nanocrystals by *Bacillus licheniformis*. Colloids Surf B: Biointerfaces 65(1):150–153

Kalishwaralal K, Deepak V, Ramkumarpandian S, Nellaiah H, Sangiliyandi G (2008) Extracellular biosynthesis of silver nanoparticles by the culture supernatant of *Bacillus licheniformis*. Mater Lett 62(29):4411–4413

Kalishwaralal K, Banumathi E, Pandian SRK (2009) Silver nanoparticles inhibit VEGF induced cell proliferation and migration in bovine retinal endothelial cells. Colloids Surf B 73(1):51–57

Kalishwaralal K, Deepak V, Ram S (2010) Biosynthesis of silver and gold nanoparticles using *Brevibacterium casei*. Colloids Surf B: Biointerfaces 77(2):257–262

Kashefi K, Lovley DR (2000) Reduction of Fe(III), Mn(IV), and toxic metals at 100°C by *Pyrobaculum islandicum*. Appl Environ Microbiol 66(3):1050–1056

Khusro A, Baraathikannan K, Aarti C, Agastian P (2017) Optimization of thermo-alkali stable amylase production and biomass yield from bacillus sp. under submerged cultivation. Ferment 3(7). https://doi.org/10.3390/fermentation3010007

Klaus T, Joerger R, Olsson E, Granqvist CG (1999) Silver based crystalline nanoparticles, microbially fabricated. Proc Natl Acad Sci U S A 96(24):13611–13614

Kohli U, Nigam P, Singh D, Chaudhary K (2001) Thermostable, alkalophilic and cellulase free xylanase production by *Thermonoactinomyces thalophilus* subgroup C. Enzym Microb Technol 28:606–610

Konishi Y, Tsukiyama T, Tachimi T, Saitoh N, Nomura T, Nagamine S (2007a) Microbial deposition of gold nanoparticles by the metal-reducing bacterium Shewanella algae. Electrochim Acta 53(1):186–192

Konishi Y, Ohno K, Saitoh N (2007b) Bioreductive deposition of platinum nanoparticles on the bacterium Shewanella algae. J Biotechnol 128(3):648–653

Kowshik M, Deshmuke N, Vogal W (2002) Microbial synthesis of semiconductor CdS nanoparticles, their characterization, and their use in the fabrication of an ideal diode. Biotechnol Bioeng 78(5):583–588

Kumar GN, Srikumar K (2011) Thermophilic laccase from xerophyte species *Opuntia vulgaris*. Biomed Chromatogr 25:707–711

Kumar GN, Srikumar K (2012) Characterization of xerophytic thermophilic laccase exhibiting metal ion-dependent dye decolorization potential. Appl Biochem Biotechnol 167:662–676

Kumar CG, Takagi H (1999) Microbial alkaline proteases: from a bioindustrial viewpoint. Biotechnol Adv 17:561–594

Kumar J, Dahiya JS, Singh D, Nigam P (2000) Production of endo-1,4-β-glucanase by a biocontrol fungus *Cladorrhinum foecundissimum*. Bioresour Technol 75:95–97

Kumar SA, Ansary AA, Abroad A, Khan MI (2007) Extracellular biosynthesis of CdSe quantum dots by the fungus, *Fusarium oxysporum*. J Biomed Nanotechnol 3(2):190–194

Labrenz M, Druschel MGK, Thomsen-Ebert T (2000) Formation of sphalerite (ZnS) deposits in natural biofilms of sulfate-reducing bacteria. Science 290(5497):1744–1747

Lee JH, Han J, Choi H, Hur HG (2007) Effects of temperature and dissolved oxygen on Se(IV) removal and Se(0) precipitation by *Shewanella sp.* HN-41. Chemosphere 68(10):1898–1905

Lefevre CT, Abreu F, Lins U, Bazylinski DA (2010a) Nonmagnetotactic multicellular prokaryotes from lowsaline, nonmarine aquatic environments and their unusual negative phototactic behavior. Appl Environ Microbiol 76(10):3220–3227

Lefevre CT, Abreu F, Schmidt ML (2010b) Moderately thermophilic magnetotactic bacteria from hot springs in Nevada. Appl Environ Microbiol 76(11):3740–3743

Lengke MF, Fleet ME, Southam G (2006a) Morphology of gold nanoparticles synthesized by filamentous cyanobacteria from gold(I)-Thiosulfate and gold(III)-chloride complexes. Langmuir 22(6):2780–2787

Lengke MF, Ravel B, Fleet ME, Wanger G, Gordon RA, Southam G (2006b) Mechanisms of gold bioaccumulation by filamentous cyanobacteria from gold(III)-chloride complex. Environ Sci Tech 40(20):6304–6309

Li W, Yu L, Zhou P, Zhu M (2007) A magnetospirillum strain WM-1 from a freshwater sediment with intracellular magnetosomes. World J Microbiol Biotechnol 23(10):1489–1492

Li Y, Niu D, Zhang L, Wang Z, Shi G (2013) Purification, characterization and cloning of a thermotolerant isoamylase produced from *Bacillus sp.* CICIM 304. J Ind Microbiol Biotechnol 40:437–446

Li J, Li Q, Ma X, Tian B, Li T, Yu J, Da S, Weng Y, Hua Y (2016) Biosynthesis of gold nanoparticles by the extreme bacterium *Deinococcus radiodurans* and an evaluation of their antibacterial properties. Int J Nanomedicine 11:5931–5944

Liu B, Zhang J, Li B, Liao X, Du G, Chen J (2013) Expression and characterization of extreme alkaline, oxidation-resistant keratinase from *Bacillus licheniformis* in recombinant *Bacillus subtilis* WB600 expression system and its application in wool fiber processing. World J Microbiol Biotechnol 29:825–832

Lloyd JR, Yong P, Macaskie LE (1998) Enzymatic recovery of elemental palladium by using sulfate-reducing bacteria. Appl Environ Microbiol 64(11):4607–4609

Lover T, Henderson W, Bowmaker GA, Seakins JM, Cooney RP (1997) Functionalization and capping of a cds nanocluster: a study of ligand exchange by electrospray mass spectrometry. Chem Mater 9(8):1878–1886

Luo H, Li J, Yang J, Wang H, Yang Y, Huang H, Shi P, Yuan T, Fan Y, Yao B (2009) A thermophilic and acid stable family-10 xylanase from the acidophilic fungus *Bispora sp.* MEY-1. Extremophiles 13:849–857

Luo H, Wang K, Huang H, Shi P, Yang P, Yao B (2012) Gene cloning, expression and biochemical characterization of an alkali-tolerant b-mannanase from *Humicola insolens* Y1. J Ind Microbiol Biotechnol 39:547–555

Ma J, Wong H, Kong LB, Peng KW (2003) Biomimetic processing of nanocrystallite bioactive apatite coating on titanium. Nanotechnology 14(6):619–623

Mahdihassan S (1988) Alchemy, Chinese versus Greek, an etymological approach: a rejoinder. Am J Chin Med 16(1–2):83–86

Mahmoudi M, Sant S, Wang B, Laurent S, Sen T (2010) Superparamagnetic iron oxide nanoparticles (SPIONs): development, surface modification and applications in chemotherapy. Adv Drug Deliv Rev 63:24–46

Mamo G, Thunnissen M, Hatti-Kaul R, Mattiasson B (2009) An alkaline active xylanase: insights into mechanisms of high pH catalytic adaptation. Biochimie 91:1187–1196

Marcano L, García-Prieto A, Muñoz D, Fernández Barquín L, Orue I, Alonso J, Muela A, Fdez-Gubieda ML (2017) Influence of the bacterial growth phase on the magnetic properties of magnetosomes synthesized by *Magnetospirillum gryphiswaldense*. Biochim Biophys Acta. https://doi.org/10.1016/j.bbagen.2017.01.012

Marques S, Alves L, Ribeiro S, Girio FM, Amaralcollaco MT (1998) Characterisation of a thermotolerant and alkalotolerant xylanase from a Bacillus sp. Appl Biochem Biotechnol 73:159–172

Matsunaga T, Ueki F, Obata K (2003) Fully automated immunoassay system of endocrine disrupting chemicals using monoclonal antibodies chemically conjugated to bacterial magnetic particles. Anal Chim Acta 475(1–2):75–83

Meng C, Tian J, Li Y, Zheng S (2010) Influence of native bacterial magnetic particles on mouse immune response. Wei Sheng Wu Xue Bao 50(6):817–821

Mergeay M, Monchy S, Vallaeys T (2003) Ralstonia metallidurans, a bacterium specifically adapted to toxic metals: towards a catalogue of metal-responsive genes. FEMS Microbiol Rev 27(2–3):385–410

Mishra A, Kumar S (2009) Kinetic studies of laccase enzyme of *Coriolus versicolor* MTCC 138 in an inexpensive culture medium. Biochem Eng J 46:252–256

Mohammed Fayaz A, Balaji K, Kalaichelvan PT, Venkatesan R (2009) Fungal based synthesis of silver nanoparticles an effect of temperature on the size of particles. Colloids Surf B: Biointerfaces 74(1):123–126

Mukherjee P, Ahmad A, Mandal D (2001a) Bioreduction of AuCl4− ions by the fungus, Verticillium sp. and surface trapping of the gold nanoparticles formed. Angew Chem Int Ed 40(19):3585–3588
Mukherjee P, Ahmad A, Mandal D (2001b) Fungus-mediated synthesis of silver nanoparticles and their immobilization in the mycelial matrix: a novel biological approach to nanoparticle synthesis. Nano Lett 1(10):515–519, 2001
Mukherjee P, Senapati S, Mandal D (2002) Extracellular synthesis of gold nanoparticles by the fungus fusarium oxysporum. Chem Bio Chem 3(5):461–463
Mukherjee P, Bhattacharya R, Patra CR (2007) Nanogold in cancer therapy and diagnosis, vol 7., chapter 3. Wiley-VCH, Weinheim
Mukherjee AK, Adhikari H, Rai SK (2008) Production of alkaline protease by a thermophilic *Bacillus subtilis* under solid-state fermentation (SSF) condition using Imperata cylindrical grass and potato peel as low-cost medium: characterization and application of enzyme in detergent formulation. J Biochem Eng 39:353–361
Nair B, Pradeep T (2002) Coalescence of nanoclusters and formation of submicron crystallites assisted by lactobacillus strains. Cryst Growth Des 2(4):293–298
Narayanan KB, Sakthivel N (2010) Biological synthesis of metal nanoparticles by microbes. Adv Colloid Interf Sci 156(1–2):1–13
Nies DH (1999) Microbial heavy-metal resistance. Appl Microbiol Biotechnol 51(6):730–750
Nigam P, Pandey A (2009) Biotechnology for agro-industrial residues utilisation. Publisher Springer Science Business Media B.V., Germany, pp 1–466
Nigam P, Singh D (1995) Enzyme and microbial systems involved in starch processing. Enzym Microb Technol 17:770–778
Nigam P, Pandey A, Prabhu KA (1987a) Ligninolytic activity of two basidiomycetes moulds in the decomposition of bagasse. Biol Wastes 21:1–10
Nigam P, Pandey A, Prabhu KA (1987b) Cellulase and ligninase production by Basidiomycetes culture in solidstate fermentation. Biol Wastes 20:1–9
Orlov IA, Sankova TP, Babich PS, Sosnin IM, Ilyechova EY, Kirilenko DA, Brunkov PN, Ataev GL, Romanov AE, Puchkova LV (2016) New silver nanoparticles induce apoptosis-like process in *E. coli* and interfere with mammalian copper metabolism. Int J Nanomedicine 11:6561–6574
Panácek A, Kvítek L, Prucek R (2006) Silver colloid nanoparticles: synthesis, characterization, and their antibacterial activity. J Phys Chem B 110(33):16248–16253
Pandey A, Selvakumar P, Soccol CR, Nigam P (1999) Solid-state fermentation for the production of industrial enzymes. Curr Sci 77:149–162
Pandey A, Nigam P, Soccol CR, Soccol VT, Singh D, Mohan R (2000a) Advances in microbial amylases. Biotechnol Appl Biochem 31:135–152
Pandey A, Soccol CR, Nigam P (2000b) Biotechnological potential of agro-industrial residues, II-cassava bagasse. Bioresour Technol 74:81–87
Pandian SRK, Deepak V, Kalishwaralal K, Muniyandi J, Rameshkumar N, Gurunathan S (2009) Synthesis of PHB nanoparticles from optimized medium utilizing dairy industrial waste using *Brevibacterium casei* SRKP2: a green chemistry approach. Colloids Surf B: Biointerfaces 74(1):266–273
Pantarotto D, Partidos CD, Hoebeke J (2003) Immunization with peptide-functionalized carbon nanotubes enhances virus-specific neutralizing antibody responses. Chem Biol 10(10):961–966
Papinutti L, Dimitriu P, Forchiassin F (2008) Stabilization studies of *Fomes sclerodermeus* laccases. Bioresour Technol 99:419–424
Parak WJ, Boudreau R, Le Gros M (2002) Cell motility and metastatic potential studies based on quantum dot imaging of phagokinetic tracks. Adv Mater 14(12):882–885
Patra CR, Bhattacharya R, Mukhopadhyay D, Mukherjee P (2010) Fabrication of gold nanoparticles for targeted therapy in pancreatic cancer. Adv Drug Deliv Rev 62(3):346–361
Perez-Gonzalez T, Jimenez-Lopez C, Neal AL (2010) Magnetite biomineralization induced by *Shewanella oneidensis*. Geochim Cosmochim Acta 74(3):967–979
Piao MJ, Kang KA, Lee IK (2011) Silver nanoparticles induce oxidative cell damage in human liver cells through inhibition of reduced glutathione and induction of mitochondria-involved apoptosis. Toxicol Lett 201(1):92–100

Polizeli ML, Rizzatti AC, Monti R, Terenzi HF, Jorge JA, Amorim DS (2005) Xylanases from fungi: properties and industrial applications. Appl Microbiol Biotechnol 67:577–591

Prade RA (1996) Xylanases: from biology to biotechnology. Biotechnol Genet Eng Rev 13:100–131

Prasad R (2014) Synthesis of silver nanoparticles in photosynthetic plants. J Nanoparticles, Article ID 963961, https://doi.org/10.1155/2014/963961

Prasad K, Jha AK (2010) Biosynthesis of CdS nanoparticles: an improved green and rapid procedure. J Colloid Interface Sci 342(1):68–72

Prasad R, Pandey R, Barman I (2016) Engineering tailored nanoparticles with microbes: quo vadis. WIREs Nanomed Nanobiotechnol 8:316–330. https://doi.org/10.1002/wnan.1363

Prasad R, Bhattacharyya A, Nguyen QD (2017a) Nanotechnology in sustainable agriculture: recent developments, challenges, and perspectives. Front Microbiol 8:1014. https://doi.org/10.3389/fmicb.2017.01014

Prasad R, Pandey R, Varma A, Barman I (2017b) Polymer based nanoparticles for drug delivery systems and cancer therapeutics. In: Kharkwal H, Janaswamy S (eds) Natural polymers for drug delivery. CAB International, UK, pp 53–70

Quaratino D, Federici F, Petruccioli M, Fenice M, D'Annibale A (2007) Production, purification and partial characterisation of a novel laccase from the white-rot fungus *Panus tigrinus* CBS 577.79. Anton Leeuw Int J G 91:57–69

Rahman RNZRA, Basri M, Salleh AB (2003) Thermostable alkaline protease from *Bacillus stearothermophilus* F1; nutritional factors affecting protease production. Ann Microbiol 53:199–210

Rautaray D, Sanyal A, Adyanthaya SD, Ahmad A, Sastry M (2004) Biological synthesis of strontium carbonate crystals using the fungus fusarium oxysporum. Langmuir 20(16):6827–6833

Reddy K, Nasr M, Kumari S, Kumar S, Gupta SK, Enitan AM, Bux F (2017) Biohydrogen production from sugarcane bagasse hydrolysate: effects of PH,S/X, Fe>sup<2+>/sup<, and magnetite nanoparticles. Environ Sci Pollut Res Int. https://doi.org/10.1007/s11356-017-8560-1

Reith F, Lengke MF, Falconer D, Craw D, Southam G (2007) The geomicrobiology of gold. ISME J 1(7):567–584

Riddin T, Gericke M, Whiteley CG (2010) Biological synthesis of platinum nanoparticles: effect of initial metal concentration. Enzym Microb Technol 46(6):501–505

Robinson T, Nigam P (2008) Remediation of textile dye-waste water using a white rot fungus *Bjerkandera adusta* through solid state fermentation (SSF). Appl Biochem Biotechnol 151:618–628

Robinson T, Chandran B, Nigam P (2001a) Studies on the decolourisation of an artificial effluent through lignolytic enzyme production by white-rot fungi in N-rich and N-limited media. Appl Microbiol Biotechnol 57:810–813

Robinson T, Chandran B, Nigam P (2001b) Studies on the production of enzymes by white-rot fungi for the decolourisation of textile dyes. Enzym Microb Technol 29:575–579

Rodríguez-Carmona E, Villaverde A (2010) Nanostructured bacterial materials for innovative medicines. Trends Microbiol 18(9):423–430

Sanghi R, Verma P (2009) A facile green extracellular biosynthesis of CdS nanoparticles by immobilized fungus. Chem Eng J 155(3):886–891

Sani JK, Saini R, Tewari L (2015) Lignocellulosic agriculture wastes as biomass feedstock for second-generation bioethanol production: concepts and recent developments. 3 Biotech 5(4):337–353

Sanyal A, Rautaray D, Bansal V, Ahmad A, Sastry M (2005) Heavy-metal remediation by a fungus as a means of production of lead and cadmium carbonate crystals. Langmuir 21(16):7220–7224

Sawle BD, Salimath B, Deshpande R, Bedre MD, Prabhakar BK, Venkataraman A (2008) Biosynthesis and stabilization of Au and Au-Ag alloy nanoparticles by fungus, *Fusarium semitectum*. Sci Technol Adv Mater 9(3):1–6

Senapati S, Manda D, Ahmad A (2004) Fungus mediated synthesis of silver nanoparticles: a novel biological approach. Indian J Phys A 78(1):101–105

Senapati S, Ahmad A, Khan MI, Sastry M, Kumar R (2005) Extracellular biosynthesis of bimetallic Au-Ag alloy nanoparticles. Small 1(5):517–520

Shan G, Xing J, Zhang H, Liu H (2005) Biodesulfurization of dibenzothiophene by microbial cells coated with magnetite nanoparticles. Appl Environ Microbiol 71(8):4497–4502

Shen Z, Li Y, Kohama K, Oneill B, Bi J (2011) Improved drug targeting of cancer cells by utilizing actively targetable folic acid-conjugated albumin nanospheres. Pharmacol Res 63(1):51–58

Shimoshige H, Nakajima Y, Kobayashi H, Yanagisawa K, Nagaoka Y, Shimamura S, Mizuki T, Inoue A, Maekawa T (2017) Formation of core-shell nanoparticles composed of magnetite and samarium oxide in *magnetospirillum magnetium* strain RSS-1. PLoS One 12(1):e0170932

Shinkai M, Yanase M, Suzuki M (1999) Intracellular hyperthermia for cancer using magnetite cationic liposomes. J Magn Magn Mater 194(1):176–184

Singaravelu G, Arockiamary JS, Kumar VG, Govindaraju K (2007) A novel extracellular synthesis of monodisperse gold nanoparticles using marine alga, *Sargassum wightii* Greville. Colloids Sur B 57(1):97–101

Singh D, Dahiya JS, Nigam P (1995) Simultaneous raw starch hydrolysis and ethanol fermentation by glucoamylase from *Rhizoctonia solani* and *Saccharomyces cerevisiae*. J Basic Microbiol 35:117–121

Sinha A, Khare SK (2011) Mercury bioaccumulation and simultaneous nanoparticle synthesis by *Enterobacter sp.* cells. Bioresour Technol 102:4281–4284

Sivaramakrishnan S, Gangadharan D, Nampoothiri KM, Soccol CR, Pandey A (2006) α-amylases from microbial sources–an overview on recent developments. Food Technol Biotechnol 44:173–184

Sneha K, Sathishkumar M, Mao J, Kwak IS, Yun YS (2010) Corynebacterium glutamicum-mediated crystallization of silver ions through sorption and reduction processes. Chem Eng J 162(3):989–996

Sony IS, Potty VP (2017) Screening of protease produced by bacterial isolates of bakery industry. Int Res J Biol Sci 6(2):36–41

Southam G, Beveridge TJ (1996) The occurrence of sulfur and phosphorus within bacterially derived crystalline and pseudocrystalline octahedral gold formed in vitro. Geochim Cosmochim Acta 60(22):4369–4376

Spring S, Schleifer KH (1995) Diversity of magnetotactic bacteria. Syst Appl Microbiol 18(2):147–153

Srinivasan MC, Rele MV (1995) Cellulase free xylanase from microorganisms and their applications to pulp and paper biotechnology: an overview. Indian J Microbiol 35:93–101

Srivastava P, Kowshik M (2017) Fluorescent lead (IV) sulfide nanoparticles synthesized by *Idiomarina* sp. strain PR58-8 for bioimaging applications. Appl Environ Microbiol 83(7):e03091–e03016

Sun JB, Duan JH, Dai SL (2007) In vitro and in vivo antitumor effects of doxorubicin loaded with bacterial magnetosomes (DBMs) on H22 cells: the magnetic bionanoparticles as drug carriers. Cancer Lett 258(1):109–117

Sun JB, Wang ZL, Duan JH (2009) Targeted distribution of bacterial magnetosomes isolated from magnetospirillum gryphiswaldense MSR-1 in healthy sprague-dawley rats. J Nanosci Nanotechnol 9(3):1881–1885

Sun T, Zhang YS, Pang B, Hyun DC, Yang M, Xia Y (2014) Engineered nanoparticles for drug delivery in cancer therapy. Angew Chem 53(46):12320–12364

Suresh AK, Pelletier DA, Wang W (2011) Biofabrication of discrete spherical gold nanoparticles using the metal reducing bacterium *Shewanella oneidensis*. Acta Biomater 7(5):2148–2152

Sweeney RY, Mao C, Gao X (2004) Bacterial biosynthesis of cadmium sulfide nanocrystals. Chem Biol 11(11):1553–1559

Tanaka T, Takeda H, Ueki F (2004) Rapid and sensitive detection of 17β-estradiol in environmental water using automated immunoassay system with bacterial magnetic particles. J Biotechnol 108(2):153–159

Tang H, Yan M, Zhang H, Xia M, Yang D (2005) Preparation and characterization of water-soluble CdS nanocrystals by surface modification of ethylene diamine. Mater Lett 59(8–9):1024–1027

Thornhill RH, Burgess JG, Matsunaga T (1995) PCR for direct detection of indigenous uncultured magnetic cocci in sediment and phylogenetic analysis of amplified 16S ribosomal DNA. Appl Environ Microbiol 61(2):495–500

Tian F, Prina-Mello A, Estrada G (2008) A novel assay for the quantification of internalized nanoparticles in macrophages. Nanotoxicology 2(4):232–242

Tripathi RM, Gupta RK, Bhadwal AS, Singh P, Shrivastav A, Shrivastav BR (2015) Fungal biomolecules assisted biosynthesis of Au-Ag alloy nanoparticles and evaluation of their catalytic property. IET Nanobiotechnol 9(4):178–183

Uthandi S, Saad B, Humbard MA, Maupin-Furlow JA (2010) LccA an archaeal laccase secreted as a highly stable glycoprotein into the extracellular medium by *Haloferax volcanii*. Appl Environ Microbiol 76:733–743

Vaidyanathan R, Kalishwaralal R, Gopalram S, Gurunathan S (2009) Nanosilver the burgeoning therapeutic molecule and its green synthesis. Biotechnol Adv 27(6):924–937

Vigneshwaran N, Kathe AA, Varadarajan PV, Nachane RP, Balasubramanya RH (2006) Biomimetics of silver nanoparticles by white rot fungus, *Phaenerochaete chrysosporium*. Colloids Surf B: Biointerfaces 53(1):55–59

Vigneshwaran N, Ashtaputre NM, Varadarajan PV, Nachane RP, Paralikar KM, Balasubramanya RH (2007) Biological synthesis of silver nanoparticles using the fungus *Aspergillus flavus*. Mater Lett 61(6):1413–1418

Vijayalakshmi S, Kumar S, Thankamani V (2011) Optimization and cultural characterization of bacillus RV.B2.90 producing alkalophilic thermophilic protease. Res J Biotechnol 6:26–32

Wang T, Yang L, Zhang B, Liu J (2010) Extracellular biosynthesis and transformation of selenium nanoparticles and application in H_2O_2 biosensor. Colloids Surf B 80(1):94–102

Wang X, Li D, Watanabe T, Shigemori Y, Mikawa T, Okajima T, Mao LQ, Ohsaka T (2012) A glucose/o-2 biofuel cell using recombinant thermophilic enzymes. Int J Electrochem Sci 7:1071–1078

Wati L, Dhamija SS, Singh D, Nigam P, Marchant R (1996) Characterisation of genetic control of thermotolerance in mutants of *Saccharomyces cerevisiae*. Genet Eng Biotechnol 16:19–26

Watson JHP, Ellwood DC, Soper AK, Charnock J (1999) Nanosized strongly-magnetic bacterially-produced iron sulfide materials. J Magn Magn Mater 203(1–3):69–72

Weissleder R, Elizondo G, Wittenberg J, Rabito CA, Bengele HH, Josephson L (1990) Ultrasmall superparamagnetic iron oxide: characterization of a new class of contrast agents for MR imaging. Radiology 175(2):489–493

Xiang L, Bin W, Huali J (2007b) Bacterial magnetic particles (BMPs)-PEI as a novel and efficient non-viral gene delivery system. J Gene Med 9(8):679–690

Xiang L, Wei J, Jianbo S, Guili W, Feng G, Ying L (2007a) Purified and sterilized magnetosomes from Magnetospirillum gryphiswaldense MSR-1 were not toxic to mouse fibroblasts in vitro. Lett Appl Microbiol 45(1):75–81

Xie J, Chen K, Chen X (2009) Production, modification and bio-applications of magnetic nanoparticles gestated by magnetotactic bacteria. Nano Res 2(4):261–2782

Xue Y, Wang X, Chen X, Hu J, Gao MT, Li J (2017) Effect of different cellulases on the release of phenolic acids from rice straw during saccharification. Bioresour Technol 234:208–2016

Yan S, He W, Sun C (2009) The biomimetic synthesis of zinc phosphate nanoparticles. Dyes Pigments 80(2):254–258

Yang H, Santra S, Holloway PH (2005) Syntheses and applications of Mn-doped II-VI semiconductor nanocrystals. J Nanosci Nanotechnol 5(9):1364–1375

Yong P, Rowson NA, Farr JPG, Harris IR, Macaskie LE (2002) Bioreduction and biocrystallization of palladium by *Desulfovibrio desulfuricans* NCIMB 8307. Biotechnol Bioeng 80(4):369–379

Zhang Y, Zhu Y, Cao Y, Li D, Zhang Z, Wang K, Ding F, Wang X, Meng D, Fan L, Wu J (2016) Size and morphology-controlled synthesis of Ni3C nanoparticles in a TEG solution and their magnetic properties. https://doi.org/10.1039/C6RA11916F

Zheng D, Hu C, Gan T, Dang X, Hu S (2010a) Preparation and application of a novel vanillin sensor based on biosynthesis of Au-Ag alloy nanoparticles. Sens Actuators B Chem 148(1):247–252

Zheng B, Qian L, Yuan H (2010b) Preparation of gold nanoparticles on eggshell membrane and their biosensing application. Talanta 82(1):177–183

Zhou W, He W, Zhong S (2009) Biosynthesis and magnetic properties of mesoporous Fe_3O_4 composites. J Magn Magn Mater 321(8):1025–1028

Zhu K, Pan H, Li J (2010) Isolation and characterization of a marine magnetotactic *Spirillum axenic* culture QH-2 from an intertidal zone of the China Sea. Res Microbiol 161(4):276–283

Chapter 8
Biological Nanoparticles: Optical and Photothermal Properties

Aditya Saran, Rajender Boddula, and Sharda Ranjan Sharan

Abstract The synthesis of metal nanoparticles (NPs) through biological approach has proved to be safer than chemical approach and environment friendly. Still the quality of the biologically synthesized metal NPs could not touch the quality standard of chemically synthesized metal NPs. This limits the application of biologically synthesized NPs. In the biosynthetic approach, the complications lie in complicated biochemical pathways, cellular morphology, physiology and other biological and environmental limitations. On the other hand, chemical synthesis process is simple because of limited chemical reactions and has ease to control and change the required physical and chemical parameters like pH, temperature, humidity, etc. as per need. Thus, chemically synthesized NPs are more uniform in shape and size, have defined surface modification and are stable than its biological counterpart. Here we will focus on the limitations of biological synthesis of metal NPs through fungus and discuss the various approaches to counter these limitations to enhance the quality of metal NPs so that biologically synthesized NPs can be used practically on a wider scale.

A. Saran (✉)
ORCID ID: 0000-0001-5433-8769; CAS Key Laboratory of Standardization and Measurement for Nanotechnology, National Center for Nanoscience and Technology, No.11 ZhongguancunBeiyitiao, Beijing 100190, People's Republic of China
e-mail: aditya8march@yahoo.com; aditya8march@gmail.com

R. Boddula
ORCID ID: 0000-0001-8533-3338; CAS Key Laboratory for Nanosystem and Hierarchical Fabrication, National Center for Nanoscience and Technology, No.11 ZhongguancunBeiyitiao, Beijing 100190, People's Republic of China

S.R. Sharan
Department of Physics, Deoghar College, S K M University, Dumka 814101, India

R. Prasad (ed.), *Fungal Nanotechnology*, Fungal Biology,
DOI 10.1007/978-3-319-68424-6_8

8.1 Introduction: Nanoparticles, Nanostructures and Nanomaterials

Nanotechnology deals with various structures of matter having dimensions of the order of a 10^{-9} of a metre. Nanoparticles are generally considered to be a number of atoms or molecules bonded together with a radius of <100 nm. The structure which comes under the dimension of a nanometre is also considered as nanostructures. Hence, a haem molecule which is incorporated into haemoglobin molecule can be considered as nanoparticle, and most of the viruses are nanostructures. Thus nanoparticles and nanostructures can be built by assembling individual atoms or subdividing bulk materials.

The physical and chemical properties change at dimensions of nanometre scale. Mostly the crystal structure of large-sized nanoparticles remains the same as that of its bulk, but lengths of primitives are somewhat different. It has been verified from X-ray diffraction experiment that 80 nm aluminium particle has face-centred cubic (fcc), which the bulk aluminium has. Particles of sizes smaller than 5 nm may have different structures. Three to 5 nm gold particles have an icosahedral structure, whereas its bulk has fcc structure. If one geometrical dimension has a nanosize length, the structure is called a quantum well. If the electronic structure is quite different from the arrangement where two dimensions in three-dimension are of nanometre length, it is called a quantum wire. But if all the lengths in all three geometrical dimensions are in the nano-range, then it is called a quantum dot (Poole and Owens 2003).

Nanomaterials are made up of cluster of atoms which can be assumed as superatoms (Khanna and Jena 1995). The number of atoms which makes the cluster stable is called as structural magic number (Poole and Owens 2003). In an atom the positive charge of the nucleus is localized at a point and the electrons respond to 1/r potential. On the other hand, in a cluster, the positive charge is assumed to be distributed over a positive ion core of the size of the cluster, and the electrons respond to a square-well-type potential that is attractive to the core. The electron energy levels in clusters are discrete and can be labelled by principal and angular momentum quantum numbers, but the ordering of levels is different from electron's energy levels in an atom. The electronic structure of an atom is like $1s^2$, $2s^2$, $2p^6$, $3s^2$, $3p^6$, $3d^{10}$, $4s^2$ and so on, but in case of cluster, it is changed to $1s^2$, $1p^6$, $1d^{10}$, $2s^2$, $1f^{14}$, $2p^6$ and so on. For example, electronic structure of aluminium is $1s^2$, $2s^2$, $2p^6$ and $3s^1$. Since Al is trivalent, an Al_{13} cluster has 39 electrons, which, in a jellium picture, would correspond to an electronic configuration of $1s^2$, $1p^6$, $1d^{10}$, $2s^2$, $1f^{14}$ and $2p^5$. Note that chemical property is determined by the electrons in the outermost orbit. Cluster also has the similar electronic features in ionization potential, electron affinity and reactivity, but uniquely their nuclear potential is fixed, and attractive potential can be easily modulated. The attractive potential can be modulated by changing its composition, the range can be controlled by its size, and its shape can be controlled by the overall topology. Further, the electronic nature can be controlled by manipulating the number of valence electrons by doping with foreign atoms, and the nature of the electronic degeneracy can be controlled by changing the geometry. This allows much better control of the electronic spectrum and thereby on its properties (Rao et al. 1999). This is the reason why nanomaterial shows different properties from its bulk material (Fig. 8.1).

8.2 Metal Nanoparticles

This chapter deals with metal nanoparticles which may or may not be nanomaterials. The optical and photothermal properties of metallic nanoparticle are mainly dependent upon its size and surface area. The nobility of a metal is decided by the ability of metal surface to be oxidized. Gold is the only metal which has endothermic chemisorption energy. It means that it will not make bond to oxygen at all, and hence, it remains inert in an oxygen atmosphere. The d-state orbital of Au is so low in energy that the interaction with oxygen 2p states is net repulsive. So it is unlikely that Au should be a good catalyst for an oxidation reaction. Uniquely in oxidation

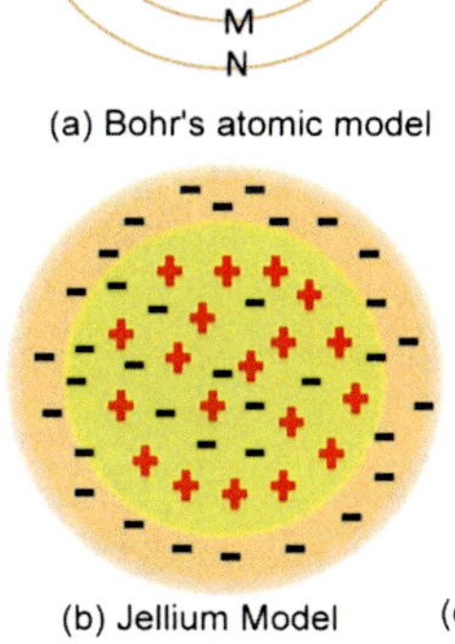

(a) Bohr's atomic model

(b) Jellium Model

Orbital Energy	Orbitals	One Electron Sphere	Interacting Electron Sphere	Energy Shell
↑	1h, 2f, 3p	112	92	3h
				1h
				2d
	1g, 2d, 3s	70		
	1f, 2p	40	58	1g
			40	2p
			34	1f
	1d, 2s	20	20	2s
			18	1d
	1p	8	8	1p
	1s	2	2	1s

(c) Distribution of electrons in different orbitals in a Jellium model

Fig. 8.1 A comparison between Bohr's atomic model and jellium model; (**a**) Bohr's atomic model; (**b**) positive charge is assumed to be distributed over a positive ion core of the size of the cluster, and the electrons respond to a square-well-type potential that is attractive within the ion core and zero outside, shown in jellium model; (**c**) distribution of electrons in different orbitals in a jellium model

of CO, Au nanoparticles are very good catalyst even at room temperature. This activity of Au nanoparticle depends upon the size (Au <5 nm shows catalytic activity) (Hvolbæk et al. 2007).

8.3 Excitation and Relaxation Processes of Metal Nanoparticles

Generally an excitation due to intersystem crossing, internal conversions, fluorescence, etc. has a defined time scale for the average lifetime of an excited state (Turro et al. 2012). In the case of metals, which have a relatively high number of electrons close in energy to a large number of available empty states, their electrons can freely transfer between states at room temperature. So in these cases, the time scale for excitation is not limited (Hartland 2011). As per Mie theory, the choice of metal, as well as size, shape, surrounding matrix, surface-bound molecules and degree of aggregation of the particles, determines the frequency of light that can excite plasmons (Kelly et al. 2003; Stamplecoskie and Scaiano 2010). Light is an electromagnetic wave and has both sinusoidal electric and magnetic fields mutually transverse and always in phase, and these are transverse to the direction of propagation of light wave. When light of a suitable wavelength or frequency is made incident on the surface of the metal, the electric field in the incident light ray exerts force on electrons and thus causes their displacements and hence changes the volume density of electron, i.e. number of electrons per unit volume. Electric field of light expenses energy, that is, light loses its intensity and it is absorbed. A periodic fluctuation of positive and negative charges occurs which is called 'surface plasmon polaritons'. The nuclei being heavy provide a restoring force on the dynamic lighter electrons which have moved away. This is similar to plasma oscillations in physics. The oscillating charges create electromagnetic radiation.

For this reason, the frequency that can be used to excite plasmon absorptions is a function of both the metal and the dielectric medium surrounding it. Some electrons of metal nanoparticles get instantaneous excitation due to incident light. Moreover the surface curvature is not uniform with respect to wavelength of light. For this reason, the plasmon absorptions by nanoparticles occur in the visible region of the electromagnetic spectrum, giving rise to the multitude of colours displayed by metal nanoparticles. The plasmons of nanoparticles cannot move away because of their confinement to the particle. Hence they are called as localized surface plasmons (LSP).

Kasha's rule states that the rate of an electronic relaxation is inversely proportional to the difference in energy between energy states. This means that, for metals with overlapping filled and unfilled energy bands, the rate of relaxation must be very fast. The coherent oscillating electrons collide with one another causing the electrons to rapidly go out of phase with one another. This dissimilarity in electron phases causes a non-Fermi distribution of electrons, and it occurs within the ~10 fs of excitation. The excited electrons further scatter with each other leading to a more random or chaotic distribution. The 'electron-electron' scattering occurs within the first ~100 fs after excitation. Most spectroscopic techniques use lasers with greater than 100 fs pulse widths, so these first electron relaxation steps are very rarely observed experimentally. The final process in plasmon relaxation dissipates heat to its surrounding. The heat transfer depends upon the excitation and thermal conductivity of the medium. Heat transfer to the surroundings occurs in hundreds of picoseconds to nanoseconds, following excitation. The ability to absorb a lot of light, and release that energy locally in a short amount of time (<1 ns), is a unique property of metal nanoparticles (Alarcon et al. 2015).

8.4 Plasmonics and Surface Plasmon Resonance

Plasmonics is a phenomenon of electromagnetic resonance encountered in nanostructured metals, due to collective oscillation of conduction electrons called plasmons. Under certain conditions plasmons are excited by light that leads to strong light scattering, absorption and local electromagnetic field (Fig. 8.2). The effect can be manipulated by doing alteration and changes in the defined conditions (Fu et al. 2012). At nanoscale we are in the middle between classical and quantum effects, macroscopic and microscopic properties of the matter (Wolf 2015).

The phenomenon of surface plasmon resonance (SPR) was first observed by Wood in 1902. He made polarized light incident on a reflection grating which causes final diffraction of light producing a pattern of 'anomalous' dark and light bands

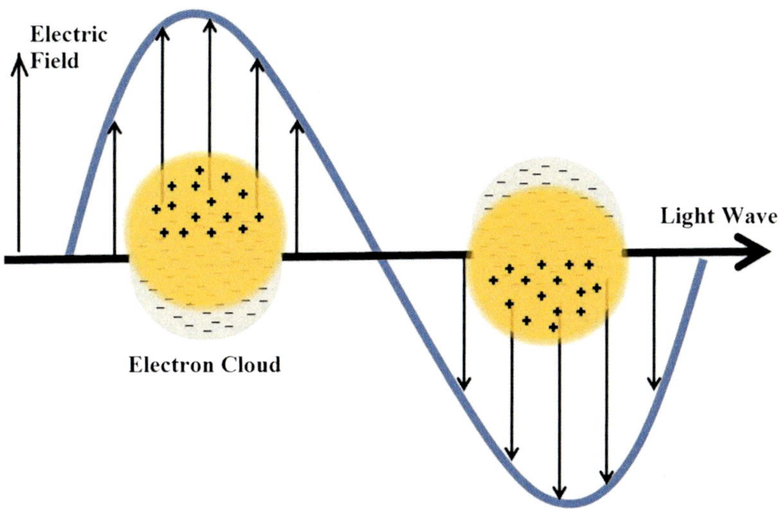

Fig. 8.2 Schematic illustration of plasmon excitation causing an instantaneous collective oscillation of electrons

(Wood 1902, 1912). Lord Rayleigh initiated the physical interpretation of this phenomenon (Rayleigh 1907), which was further refined by Fano (1941). Finally a detailed explanation came in 1968 by Otto (1968) and independently by Kretschmann and Raether (1968).

Experimentally a dip in the intensity of the reflected light is observed when a polarized light is incident from the light source on the sensor chip with a gold coating on a prism. At a certain angle of incidence (ϕ), excitation of surface plasmons occurs which results in a dip in the intensity of the reflected light. This dip in intensity is caused by the oscillation of the free electrons which is generated because of the interaction of photons of plane-polarized light with the free electrons of the metal layer. The angle at which the maximum loss of the reflected light intensity occurs is called resonance angle or SPR angle. The SPR angle is dependent on the optical characteristics of the system such as on the refractive indices of the media at both sides of the metal, usually gold. When an accumulated mass adsorbs (protein, DNA, etc.) on the surface of metal in its vicinity and refractive index at the prism side remains constant, the surface plasmon resonance condition changes, and the shift of SPR angle provides the information on kinetics.

8.4.1 SPR Assay

To construct a SPR assay, the surface of the sensor needs to be modified to allow selective capturing and thus selective measurement of the target compound. Often SPR measurements are carried out to determine the kinetics of a binding process. Recently SPR signals are also utilized in the detection of biomolecules through ELISA (de la Rica and Stevens 2012, 2013; Rodriguez-Lorenzo et al. 2012; Cecchin et al. 2014; Nie et al. 2014; Peng et al. 2014; Xianyu et al. 2014; Liang et al. 2015; Zhang et al. 2015; Huang et al. 2016; Zhao et al. 2016; Hu et al. 2014).

8.4.2 Surface Plasmon Resonance (SPR) Biosensors

The term biosensor was introduced around 1975. Literally all devices capable of reporting parameters of the living body are biosensors including a thermometer. As per present definition, a biosensor should have a recognition element comprised of biological origin and a physiochemical transducer.

The transducer detects a physiochemical change because of specific interaction between the target analyte and the biological material. Further the transducer then yields an analogue electronic signal proportional to the concentration of a specific analyte/analytes. Application of SPR-based sensors to biomolecular interaction monitoring was first demonstrated as physical methods for label-free, real-time detection of biomolecules by Lundstrom in 1983(Liedberg et al. 1983). In 1990, Pharmacia Biosensor AB launched the first commercial SPR product, the Biacore

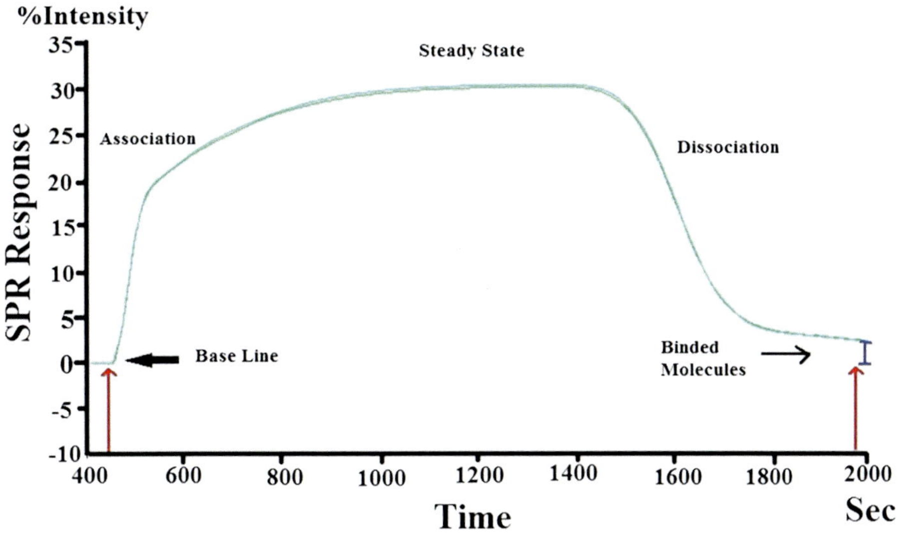

Fig. 8.3 A typical SPR signal of SPR biosensor detecting the binding of N-succinimidyl 3-(2-pyridyldithio)propionate (SPDP) to human immunoglobulin G (IgG). The red arrow is showing the baseline of human IgG. The *X-axis* is showing the time line in seconds for the association, steady state and dissociation of SPDP molecule. The blue bar is the SPR response in terms of intensity for SPDP molecule. Higher SPR response corresponds to higher binding

instrument, which was the most advanced, sensitive, accurate, reliable and reproducible direct biosensor technique (Jonsson et al. 1991). A typical binding of SPDP to human IgG is demonstrated and analysed by SPR as shown in Fig. 8.3.

8.5 Surface-Enhanced Raman Spectroscopy (SERS)

Raman spectra usually contain many sharp peaks that correspond to specific molecular vibrational frequencies, and these can provide a clear signature defining the presence of specific molecules in a sample. Accordingly, Raman spectra can be used to qualitatively and quantitatively discriminate between chemical species in materials (Motz et al. 2006).

Raman scattering is an inelastic phenomenon that occurs when a sample is illuminated by a laser light, as first demonstrated by Raman and Krishnan in 1928 (Raman and Krishnan 1998). In Raman scattering the frequency or wavelength of scattered light remains mostly unchanged except very weak lines with increase or decrease in wavelength. This increase in wavelength is Stokes Raman scattering, and decrease in wavelength is anti-Stokes Raman scattering.

The signals are weak and to enhance it, high-power lasers and long acquisition times are required which are unsuitable for biological samples. In order to enhance the intensity of Raman spectral lines, a widely used method is SERS. Finally Fleischmann in 1974 (Fleischmann et al. 1974) observed the reliability on absorption effect on Raman spectral lines. The absorption function of analyte(s) relies on the surface of metal structures. The intensity of absorbed Raman lines can be 106–108 times more. This is indicative of the importance of SERS. It is being discussed in the following section.

Surface-enhanced Raman scattering (SERS) bypasses this limitation (Alarcon et al. 2015). SERS evolves as the most promising label-free technique for molecular sensing (Campion and Kambhampati 1998). SERS, used for extremely low concentrations of molecules, overcome the low Raman cross-sectional barrier by exploiting the large field enhancement caused by electromagnetic coupling between the nanoparticles (Alarcon et al. 2015). Under such conditions, the Raman signal of the target molecule is enhanced by several orders of magnitude, enabling detection down to a single molecule (Qian and Nie 2008; Abalde-Cela et al. 2010; Oh et al. 2011; Choi et al. 2010). Generally SERS is performed on gold, silver and copper. The absorbance is mainly in the visible and near-IR wavelengths. Raman signal is amplified because of two reasons: (1) the strong enhancement of the electromagnetic field by the localized surface plasmon resonance (SPR) of metallic nanostructures and (2) surface chemical enhancement mechanism (Schatz and Van Duyne 2002; Faulds et al. 2004).

SERS has been widely used as an analytical technique in the fields of biochemistry, forensics, food safety, threat detection and medical diagnostics (Qian and Nie 2008; Ewená Smith 2002; McNay et al. 2011; Graham and Goodacre 2008). The reason behind the wide acceptance of SERS is that it is a label-free molecular fingerprint down to the level of single-molecule detection with multiplex detection which works in both in situ and in vitro detection with minimum sample requirements (Alarcon et al. 2015; Siddhanta et al. 2017).

8.6 Photothermal Effect

Photothermal therapy using metal nanoparticles exploits the infrared light to heat conversion of metal nanoparticles for selective destruction of tumours. The larger wavelength has longer penetrating depth than the shorter wavelength; this is the reason some of the red light makes it through the skin or mouth tissue. In photothermal therapy, the wavelength of excitation is an important factor then, since only red light can penetrate deep into the tissue to excite the absorbing species (that subsequently releases heat). The multiphoton cross section for metal nanoparticles is very high in comparison to dyes. This becomes even more relevant in PPTT treatments when it is desirable to use near-infrared laser excitation. Metal nanoparticles are particularly effective at absorbing near-infrared (NIR) light and converting that light into heat (Alarcon et al. 2015).

Silver nanoparticles have large absorption cross section, and it almost absorbs visible light and produces heat, thereby increasing its temperature. This makes it an important candidate with regard to plasmonic photothermal therapy (PPTT) (El-Sayed et al. 2006). By controlling the surface functionalization of nanoparticles, they can be tailored to accumulate in tumour cells.

Most photothermal therapy strategies presently use gold nanoparticles due to their superior stability under many different conditions, in biological relevant mediums. Biomolecule conjugated AgNP synthesis, with good stability, is rapidly developing and opening the door to using AgNP in more PPTT strategies (Alarcon et al. 2012, 2013). The superior optical properties of AgNP in comparison to gold make it even more attractive for higher efficiencies of light to heat conversion in PPTT.

8.7 Application of Optical and Photothermal Effect in Biology

Day by day new methods are being introduced based on the optical and photothermal effects for the detection, diagnostics (in vitro and in vivo) and therapy.

8.7.1 *In Vitro Detection and Diagnostics*

In the field of in vitro detection of biomolecules, a new type of ELISA strategy is developed and introduced by Steven's group called plasmonic ELISA. They developed a plasmonic nanosensor with inverse sensitivity via enzyme-guided crystal growth (Rodriguez-Lorenzo et al. 2012). Gold nanostars were used as the nanosensors. Glucose oxidase (GOx) enzyme-linked detection antibody controls the generation of hydrogen peroxide (H_2O_2), which reduces silver ions to atoms. At low concentrations of GOx, conformal coating of silver on the gold nanostar is favoured due to low Ag self-nucleation. It causes a large plasmon peak shift. In contrast, at high loading of GOx, competitive Ag self-nucleation dominates owing to more available H_2O_2. It reduces the deposition of silver on the gold nanostar, thus generating less peak shift. It therefore provides an inverse graph of spectral shift vs. analyte concentration. In the buffered solution, 10^{-18} g ml^{-1} of a cancer marker prostate-specific antigen (PSA) has been detected with a dynamic range between 10^{-13} and $10^{-18}gml^{-1}$.

In another work of Steven's group, catalase enzyme was linked to the immunocomplex through interactions between enzyme-decorated streptavidin and biotinylated secondary antibodies. In the presence of hydrogen peroxide, gold ions are reduced. High concentration of hydrogen peroxide favours the formation of non-aggregated, spherical nanoparticles that give rise to a red solution. When the concentration of hydrogen peroxide decreases, for example, due to the biocatalytic action of the enzyme catalase, aggregates of nanoparticles are formed, and this turns the solution blue. The range of detection was obtained 10^{-15}–10^{-19}. This is the highest possible limit of detection (de la Rica and Stevens 2012).

Inspired from this work, many researchers have followed the similar design idea and demonstrated the great potentials of plasmonic ELISA in ultrasensitive detections (Zhao et al. 2016; Xianyu et al. 2014; Peng et al. 2014; Nie et al. 2014; Cecchin et al. 2014; Huang et al. 2016; Tang et al. 2015). For instance, using antibody-tagged gold nanoclusters (AuNCs) to catalyse the decomposition of hydrogen peroxide, Zhao et al. showed sensitive detection of 1×10^{-20} M protein avidin, 7.52×10^{-14} U/ml breast cancer antigen 15-3, 2×10^{-15} mg/ml triiodothyronine and 2.3×10^{-18} mg/ml methamphetamine (Zhao et al. 2016). Using alkaline phosphatase (ALP) linking antibody in combination with iodine-mediated etching of gold nanorods, Zhang et al. exhibited sensitive detection of IgG with a LOD of 100 pg/ml (Zhang et al. 2015).

Based on the chiroplasmonic properties of nanoparticle dimers, Wu et al. (2013) demonstrated the ultralow LOD of 5×10^{-10} ng/ml for PSA. Li et al. fabricated a three-dimensional (3D) hierarchical plasmonic nanostructure. The capture antigen molecules were immobilized on plasmonic gold triangle nanopatterns, while the detection antibody molecules were linked to Raman reporter-labelled gold nanostars. In presence of biomarkers, greatly enhanced electromagnetic field amplified the Raman signal remarkably. In the buffer solution for human immunoglobulin G protein, the SERS immunosensor exhibits a detection range from 0.1 pg/ml to 10 ng/ml with a LOD of 7 fg/ml (Li et al. 2013). The strategy of etching of gold nanorods through the mediation of iodine to generate a plasmonic effect can detect up to 100 pg/ml of IgG (Zhang et al. 2015). In another work for the detection of glucose, gold nanorods were etched by the gradual oxidation in presence of trace concentration of H_2O_2 through the activity of HRP assisted by halide ions by Saa et al. (2014). Silver nanoprism etching-based plasmonic ELISA can detect 100 pg/ml with limit of detection of 4.1 fg/ml (Liang et al. 2015). In the chemical etching, halide ions act as a ligand to reduce the electron potential of the gold species, which enables ferric ions to oxidize the gold nanorods and results in the etching of the gold nanorods. The redox etching leads to a significant decrease of the gold nanorods in length but little change in diameter, which could be attributed to less surface passivation or higher chemical reactivity of the tips of the gold nanorods (Zou et al. 2009).

When gold nanoparticles with about 13 nm in diameter were modified by goat antihuman IgG, the addition of human IgG could change the absorption of colloidal gold solution, and the absorption intensity at 740 nm depended on the amount of human IgG. A dynamic range of 10–500 μg/3 mL was found (Cao et al. 2009). Rod-shaped Au@PtCu nanostructures show enhanced peroxidase-like activity, and it can replace horse red peroxidase in ELISA. Through this the detection limit reached to 90 picogram per ml (Hu et al. 2014). The enzyme-like activity of NPs can vary by replacing one metal with another. Increasing the percentage of Ag in Au@PtAg decreases the catalytic activity reflected by K_{cat} (Hu et al. 2013b). Additionally Au@PtAg nanorods have a potential antibacterial activity. On increasing the Ag fraction in the alloy shell up to 80%, the antibacterial activity gradually increases, demonstrating a flexible way to tune this activity. At 80% Ag, the antibacterial activity is better than that of a pure Ag shell (Hu et al. 2013a).

$HAuCl_4$ salt solution is yellow in colour but changes to orange upon addition of CTAB. Addition of ascorbic acid results in the formation of a colourless solution forming the precursor solution for the plasmonic nanosensor. Irradiation with ionizing radiation results in the formation of coloured dispersions of gold nanoparticle. The colour of gold nanoparticle dispersions (AuNPs) can vary depending on the size of the nanoparticles (Pérez-Juste et al. 2004). This mechanism was utilized by Pushpavanam et al. (2015) for the development of plasmonic nanosensor for dosimetry of therapeutic levels of ionizing radiation. They accomplished this by employing ionizing radiation, in concert with templating lipid surfactant micelles, in order to convert colourless salt solutions of univalent gold ions (Au^{I}) to maroon-coloured dispersions of plasmonic gold nanoparticles (Pushpavanam et al. 2015).

8.7.2 *In Vivo Monitoring, Detection and Photothermal Effect*

Because of biocompatibility and unique optical properties, gold nanorods (AuNRs) have use in various applications in biomedical research such as imaging, drug and gene delivery and thermal therapy (Weissleder 2001; Huang et al. 2006; Bonoiu et al. 2009; von Maltzahn et al. 2009; Prasad et al. 2017). Jin et al. (2013) showed that negatively charged AuNRs, other than positively charged AuNRs, can penetrate deep into the tumour spheroids and achieve a significant thermal therapeutic benefit. In thermal therapy due to localized surface plasmon resonance, AuNRs convert luminous energy into heat when activated by laser at a specific wavelength (Dickerson et al. 2008). The LSPR maximum can be tuned to near-infrared region by controlling the aspect ratio of AuNRs (Jain et al. 2008). Tri-iodobenzene is used as clinical X-ray contrast agent. Gold has higher absorption than iodine with less bone and tissue interference achieving better contrast with lower X-ray dose. Nanoparticles clear the blood more slowly than iodine agents, permitting longer imaging times. Gold nanoparticles can be used as X-ray contrast agents with properties that overcome some significant limitations of iodine-based agents (Hainfeld et al. 2006).

8.8 Synthesis of Metal NPs

Metal nanoparticles can be synthesized through either of the two approaches, ‘top-down’ or ‘bottom-up’. Top-down approach involves the use of bulk materials and reduces them into nanoparticles by way of physical, chemical or mechanical processes, whereas bottom-up starts from molecules or atoms to obtain nanoparticles.

Top-down approach usually comprises of mechanical energy, high laser energy, lithography, atomization, annealing, arc discharge, laser ablation, electron beam evaporation, radio frequency sputtering, focused ion beam lithography, etc. Bottom-up approach can be divided into gaseous phase, liquid phase, solid phase and biological methods. Chemical vapour deposition and atomic layer deposition come under gaseous phase. Reduction of metal salts, sol-gel processes, templated synthesis and electrodeposition come under the category of liquid-phase methods (Hornyak et al. 2008). Here in this chapter, we will discuss the synthesis as nonbiological and biological synthesis. For the nonbiological method, we will more focus on bottom-up approach because biological synthesis follows only bottom-up approach.

Synthesis via chemical reduction is widely used. It involves different phenomenon/mechanism like reduction, nucleation, growth, etching and some others (Cushing et al. 2004). For any redox reaction, the values of the standard reduction potentials (E_0) determine the pairs of reactants required for successful chemical conversion. It means that the free energy change in the reaction, ΔG_0, must be negative or what is equivalent to $\Delta E_0>0$. For example, in the case of silver, the relatively large electropositive reduction potential of $Ag^+ \rightarrow Ag^0$ in water (E_0 = +0.799 V, Haynes (2016)) permits the use of several reducing agents, e.g. sodium citrate (E_0 = −0.180 V, Li et al. (2013)), sodium borohydride (E_0 = −0.481 V, Haynes (2016)), hydrazine (E_0 = −0.230 V, Cushing et al. (2004)) and hydroquinone (E_0 = −0.699 V, Ullmann (2000); Alarcon et al. (2015)).

Some other commonly used methods are photochemical and electrochemical methods. Direct photoreduction has been established as an important technique for metal NP synthesis, where Mo is formed through the direct excitation of a metal source, normally a salt. Due to the advantage of being free from reducing agents, it has been widely employed in the various mediums including polymer films, glasses, cells, etc. (Sakamoto et al. 2009).

8.8.1 Biological Methods

Biological approach can be classified on the basis of extracellular and intracellular synthesis. Chemical reduction is the main mechanism behind almost all biological methods for the synthesis of nanoparticles. A lot of reducing agents are available in both the intracellular and extracellular environments. Simultaneously a lot of reducing agents are involved in the reduction process. Hence, a biological nanoparticle produced in an intracellular or extracellular environment is the result of many reducing agents. There is a lot of work published on the biological methods of synthesis via bacteria, actinomycetes, algae, fungi, yeasts, viruses, etc. (Mukherjee et al. 2001; Shankar et al. 2004; Gericke and Pinches 2006; Prasad 2014; Prasad et al. 2016).

8.9 Factors Behind Uniformity of NPs

For a tuned optical and photothermal effect, the shape and size of a synthesized nanoparticle must be uniform. A more controlled and regulated process of synthesis yields uniform nanoparticles accordingly. For a controlled and regulated synthetic process, the reduction rate, reaction kinetics and physical parameters are the key factor. Further identification of chemicals which takes part in reaction or affects the reaction or attached to the surface of nanoparticles is necessary for a controlled synthesis process which yields desired uniform nanoparticles in turn to get specific optical or photothermal properties.

The redox potential is a characteristic of the chemical species to undergo an oxidation-reduction reaction. It is a stored energy that has the ability to do work and is measured in volt; thus, the greater the voltage, the greater the ability and propensity to undergo a redox reaction (Harris 2010). The stronger the reducing agent, the faster is the reaction. Mild reducing agent has a great role in achieving desired shape and size of NPs. At many instances, synthesis of bimetallic or tri-metallic NPs is achieved through mild reducing agents. It might be possible that during synthesis of multimetallic NPs, strong reducing agent reduces one metal completely and other metals may not reduce or partially reduce.

Kinetics of a reaction plays a crucial role in the determination of uniformity of the product (NPs). A slow reaction is easy to monitor and control. A slow reaction provides sufficient time to add other ingredients during different steps of a reaction. A very fast reaction can provide non-uniform NPs. Taking silver nucleation as an example, at too much basic condition, the reaction kinetics increases and yields non-uniform AgNPs. When the same reaction is controlled and slowed down by adding dispersing agent, the yield is almost uniform AgNPs with a sharp absorption peak at 400 nm. Kinetics of a reaction is affected by parameters like temperature, pH and concentration of ingredients and presence of poisons, drugs and inhibitors. Catalytic activity of gold nanoparticles is poisoned by sulphur compounds, such as sulphides and sulphites, and inhibited by protecting molecules, such as polyvinyl alcohol (Comotti et al. 2004). There are a variety of types of inhibitors such as irreversible and reversible ones. Reversible inhibition can be further differentiated as competitive, uncompetitive or mixed inhibition, each affecting the Michaelis-Menten constant (Km) and maximal reaction velocity (V_{max}) in a specific fashion. NaN_3 and Fe^{2+} ions show inhibition behaviours to the oxidase-like activity of Au@ Pt NRs (Liu et al. 2011).

Surfactants and some other directional agents are used for the growth of NPs in a shape and aspect ratio. Dispersing agents prevent the NPs from aggregation and allow NPs to grow independently to a uniform shape and size. CTAB is a detergent which acts as both surfactant and dispersing agent. Seed-mediated growth is a good approach for the synthesis of uniform-shaped and uniform-sized NPs. Here seed acts as nuclei which grow further. The directional agent directs the growth of a seed to a specific shape. Similarly size and aspect ratio are determined by other constituents of the reaction such as Ag+ in case of AuNRs. The gold nanorods are synthesized through seed-mediated and surfactant-assisted method. The seed which is a ~ 1.5 nm diameter gold nanoparticle is synthesized by mixing aqueous solutions of hexadecylcetyltrimethylammonium bromide (CTAB) , hydrogen tetrachloroaurate (III) hydrate and sodium borohydride in a specific ratio (Orendorff and Murphy 2006). These fresh gold seeds are then added to a growth solution of concentrated CTAB, silver nitrate, hydrogen tetrachloroaurate (III) hydrate and ascorbic acid (Nikoobakht and El-Sayed 2003). Ascorbic acid is a weak/mild reducing agent that induces heterogeneous gold deposition at the surface of the seed particles (Murphy et al. 2005). Anisotropic nanorod growth results from facet-selective gold deposition promoted by the silver ions, which get adsorbed on the gold surfaces by an underpotential deposition (UPD) mechanism as elucidated by Liu and Guyot-Sionnest (Liu and Guyot-Sionnest 2005). The nanorod aspect ratio can be increased to a certain extent, up to 4.5, by increasing the silver concentration (Nikoobakht and El-Sayed 2003), and the absence of Ag^+ from the reactions leads to only a very low yield of Au nanorods. CTAB coats the nanorod surface as a bilayer that prevents aggregation (Nikoobakht and El-Sayed 2001). CTAB is also believed to aid nanorod growth by facet-sensitive surface adsorption (Smith and Korgel 2008). The synthesis process needs specific ingredients with purity, controlled and regulated reaction for the yield of uniform gold nanorods and reproducible results. So a controlled and regulated synthesis procedure with known ingredients and reaction mechanism results in the yield of uniform NPs and reproducible results.

8.10 How to Get Uniform Biological NPs

To get uniform-shaped and uniform-sized biological NPs, the ingredients, reducing agents, reaction mechanism and all other factors which can promote or inhibit the synthesis must be known. Usually in an extracellular or intracellular environment, several reducing agents are present in different concentrations from mild to strong. The pH of the cellular environment is controlled by the cell and cannot be altered too much. Inhibitors may also be present in the biological media which can inhibit fully or partially the synthesis process. All the above-mentioned factors can vary in concentrations with respect to growth rate or growth conditions and with change in strains of bacteria or fungus. This is applicable to both types of extracts whether it is intracellular or extracellular.

Mukherjee et al. (2012) tried to find out the mechanism of biosynthesis of gold nanoparticle. It has been found that low (10–15 kDa) and high molecular weight protein molecules (100–200 kDa), amino acids, phenolic compounds (wedelolactone, desmethylwedelolactone, stigmasterol, etc.), starch, polysaccharides, alkaloids, alcoholic compounds, vitamins, enzymes, etc. present in *Eclipta alba* leaf help in the formation and stabilization of gold nanoparticles (AuNPs) (Shukla et al. 2008; Wilson et al. 1986; Jadhav et al. 2009; Chaudhary et al. 2011; Xie et al. 2007; Zhu et al. 2008; Iravani 2011; Bhargava et al. 2005; Sarma and Chattopadhyay 2004). There may be several types of molecules present in a biological extract.

A few can act as reductant and others may act as stabilizing agent. Many biologically synthesized NPs show good stability at different pH conditions. This is due to many naturally present stabilizers which form mono- or multilayer over the surface of NPs. Some compound acts as directional agents for growth process. The presence of different directional agents causes the yield of different-shaped NPs from the same synthesis method with different concentrations. Because of many directional agents in the same reaction, the uniformity is not maintained.

Ascorbic acid is a naturally occurring mild reducing agent widely used for the synthesis of NPs. CTAB is the surfactant, directional agent, dispersing agent and stabilizer for NPs. It is widely used for the synthesis of AuNRs. A few reports have been published about the biological synthesis of AuNRs (Parial et al. 2012). Till date we don't have any biological equivalent of CTAB which can replace it. For defined NP synthesis, if the ingredients from biological origin such as reducing agents, surfactants, directional agents, inhibitors and stabilizers are known, then green NPs can be produced in vitro. Unknown biological compounds involved in the synthesis of NPs can be known through different characterization techniques like HPLC, gas chromatography, NMR, etc.

The spectra of nanoparticles provide information about the uniformity in terms of shape and size. A sharp-pointed peak in spectra denotes specific information about nanoparticle. A broad peak denotes nanoparticles of different shapes and sizes. Spherical nanoparticles show a single peak, while nanorods show two peaks, one longitudinal and the other transverse. It also provides information of other chemicals present in the sample. For example, ascorbic acid which is widely used in the synthesis of nanoparticles has a specific absorbance peak at 265 nm. A typical spectrum of silver nanoparticles is shown in Fig. 8.4.

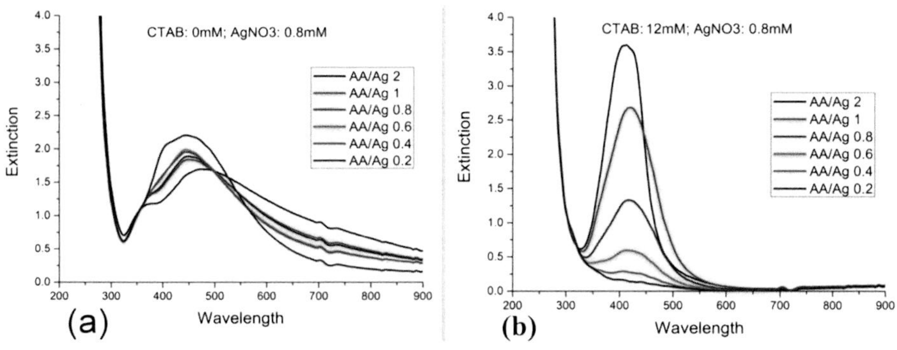

Fig. 8.4 Spectra of silver nanoparticles synthesized via ascorbic acid as reducing agent and CTAB as dispersing agent. Silver concentration is constant. Here CTAB controls the reaction rate. (**a**) In absence of dispersing agent CTAB, the reaction is very fast that leads in non-uniform silver nanoparticles having wide distribution in their shape and size. (**b**) At high concentration of CTAB, the reaction rate is slow and thus controlled. The shape is uniform. With increase in reducing agent AA, more silver reduced and results in the increase of size which is reflected from the increasing extinction

8.11 Conclusion

To obtain biological nanoparticles with defined optical and photothermal properties, uniformity in shape and size is a must. Uniformity can be achieved by understanding the mechanism of reaction, and one should have a control over it. Identification of enzymes, reducing agents, dispersing agents, directional agents and stabilizers of biological origin is necessary in order to synthesize uniform nanoparticles in vitro. In vitro synthesis of nanoparticles has always better control over in vivo synthesis. Additionally the purification of nanoparticles is easy and saves time.

References

Abalde-Cela S, Aldeanueva-Potel P, Mateo-Mateo C, Rodríguez-Lorenzo L, Alvarez-Puebla RA, Liz-Marzán LM (2010) Surface-enhanced Raman scattering biomedical applications of plasmonic colloidal particles. J R Soc Interface 7(Suppl 4):S435–S450

Alarcon EI, Udekwu K, Skog M, Pacioni NL, Stamplecoskie KG, Gonzalez-Bejar M, Polisetti N, Wickham A, Richter-Dahlfors A, Griffith M, Scaiano JC (2012) The biocompatibility and antibacterial properties of collagen-stabilized, photochemically prepared silver nanoparticles. Biomaterials 33(19):4947–4956. https://doi.org/10.1016/j.biomaterials.2012.03.033

Alarcon EI, Bueno-Alejo CJ, Noel CW, Stamplecoskie KG, Pacioni NL, Poblete H, Scaiano JC (2013) Human serum albumin as protecting agent of silver nanoparticles: role of the protein conformation and amine groups in the nanoparticle stabilization. J Nanopart Res 15(1):1374. https://doi.org/10.1007/s11051-012-1374-7

Alarcon E, Griffith M, Udekwu KI (2015) Silver nanoparticle applications: in the fabrication and design of medical and biosensing devices. Springer, Cham

Bhargava SK, Booth JM, Agrawal S, Coloe P, Kar G (2005) Gold nanoparticle formation during bromoaurate reduction by amino acids. Langmuir 21(13):5949–5956

Bonoiu AC, Mahajan SD, Ding H, Roy I, Yong K-T, Kumar R, Hu R, Bergey EJ, Schwartz SA, Prasad PN (2009) Nanotechnology approach for drug addiction therapy: gene silencing using delivery of gold nanorod-siRNA nanoplex in dopaminergic neurons. Proc Natl Acad Sci 106(14):5546–5550

Campion A, Kambhampati P (1998) Surface-enhanced Raman scattering. Chem Soc Rev 27(4):241–250

Cao QH, Yuan H, Cai RX (2009) Homogeneous detection of human IgG by gold nanoparticle probes. J Wuhan Univ Technol 24(5):772–775. https://doi.org/10.1007/s11595-009-5772-3

Cecchin D, de la Rica R, Bain RES, Finnis MW, Stevens MM, Battaglia G (2014) Plasmonic ELISA for the detection of gp120 at ultralow concentrations with the naked eye. Nanoscale 6(16):9559–9562. https://doi.org/10.1039/c3nr06167a

Chaudhary H, Dhuna V, Singh J, Kamboj SS, Seshadri S (2011) Evaluation of hydro-alcoholic extract of Eclipta alba for its anticancer potential: an in vitro study. J Ethnopharmacol 136(2):363–367

Choi CJ, Xu Z, Wu H-Y, Liu GL, Cunningham BT (2010) Surface-enhanced Raman nanodomes. Nanotechnology 21(41):415301

Comotti M, Della Pina C, Matarrese R, Rossi M (2004) The catalytic activity of "naked" gold particles. Angew Chem Int Ed Eng 43(43):5812–5815. https://doi.org/10.1002/anie.200460446

Cushing BL, Kolesnichenko VL, O'Connor CJ (2004) Recent advances in the liquid-phase syntheses of inorganic nanoparticles. Chem Rev 104(9):3893–3946. https://doi.org/10.1021/cr030027b

Dickerson EB, Dreaden EC, Huang X, El-Sayed IH, Chu H, Pushpanketh S, McDonald JF, El-Sayed MA (2008) Gold nanorod assisted near-infrared plasmonic photothermal therapy (PPTT) of squamous cell carcinoma in mice. Cancer Lett 269(1):57–66

El-Sayed IH, Huang X, El-Sayed MA (2006) Selective laser photo-thermal therapy of epithelial carcinoma using anti-EGFR antibody conjugated gold nanoparticles. Cancer Lett 239(1):129–135. https://doi.org/10.1016/j.canlet.2005.07.035

EwenáSmith W (2002) Selective functionalisation of TNT for sensitive detection by SERRS. Chem Commun 6:580–581

Fano U (1941) The theory of anomalous diffraction gratings and of quasi-stationary waves on metallic surfaces (Sommerfeld's Waves). J Opt Soc Am 31(3):213–222. https://doi.org/10.1364/JOSA.31.000213

Faulds K, Littleford RE, Graham D, Dent G, Smith WE (2004) Comparison of surface-enhanced resonance Raman scattering from unaggregated and aggregated nanoparticles. Anal Chem 76(3):592–598

Fleischmann M, Hendra PJ, McQuillan AJ (1974) Raman spectra of pyridine adsorbed at a silver electrode. Chem Phys Lett 26(2):163–166

Fu Y, Wang J, Zhang D (2012) Modelling at nanoscale. Plasmonics – principles and applications. Intech, Rijeka. doi: 40334. https://www.intechopen.com/books/plasmonics-principles-and-applications/modelling-at-nanoscale

Gericke M, Pinches A (2006) Microbial production of gold nanoparticles. Gold Bull 39(1):22–28

Graham D, Goodacre R (2008) Chemical and bioanalytical applications of surface enhanced Raman scattering spectroscopy. Chem Soc Rev 37(5):883–884

Hainfeld JF, Slatkin DN, Focella TM, Smilowitz HM (2006) Gold nanoparticles: a new X-ray contrast agent. Brit J Radiol 79(939):248–253. https://doi.org/10.1259/Bjr/13169882

Harris DC (2010) Quantitative chemical analysis. W. H. Freeman and Company, New York

Hartland GV (2011) Optical studies of dynamics in noble metal nanostructures. Chem Rev 111(6):3858–3887. https://doi.org/10.1021/cr1002547

Haynes WM (2016) CRC handbook of chemistry and physics, 97th edn. CRC Press

Hornyak GL, Dutta J, Tibbals HF, Rao A (2008) Introduction to nanoscience. CRC Press, Boca Raton

Hu X, Zhao Y, Hu Z, Saran A, Hou S, Wen T, Liu W, Ji Y, Jiang X, Wu X (2013a) Gold nanorods core/AgPt alloy nanodots shell: a novel potent antibacterial nanostructure. Nano Res 6(11):822–835

Hu XN, Saran A, Hou S, Wen T, Ji YL, Liu WQ, Zhang H, He WW, Yin JJ, Wu XC (2013b) Au@PtAg core/shell nanorods: tailoring enzyme-like activities via alloying. RSC Adv 3(17):6095–6105. https://doi.org/10.1039/C3ra23215h

Hu X, Saran A, Hou S, Wen T, Ji Y, Liu W, Zhang H, Wu X (2014) Rod-shaped Au@PtCu nanostructures with enhanced peroxidase-like activity and their ELISA application. Chin Sci Bull 59(21):2588–2596. https://doi.org/10.1007/s11434-014-0316-4

Huang X, El-Sayed IH, Qian W, El-Sayed MA (2006) Cancer cell imaging and photothermal therapy in the near-infrared region by using gold nanorods. J Am Chem Soc 128(6):2115–2120

Huang XL, Chen R, Xu HY, Lai WH, Xiong YH (2016) Nanospherical brush as catalase container for enhancing the detection sensitivity of competitive plasmonic ELISA. Anal Chem 88(3):1951–1958. https://doi.org/10.1021/acs.analchem.5b04518

Hvolbæk B, Janssens TVW, Clausen BS, Falsig H, Christensen CH, Nørskov JK (2007) Catalytic activity of Au nanoparticles. Nano Today 2(4):14–18. https://doi.org/10.1016/s1748-0132(07)70113-5

Iravani S (2011) Green synthesis of metal nanoparticles using plants. Green Chem 13(10):2638–2650

Jadhav V, Thorat R, Kadam V, Salaskar K (2009) Chemical composition, pharmacological activities of *Eclipta alba*. J Pharm Res 2(8):1229–1231

Jain PK, Huang X, El-Sayed IH, El-Sayed MA (2008) Noble metals on the nanoscale: optical and photothermal properties and some applications in imaging, sensing, biology, and medicine. Acc Chem Res 41(12):1578–1586

Jin S, Ma X, Ma H, Zheng K, Liu J, Hou S, Meng J, Wang PC, Wu X, Liang X-J (2013) Surface chemistry-mediated penetration and gold nanorod thermotherapy in multicellular tumor spheroids. Nanoscale 5(1):143–146

Jonsson U, Fagerstam L, Ivarsson B, Johnsson B, Karlsson R, Lundh K, Lofas S, Persson B, Roos H, Ronnberg I et al (1991) Real-time biospecific interaction analysis using surface plasmon resonance and a sensor chip technology. BioTechniques 11(5):620–627

Kelly KL, Coronado E, Zhao LL, Schatz GC (2003) The optical properties of metal nanoparticles: the influence of size, shape, and dielectric environment. J Phys Chem B 107(3):668–677. https://doi.org/10.1021/jp026731y

Khanna S, Jena P (1995) Atomic clusters: building blocks for a class of solids. Phys Rev B 51(19):13705

Kretschmann E, Raether H (1968) Radiative decay of nonradiative surface plasmons excited by light. Z Naturforsch A 23:2135. doi:citeulike-article-id:3901347

Li M, Cushing SK, Zhang J, Suri S, Evans R, Petros WP, Gibson LF, Ma D, Liu Y, Wu N (2013) Three-dimensional hierarchical plasmonic nano-architecture enhanced surface-enhanced Raman scattering immunosensor for cancer biomarker detection in blood plasma. ACS Nano 7(6):4967–4976. https://doi.org/10.1021/nn4018284

Liang JJ, Yao CZ, Li XQ, Wu Z, Huang CH, Fu QQ, Lan CF, Cao DL, Tang Y (2015) Silver nanoprism etching-based plasmonic ELISA for the high sensitive detection of prostate-specific antigen. Biosens Bioelectron 69:128–134. https://doi.org/10.1016/j.bios.2015.02.026

Liedberg B, Nylander C, Lunström I (1983) Surface plasmon resonance for gas detection and biosensing. Sensors Actuators 4:299–304. https://doi.org/10.1016/0250-6874(83)85036-7

Liu M, Guyot-Sionnest P (2005) Mechanism of silver(I)-assisted growth of gold nanorods and bipyramids. J Phys Chem B 109(47):22192–22200. https://doi.org/10.1021/jp054808n

Liu JB, Hu XN, Hou S, Wen T, Liu WQ, Zhu X, Wu XC (2011) Screening of inhibitors for oxidase mimics of Au@Pt nanorods by catalytic oxidation of OPD. Chem Commun 47(39):10981–10983. https://doi.org/10.1039/c1cc14346h

von Maltzahn G, Park J-H, Agrawal A, Bandaru NK, Das SK, Sailor MJ, Bhatia SN (2009) Computationally guided photothermal tumor therapy using long-circulating gold nanorod antennas. Cancer Res 69(9):3892–3900

McNay G, Eustace D, Smith WE, Faulds K, Graham D (2011) Surface-enhanced Raman scattering (SERS) and surface-enhanced resonance Raman scattering (SERRS): a review of applications. Appl Spectrosc 65(8):825–837

Motz JT, Fitzmaurice M, Miller A, Gandhi SJ, Haka AS, Galindo LH, Dasari RR, Kramer JR, Feld MS (2006) In vivo Raman spectral pathology of human atherosclerosis and vulnerable plaque. J Biomed Opt 11(2):021003–021003. -021009

Mukherjee P, Ahmad A, Mandal D, Senapati S, Sainkar SR, Khan MI, Ramani R, Parischa R, Ajayakumar PV, Alam M, Sastry M, Kumar R (2001) Bioreduction of AuCl(4)(−) ions by the fungus, *Verticillium* sp. and surface trapping of the gold nanoparticles formed. Angew Chem Int Ed Eng 40(19):3585–3588

Mukherjee S, Sushma V, Patra S, Barui AK, Bhadra MP, Sreedhar B, Patra CR (2012) Green chemistry approach for the synthesis and stabilization of biocompatible gold nanoparticles and their potential applications in cancer therapy. Nanotechnology 23(45):455103

Murphy CJ, Sau TK, Gole AM, Orendorff CJ, Gao J, Gou L, Hunyadi SE, Li T (2005) Anisotropic metal nanoparticles: synthesis, assembly, and optical applications. J Phys Chem B 109(29):13857–13870. https://doi.org/10.1021/jp0516846

Nie XM, Huang R, Dong CX, Tang LJ, Gui R, Jiang JH (2014) Plasmonic ELISA for the ultrasensitive detection of *Treponema pallidum*. Biosens Bioelectron 58:314–319. https://doi.org/10.1016/j.bios.2014.03.007

Nikoobakht B, El-Sayed MA (2001) Evidence for bilayer assembly of cationic surfactants on the surface of gold nanorods. Langmuir 17(20):6368–6374. https://doi.org/10.1021/la010530o

Nikoobakht B, El-Sayed MA (2003) Preparation and growth mechanism of gold nanorods (NRs) using seed-mediated growth method. Chem Mater 15(10):1957–1962. https://doi.org/10.1021/cm020732l

Oh YJ, Park SG, Kang MH, Choi JH, Nam Y, Jeong KH (2011) Beyond the SERS: Raman enhancement of small molecules using nanofluidic channels with localized surface plasmon resonance. Small (Weinheim an der Bergstrasse, Germany) 7(2):184–188

Orendorff CJ, Murphy CJ (2006) Quantitation of metal content in the silver-assisted growth of gold nanorods. J Phys Chem B 110(9):3990–3994. https://doi.org/10.1021/jp0570972

Otto A (1968) Excitation of nonradiative surface plasma waves in silver by the method of frustrated total reflection. Zeitschrift für Physik A Hadrons and nuclei 216(4):398–410. https://doi.org/10.1007/bf01391532

Parial D, Patra HK, Roychoudhury P, Dasgupta AK, Pal R (2012) Gold nanorod production by cyanobacteria-a green chemistry approach. J Appl Phycol 24(1):55–60. https://doi.org/10.1007/s10811-010-9645-0

Peng CF, Duan XH, Khamba GW, Xie ZJ (2014) Highly sensitive "signal on" plasmonic ELISA for small molecules by the naked eye. Anal Methods-Uk 6(24):9616–9621. https://doi.org/10.1039/c4ay01993h

Pérez-Juste J, Liz-Marzan L, Carnie S, Chan DY, Mulvaney P (2004) Electric-field-directed growth of gold nanorods in aqueous surfactant solutions. Adv Funct Mater 14(6):571–579

Poole CP, Owens FJ (2003) Introduction to nanotechnology. Wiley, Hoboken

Prasad R (2014) Synthesis of silver nanoparticles in photosynthetic plants. J Nanoparticles, Article ID 963961, https://doi.org/10.1155/2014/963961

Prasad R, Pandey R, Barman I (2016) Engineering tailored nanoparticles with microbes: quo vadis. WIREs Nanomedicine Nanobiotechnology 8:316–330. https://doi.org/10.1002/wnan.1363

Prasad R, Bhattacharyya A, Nguyen QD (2017) Nanotechnology in sustainable agriculture: recent developments, challenges, and perspectives. Front Microbiol 8:1014. https://doi.org/10.3389/fmicb.2017.01014

Pushpavanam K, Narayanan E, Chang J, Sapareto S, Rege K (2015) A colorimetric plasmonic nanosensor for dosimetry of therapeutic levels of ionizing radiation. ACS Nano 9(12):11540–11550. https://doi.org/10.1021/acsnano.5b05113

Qian X-M, Nie S (2008) Single-molecule and single-nanoparticle SERS: from fundamental mechanisms to biomedical applications. Chem Soc Rev 37(5):912–920

Raman C, Krishnan K (1998) A new type of secondary radiation (reprinted from nature, vol 121, pg 501–502, 1928). Curr Sci 74(4):381–381

Rao B, Khanna S, Jena P (1999) Designing new materials using atomic clusters. J Clust Sci 10(4):477–491

Rayleigh L (1907) On the dynamical theory of gratings. Proc R Soc London Ser A 79(532):399–416. https://doi.org/10.1098/rspa.1907.0051

de la Rica R, Stevens MM (2012) Plasmonic ELISA for the ultrasensitive detection of disease biomarkers with the naked eye. Nat Nanotechnol 7(12):821–824. https://doi.org/10.1038/nnano.2012.186

de la Rica R, Stevens MM (2013) Plasmonic ELISA for the detection of analytes at ultralow concentrations with the naked eye. Nat Protoc 8(9):1759–1764. https://doi.org/10.1038/nprot.2013.085

Rodriguez-Lorenzo L, de la Rica R, Alvarez-Puebla RA, Liz-Marzan LM, Stevens MM (2012) Plasmonic nanosensors with inverse sensitivity by means of enzyme-guided crystal growth. Nat Mater 11(7):604–607. https://doi.org/10.1038/Nmat3337

Saa L, Coronado-Puchau M, Pavlov V, Liz-Marzan LM (2014) Enzymatic etching of gold nanorods by horseradish peroxidase and application to blood glucose detection. Nanoscale 6(13):7405–7409. https://doi.org/10.1039/C4nr01323a

Sakamoto M, Fujistuka M, Majima T (2009) Light as a construction tool of metal nanoparticles: synthesis and mechanism. J Photochem Photobiol C: Photochem Rev 10(1):33–56. https://doi.org/10.1016/j.jphotochemrev.2008.11.002

Sarma TK, Chattopadhyay A (2004) Starch-mediated shape-selective synthesis of Au nanoparticles with tunable longitudinal plasmon resonance. Langmuir 20(9):3520–3524

Schatz GC, Van Duyne RP (2002) Electromagnetic mechanism of surface-enhanced spectroscopy. Handbook of vibrational spectroscopy. Wiley, Chichester. http://onlinelibrary.wiley.com/book/10.1002/0470027320

Shankar SS, Rai A, Ankamwar B, Singh A, Ahmad A, Sastry M (2004) Biological synthesis of triangular gold nanoprisms. Nat Mater 3(7):482–488. https://doi.org/10.1038/nmat1152

Shukla R, Nune SK, Chanda N, Katti K, Mekapothula S, Kulkarni RR, Welshons WV, Kannan R, Katti KV (2008) Soybeans as a phytochemical reservoir for the production and stabilization of biocompatible gold nanoparticles. Small (Weinheim an der Bergstrasse, Germany) 4(9):1425–1436. https://doi.org/10.1002/smll.200800525

Siddhanta S, Paidi SK, Bushley K, Prasad R, Barman I (2017) Exploring morphological and biochemical linkages in fungal growth with label-free light sheet microscopy and Raman spectroscopy. ChemPhysChem 18(1):72–78

Smith DK, Korgel BA (2008) The importance of the CTAB surfactant on the colloidal seed-mediated synthesis of gold nanorods. Langmuir 24(3):644–649. https://doi.org/10.1021/La703625a

Stamplecoskie KG, Scaiano JC (2010) Light emitting diode irradiation can control the morphology and optical properties of silver nanoparticles. J Am Chem Soc 132(6):1825–182+. https://doi.org/10.1021/ja910010b

Tang Y, Zhang W, Liu J, Zhang L, Huang W, Huo FW, Tian DB (2015) A plasmonic nanosensor for lipase activity based on enzyme-controlled gold nanoparticles growth in situ. Nanoscale 7(14):6039–6044. https://doi.org/10.1039/c4nr07579j

Turro NJ, Ramamurthy V, Scaiano JC (2012) Modern molecular photochemistry of organic molecules. Wiley Online Library

Ullmann F (2000) Ullmann's encyclopedia of industrial chemistry. Wiley. http://onlinelibrary.wiley.com/doi/10.1002/14356007.a27_443/abstract

Weissleder R (2001) A clearer vision for in vivo imaging. Nat Biotechnol 19(4):316–316

Wilson KA, Rightmire BR, Chen JC, Tan-Wilson AL (1986) Differential proteolysis of glycinin and β-conglycinin polypeptides during soybean germination and seedling growth. Plant Physiol 82(1):71–76

Wolf EL (2015) Nanophysics and nanotechnology: an introduction to modern concepts in nanoscience. Wiley

Wood RW (1902) XLII. On a remarkable case of uneven distribution of light in a diffraction grating spectrum. Philos Mag Ser 6 4(21):396–402. https://doi.org/10.1080/14786440209462857

Wood RW (1912) XXVII. Diffraction gratings with controlled groove form and abnormal distribution of intensity. Philos Mag Ser 6 23(134):310–317. https://doi.org/10.1080/14786440208637224

Wu X, Xu L, Liu L, Ma W, Yin H, Kuang H, Wang L, Xu C, Kotov NA (2013) Unexpected chirality of nanoparticle dimers and ultrasensitive chiroplasmonic bioanalysis. J Am Chem Soc 135(49):18629–18636. https://doi.org/10.1021/ja4095445

Xianyu YL, Wang Z, Jiang XY (2014) A plasmonic nanosensor for immunoassay via enzyme-triggered click chemistry. ACS Nano 8(12):12741–12747. https://doi.org/10.1021/nn505857g

Xie J, Lee JY, Wang DI, Ting YP (2007) Identification of active biomolecules in the high-yield synthesis of single-crystalline gold nanoplates in algal solutions. Small (Weinheim an der Bergstrasse, Germany) 3(4):672–682

Zhang ZY, Chen ZP, Wang SS, Cheng FB, Chen LX (2015) Iodine-mediated etching of gold nanorods for plasmonic ELISA based on colorimetric detection of alkaline phosphatase. ACS Appl Mater Interfaces 7(50):27639–27645. https://doi.org/10.1021/acsami.5b07344

Zhao Q, Huang HW, Zhang LY, Wang LQ, Zeng YL, Xia XD, Liu FP, Chen Y (2016) Strategy to fabricate naked-eye readout ultrasensitive plasmonic nanosensor based on enzyme mimetic gold nanoclusters. Anal Chem 88(2):1412–1418. https://doi.org/10.1021/acs.analchem.5b04089

Zhu X, Yang Q, Huang J, Suzuki I, Li G (2008) Colorimetric study of the interaction between gold nanoparticles and a series of amino acids. J Nanosci Nanotechnol 8(1):353–357

Zou R, Guo X, Yang J, Li D, Peng F, Zhang L, Wang H, Yu H (2009) Selective etching of gold nanorods by ferric chloride at room temperature. Cryst Eng Comm 11(12):2797. https://doi.org/10.1039/b911902g

Chapter 9
Biogenic Synthesis of Silver Nanoparticles and Their Applications in Medicine

Manjit K. Selwal and Krishan K. Selwal

Abstract The synthesis of metal nanoparticles is gaining momentum because of their novel chemical, optical, electrical, and magnetic properties. Several methods have been used for biofabrication of metal nanoparticles such as chemical reduction, photochemical or radiation reduction, metallic wire explosion, sonochemical and polyol methods. In order to eliminate toxic chemical emanation during these processes, green synthesis of nanoparticles using bacteria, fungi, enzyme, and plant extract is an ecofriendly alternative to the abovementioned processes.

The nanoparticles produced through this technique have attained great interest due to changes in optical, chemical, and photoelectrochemical properties and their composition, shape, and size. The nanoparticles have several important applications in the field of bio-labeling, sensors, antimicrobial agents, filters purifying drinking water, degrading pesticides, and killing human pathogenic bacteria. Furthermore, the development of nanoparticle structures can be made by employing different substrates for delivery of drugs or therapeutic compounds.

9.1 Introduction

The concept of nanotechnology though considered to be a modern science has its history dating to as back as the ninth century. Nanoparticles of gold and silver were used by the artisans of Mesopotamia to generate a glittering effect to pots. The first scientific description of the properties of nanoparticles was reported in 1857 by Michael Faraday in his famous article "Experimental Relations of Gold (and Other Metals) to Light" (Faraday 1857). The nano-revolution conceptually started in the early 1980s with the first paper on nanotechnology being published in 1981 by K. Eric Drexler of Space Systems Laboratory, Massachusetts Institute of Technology. Nanobiotechnology is rapidly growing by generating various nanoproducts such as nanotubes, biosensor, dye-resistant clothing, and gene and drug delivery vehicles

M.K. Selwal • K.K. Selwal (✉)
Department of Biotechnology, Deenbandhu Chhotu Ram University of Science & Technology, Murthal, Sonepat, Haryana, India
e-mail: krishan.kselwal@gmail.com

R. Prasad (ed.), *Fungal Nanotechnology*, Fungal Biology,
DOI 10.1007/978-3-319-68424-6_9

and promises to be critical in advances in other related fields. Nanoparticles are particulate dispersions or solid particles with a size in the range of 10–1000 nm. Nanotechnology is being utilized in offering many new developments in the fields of biosensors, biomedicine, bio-nanotechnology diagnosis, and therapeutic drug delivery and in the development of treatments for many diseases and disorders (Rudge et al. 2001; Prasad et al. 2014, 2016, 2017a, b).

Synthesis of silver nanoparticles (AgNPs) has great revolution in the field of nanotechnology in the last 10 years and proves its potentials by new and varied applications due to its unique physical, chemical, and biological properties. Various synthesis methods are used that give different shapes, sizes, and voids. Recently, there is an increasing trend of using biological materials such as plant, fungi, and bacterial extract to reduce silver ions to elemental state. The unique properties of NPs synthesized by biological materials have many advantages over other methods. This chapter is devoted to preparation and applications of silver nanoparticles using biological materials.

9.2 Approaches for Silver Nanoparticle Synthesis

The development of consistent processes for the synthesis of silver nanoparticles is an important aspect of current nanotechnology research. Generally, physical and chemical methods have been used in both for AgNP synthesis with particular shape and size depending on specific requirements (Fig. 9.1). These methods for nanoparticle synthesis used either the bottom-up or top-down approach. In the former approach, production of silver nanoparticles entails mechanical grinding of a bulk piece of the material, whereas the latter approach comprising chemical and

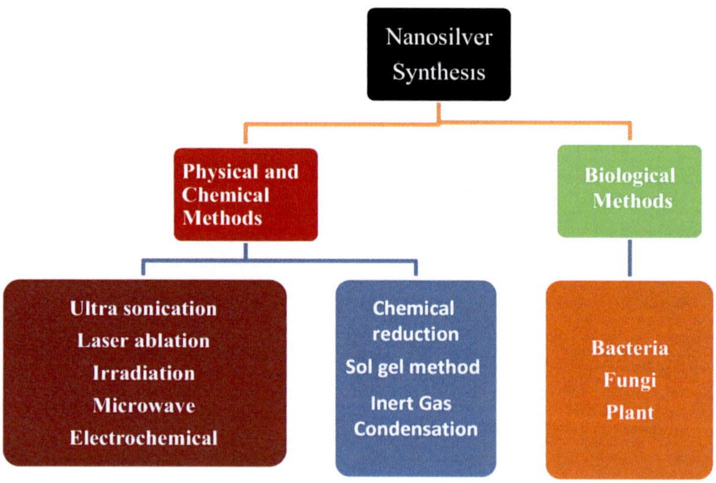

Fig. 9.1 Different methods for silver nanoparticle synthesis

biological means to make nanostructures. These processes cause controlled condensation of solute molecules that are formed during a chemical reaction. The restriction of the condensation or growth leads to the formation of particles of desired size and shape. Some of the commonly used methods are ion sputtering, thermal synthesis, inert gas condensation, reduction, and sol-gel technique as described below.

9.2.1 *Evaporation-Condensation*

Various metal nanoparticles such as Ag, Au, PbS, and fullerene have previously been produced using the evaporation/condensation method (Kruis et al. 2000). In this method the inorganic or organic material is vaporized in the furnace followed by homogeneous nucleation in gas phase and grows by coalescence of atoms. The size and stability of NPs depend on gas pressure and vapor pressure of inert gas (Gurav et al. 1994). However, the synthesis of silver nanoparticles using a tube furnace has several drawbacks, because a tube furnace occupies a large space and requires power consumption of more than several kilowatts and a long preheating time to attain a stable operating temperature.

9.2.2 *Laser Ablation*

Laser ablation is the process of removing material from a solid surface by irradiating with a laser beam. At low laser flux, the material is heated by absorbed laser energy and evaporates or sublimates. At higher flux, the material is converted to plasma. The depth over which laser energy is absorbed and the amount of material removed by single laser pulse depend on the material's optical properties and laser wavelength (Mafune et al. 2000). Carbon nanotubes can be produced by this method. Various workers used this technique for the synthesis of silver nanoparticles from metallic bulk materials and carbon nanotube in solution (Chen and Yeh 2002; Kabashin and Meunier 2003). The advantage of laser ablation compared to other conventional method for preparing metal colloids is the absence of chemical reagents in solutions (Sylvestre et al. 2004).

9.2.3 *Inert Gas Technique*

This technique was developed in Japan in 1960, a very common method for nanoparticle (NP) synthesis. The method synthesizes NPs by evaporation of substrate and then subsequently condensation to produce required size particles normally in the range of 10–100 nm. Inert gas such as helium or argon is introduced in the chamber with a pressure of 1–40 torr leading to formation of NPs.

9.2.4 *Sputtering*

This is an another version of evaporation and condensation technique which comprises vaporizing materials from a solid surface by bombarding a target with high-velocity ions of an inert gas, which causes ejection of atoms and clusters. Gunther and Kumpmann (1992) applied an electron beam in an inert gas atmosphere with pressures up to 5 mbar in order to produce 5 nm Al_2O_3 and SiO_2 particles. The advantage of sputtering is that it is mainly the target material which is heated and that the composition of the sputtered material is the same as that of the target.

9.3 Chemical Method

Chemical reduction is the most frequently applied method for the preparation of AgNPs as stable, colloidal dispersions in water or organic solvents. The process requires reaction between substrate, reducing agent, and stabilizing agents resulting in the formation of uniform shape and size and stable silver nanoparticles. The commonly used reducing agents are borohydride, citrate, ascorbate and elemental hydrogen, polyvinylpyrrolidone (PVP) with ethylene glycol, and oleylamine-liquid paraffin system. The reduction of silver ions (Ag^+) in aqueous solution generally yields colloidal silver with particle diameters of several nanometers. Initially, the reduction of various complexes with Ag^+ ions leads to the formation of silver atoms (Ag0), which is followed by agglomeration into oligomeric clusters. These clusters eventually lead to the formation of colloidal Ag particles ranging from monodisperse to spherical in nature.

9.4 Disadvantages of Physical and Chemical Methods

The synthesis of nanoparticles including physical and chemical processes is costly and laborious. The use of these synthesis methods requires both strong and weak reducing agents and protective agents like sodium borohydride, sodium citrate, and alcohols. These mostly used agents are toxic, flammable, and cannot be easily degraded which pose serious environmental issues. Chemical methods produced nanoparticles of desired size; however, they require stabilizing agents to prevent the AgNPs from agglomeration. So there is a need to find out a green technology comprising microorganisms or plant extracts for synthesis of NPs.

9.5 Biological Synthesis of Silver Nanoparticles

9.5.1 Bacterial-Mediated Synthesis

Microorganisms are considered as a potential biofactory for the synthesis of nanoparticles like gold, silver, and cadmium sulfide. Bacteria are known to produce inorganic materials either intracellularly or extracellularly. Some well-known examples of bacteria synthesizing inorganic materials include magnetotactic bacteria (synthesizing magnetic nanoparticles) and S-layer bacteria which produce gypsum and calcium carbonate layers. The probability of the bacterial-mediated synthesis of silver nanoparticles has been successfully reported by various workers. The first evidence of bacteria synthesizing silver nanoparticles was established using the *Pseudomonas stutzeri* AG259 strain that was isolated from silver mine (Haefeli et al. 1984). The research took momentum afterward, and various investigators reported AgNP synthesis using different bacterial species. The process involves the preparation of bacterial extract and reaction between the extract and substrate for reduction of silver ions (Fig. 9.2). For preparation of bacterial extract, the desired culture needs to grow in liquid medium and separation of supernatant by centrifugation (Fig. 9.3). The presence of nitrate reductase, NADH, electron donor, and acceptor in bacterial cell membrane facilitates synthesis of AgNPs (Prasad et al. 2016). Shahverdi et al. (2007) reported the rapid synthesis of metallic nanoparticles of silver using the reduction of aqueous Ag+ ion using the culture supernatants of *Klebsiella pneumoniae*, *Escherichia coli*, and *Enterobacter cloacae*. Silver nanoparticles thus produced were then characterized by UV-vis spectroscopy, Fourier transform infrared spectroscopy (FTIR), transmission electron microscopy (TEM), and confirmed metallic nature of particles. Saifuddin et al. (2009) have described a novel combinational approach for green biosynthesis using a combination of culture supernatant of *Bacillus subtilis* and microwave irradiation in water. They reported the extracellular biosynthesis of monodispersed silver NPs in the size range of 5–50 nm using culture supernatants and microwave radiation which help in heating around the NPs and thus prevent aggregation. Similarly, Jain et al. (2010) used the spore crystal mixture of *Bacillus thuringiensis* for synthesis of silver nanoparticles. Nanoparticles were characterized using UV-vis absorption spectroscopy, X-ray diffraction (XRD), and scanning electron microscopy (SEM). X-ray diffraction and TEM analysis showed the average particle size of 15 nm with cubic and hexagonal structure.

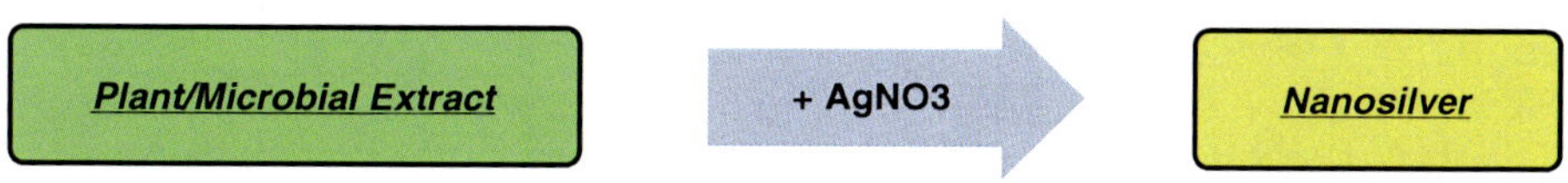

Fig. 9.2 Plant- and microbial-mediated synthesis of silver nanoparticles

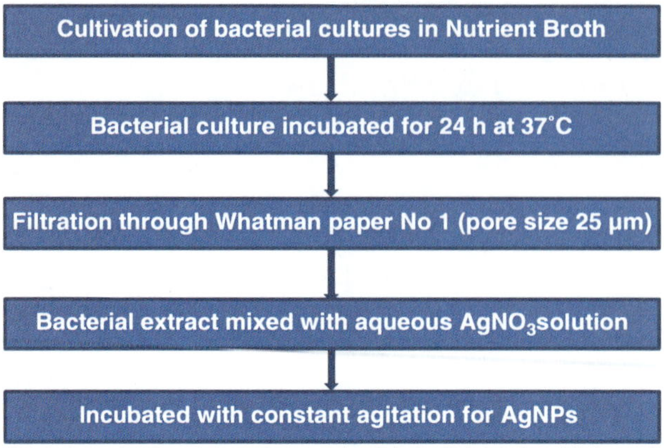

Fig. 9.3 Flow chart for biogenic synthesis of AgNPs using bacteria

The synthesis and stability of nanoparticles appeared to depend on the temperature, pH, or species of bacteria from which the supernatant was used. It was observed that the AgNP synthesis by *Arthrobacter kerguelensis* supernatant was not influenced by temperature, pH of media, and growth phase of culture, whereas *Penicillium antarctica*, *B. subtilis*, and *B. licheniformis* culture extract produced stable AgNPs at different growth conditions (Shivaji et al. 2011). Wei et al. (2012) obtained silver NPs using solar irradiation of cell-free extracts of *Bacillus amyloliquefaciens* and $AgNO_3$. They demonstrated that light intensity, extract concentration, and NaCl addition influenced the synthesis of AgNPs. TEM and XRD analysis confirmed that circular and triangular crystalline nanoparticles with a mean diameter of 14.6 nm were synthesized. Nanda et al. (2015) exhibited synthesis of stable, spherical silver nanoparticles with an average size of 35.42 nm by exposing aqueous silver ions to extracellular exudates of *B. subtilis* A1 under optimized laboratory conditions.

9.5.2 Fungus-Mediated Synthesis of AgNPs

Fungi have been broadly used for the biosynthesis of nanoparticles due to better control on size, shape, ease to downstream process, and stability of nanoparticles (Prasad et al. 2016). Furthermore, monodisperse nanoparticles with well-defined dimensions and larger amount can be obtained using fungi. Most fungi have a very high wall-binding capacity as well as intracellular metal uptake capacities (Volesky and Holan 1995). They are easy to culture on a large scale by solid substrate fermentation, thus making a large amount of biomass available for processing. Fungi can grow over the surface of inorganic substrate which leads to the metal being distributed in a more efficient way as a catalyst. The fungal biomass is generally obtained aerobically in medium like potato dextrose broth or Czapek Dox broth incubated at 25 °C. After incubation, the fungal biomass was separated and thoroughly washed with double deionized water. The biomass is suspended in distilled water for further incubation at 25 °C for 24–72 h. Finally the supernatant is obtained by passing the content through Whatman paper no. 1 and the same mixed with $AgNO_3$ solution to facilitate nanoparticle formation. The change in color of the reaction mixture from colorless to brown indicates formation of metallic silver nanoparticles. *Verticillium* sp. fungal biomass when exposed to aqueous $AgNO_3$ solution resulted in the intracellular formation of silver nanoparticles, while *Fusarium oxysporum* biomass resulted in the extracellular silver nanoparticles (Senapati et al. 2004). Duran et al. (2005) carried out the extracellular production of silver nanoparticles using *F. oxysporum* that leads to formation of silver nanoparticles in the range of 20–50 nm in dimensions. Kumar et al. (2007) directly used the purified nitrate reductase from the organism *F. oxysporum* for the synthesis of silver nanoparticle in test tube. Their reaction mixture contained only the enzyme nitrate reductase, silver nitrate, and NADPH. Slowly, the reaction mixture turned brown indicating the formation of silver nanoparticles. This was the first direct evidence for the involvement of nitrate reductase in the synthesis of silver nanoparticles. The mechanism underlying synthesis of AgNPs is not clearly understood till date, nevertheless reduction of precursor ions to metallic form occurs either by cysteine/amine group of cell wall proteins or via electrostatic interaction of Ag^+ ions with carboxylate group of fungal enzyme could be the possible rational behind AgNPs synthesis (Kathiresan et al. 2009). Various workers have reported biological synthesis of AgNPs using different fungal species such as *Aspergillus fumigatus*, *Penicillium fellutanum*, *Amylomyces rouxii*, *Fusarium solani*, and *Mucor hiemalis* (Bhainsa D'Souza 2006; Sadowski et al. 2008; Musarrat et al. 2010; Vahabi et al. 2011; El-Rafi et al. 2012; Aziz et al. 2016). Recently, our study explored stable silver nanoparticle synthesis with the size range of 5–25 nm using *Penicillium atramentosum* KM. The process parameters were optimized to produce stable AgNPs and showed antimicrobial activity against various enteric pathogens. The XRD pattern depicted in Fig. 9.4 confirmed the crystalline nature of AgNPs (Sarsar et al. 2015).

9.5.3 Plant-Mediated Synthesis of Silver Nanoparticles

The plant extract is also important for the synthesis of silver nanoparticles. It is easily prepared, safe, nontoxic to use that led to rapid synthesis of AgNPs in comparison to microbial and fungal extract (Mude et al. 2009). Various plants are used for AgNPs ranging from algae, medicinal plants, and angiosperm. The process of reduction of $AgNO_3$ is accomplished by plant extract prepared in similar manner as that of microbes and fungal extract. For preparation of plant extract, any plant part such as leaves, bark, stem, fruit, and flower can be utilized. The thoroughly washed plant materials are finely chopped (10–20 g) and mixed with 100 ml of double-distilled water. The suspension is allowed to boil at 60 °C for 30 min followed by filtration through appropriate membrane, and the resultant extract solution would be used for reduction of Ag^+ ions to Ag^0. The bioreduction of Ag^+ ions was monitored

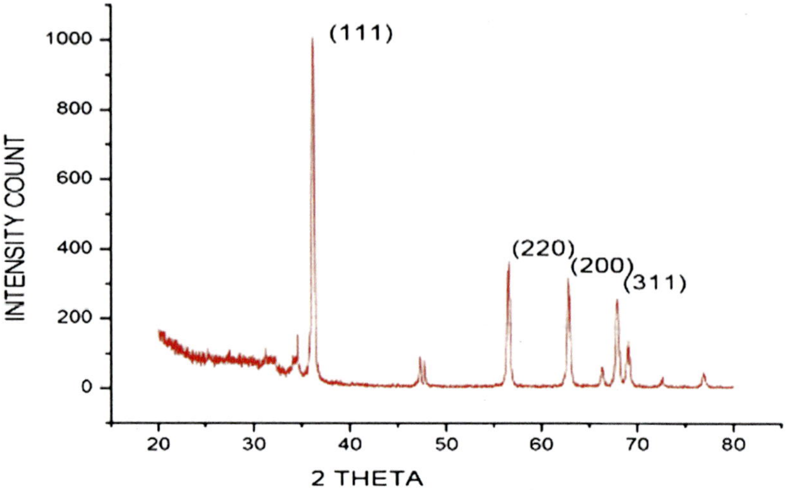

Fig. 9.4 XRD peaks of 111, 200, 220, and 311 nanoparticles implying crystalline nature of AgNPs (Sarsar et al. 2015)

by UV-vis spectra of the solution (Ahmad et al. 2011; Prasad and Swamy 2013; Sarsar et al. 2014).

Phytochemicals are assumed to be responsible for the formation of silver nanoparticles. The main phytochemicals involved are terpenoids, flavones, ketones, aldehydes, amides, and carboxylic acids (Prasad 2014). Flavones, organic acids, and quinones are water-soluble phytochemicals that are responsible for the immediate reduction of the ions (Jha et al. 2009; Karthikeyan et al. 2012). Li et al. (2007) reported green synthesis of silver nanoparticles using *Capsicum annuum* leaf extract. The results indicated that the proteins, which have amine groups, played a reducing and controlling role during the formation of silver NPs in the solutions and that the secondary structure of the proteins changed after reaction with silver ions.

Balaprasad et al. (2005) reported biosynthesis of gold and silver nanoparticles using *Emblica officinalis* fruit extract mixed with aqueous silver sulfate and chloroauric acid solutions, rapid reduction of silver, and chloroaurate ions is observed leading to the formation of highly stable silver and gold nanoparticles in solution. Chandran et al. (2006) reported synthesis of gold and AgNPs using *Aloe vera* extract. Reduction of silver ions by *Aloe vera* extract led to the formation of spherical silver nanoparticles of 15.2 ± 4.2 nm size. Detection and characterization of synthesized silver nanoparticles are carried out by FTIR and SEM. The former showed peaks between 1000 and 2000 cm^{-1} which confirmed the presence of proteins and other ligands required for the synthesis and stabilization of silver nanoparticles, whereas the latter micrograph exhibited the size of silver nanoparticles. Jha et al. (2009) reported confirmation of silver nanoparticles using *Eclipta* leaf by X-ray and TEM analyses which revealed spherical NPs with a size range of 2–6 nm. Similarly, Singhal et al. 2010 used UV-Vis, atomic absorption spectroscopy (AAS), dynamic light scattering (DLS), XRD, FTIR, and TEM techniques for characterization of AgNPs using *Ocimum sanctum* (tulsi) leaf extract.

Saxena et al. (2010) reported biological synthesis of silver nanoparticles by using onion (*Allium cepa*) extract and their antibacterial activity. The morphology of silver nanoparticles was confirmed by TEM. The antibacterial activity of these nanoparticles was studied against *E. coli* and *Salmonella typhimurium*. The bactericidal property of nanoparticles was analyzed by measuring the growth curve of bacterial population against 50 μg/mL concentration of silver nanoparticles.

9.6 Factors Affecting Green Synthesis of Silver Nanoparticles

There is always a continuous interaction between living organisms and the environment in which they live. The environmental conditions exert an influence on growth and development of organisms. The metabolite production by bacteria, fungi, and plant is influenced by the conditions in which the organisms are cultivated (Fig. 9.5). Therefore, optimization studies will not only support good growth but also enhance product yield. Our previous studies also suggested that optimization of process parameters such as incubation temperature, pH, incubation time, concentration of the metal ions, the concentration ratio of the extract and metal ions played an important role that increased productivity of AgNPs (Sarsar et al. 2014, 2015, 2016). Deva et al. (2016) demonstrated that 1 mM precursor salt concentration, 1:9 ratio of bacterial extract and silver nitrate, 72 h of incubation at pH 7, and temperature 45 °C resulted in higher yield of NPs using *Microbacterium* sp. Similarly, Nayak et al.

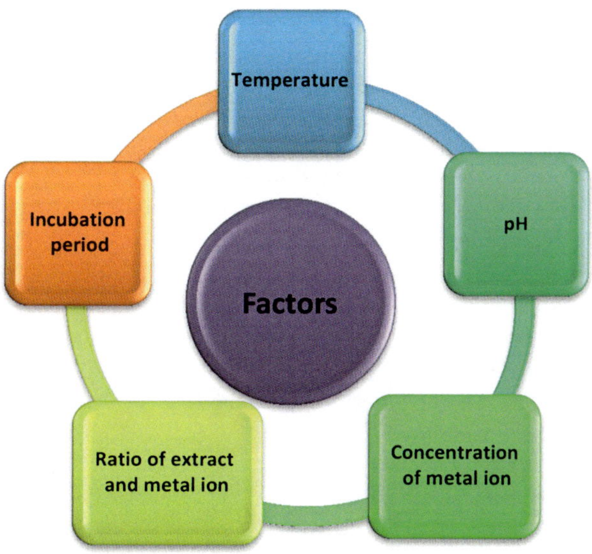

Fig. 9.5 Factors affecting biogenic synthesis of AgNPs

(2011) demonstrated that increase in concentration of $AgNO_3$ increases the formation of silver nanoparticle using fungal extract of *Penicillium purpurogenum* NPMF. At 5 mM concentration, highly populated polydispersed nanoparticles were formed. Furthermore, change in pH of the reaction mixture leads to changes in shape and size of silver nanoparticles. At lower pH two peaks were observed in the absorption spectra showing polydispersity of nanoparticles. However, highly monodispersed spherical nanoparticles of 8–10 nm size form with 1 mM $AgNO_3$ concentration at pH 8. Dwivedi and Gopal (2010) also achieved higher production of silver nanoparticles at a concentration of 1 mM $AgNO_3$, incubation period of 25 min at 37 °C, and pH 7 using leaf extract of *Chenopodium album*.

9.7 Applications of AgNPs

9.7.1 Anticancer Activity of AgNPs

AgNPs attract the attention of researchers because of their extensive applications in the field of medicine and cancer therapy. Drugs available for cancer treatment, such as doxorubicin, cisplatin, vinblastine, bleomycin, and daunorubicin, are limited with poor specificity, high toxicity, high cost, and emergence of drug-resistant diseases. It is imperative to discover alternative diagnosis method and drug therapies to solve the problems at an early stage of disease. Nowadays, the development of nano-formulations is underway to fight against life-threatening diseases like glioblastoma, HIV, TB, and MDR bacterial infections (Locatelli et al. 2014). AgNPs have been used in the screening, detection, and imaging of tumors studded with biomarkers on nanoparticles for targeting tumor, drug delivery, and hyperthermia therapy (Ong et al. 2013). Researchers have extensively exploited chemically as well as biologically synthesized AgNPs to evaluate the in vitro as well as in vivo anticancer and antitumor activities against different cancerous cell lines such as MCF-7 human breast cancer cell line, B10F17 melanoma cell, A549 lung epithelial adenocarcinoma cell line, SiHa cervical cancer cell line, and HeLa cell line (Selvi et al. 2011; Jannathul and Lalitha 2015).

Coulter et al. (2013) proposed a mechanism to treat tumor cell sensitization. They suggested generation of an oxidizing agent by AgNPs, reducing the production of ATP in tumor cells and thus increasing the production of intracellular reactive oxygen species (ROS). The action of AgNPs on tumor cells can also occur via DNA damage, because AgNPs have been shown to negatively regulate the activity of DNA-dependent protein kinase, a key enzyme involved in DNA repair. Jeyaraj et al. (2013) studied the caspase-mediated apoptotic cell death on treatment of AgNPs synthesized from *Podophyllum hexandrum* leaf extract. The AgNPs of size ranging from 12 to 40 nm showed the 50% cell death at a concentration of 20 mg/mL. Similarly, Coccini et al. (2014) demonstrated that the AgNPs induced possible changes in the expression of oxidative gene in the testis and liver and induced oxidative stress gene, i.e., Gpx1, SOD, FMO2, and GAPDH. The upregulation of gene SOD and Gpx1 is a protective mechanism in oxidative stress indicating the AgNP-induced cytotoxic effect.

Recently, Urbanska et al. (2015) explored antiproliferative properties of AgNPs against human glioblastoma multiforme (GBM) cells and indicated that silver nanoparticles showed proapoptotic properties against GBM cells. He et al. (2010) observed cytotoxic effect of green-synthesized AgNPs on human lung cancer H1299 cells by MTT and trypan blue assays.

9.7.2 *Silver Nanoparticles as Antiviral*

AgNPs act as an excellent therapeutic agent in medical field as it displays antiviral activity against monkey poxvirus (Rogers et al. 2008), herpes simplex virus (Baram-Pinto et al. 2010), and hepatitis B virus (Lu et al. 2008). AgNPs synthesized from *Fusarium oxysporum*, *Chaetomium indicum*, and *Curvularia* species showed cytotoxicity up to a concentration of 50 mg/mL, whereas antiviral concentration ranged from 0.5 to 5 mg/mL. Silver NPs showed inhibitory activity against the influenza A H1N1 virus, which was proved through hemagglutination-inhibition tests and embryo inoculation assays (Xiang et al. 2011; Gaikwad et al. 2013). The use of the combination of monoclonal antibodies and AgNPs was described by Lara et al. (2011) in which the former act on the structure of the viral envelope and prevent binding and entry of HIV viruses. Tefry and Wooley (2012) reported the development of a method to evaluate the activity of AgNPs against pseudotyped HIV-1-based viruses.

In another method, the infectivity of adenovirus type inhibited when viral aliquots were incubated with AgNPs for 2 h at 37 °C (Chen et al. 2013). They found the cytotoxicity of chemically synthesized AgNPs at 50 mg/mL strongly damaged not only the entire virus particles but DNA also. The particles damaged the capsid protein which inhibits the virus attachment to host cell.

9.7.3 *Silver Nanoparticles as Antibacterial/Antifungal*

Silver nanoparticles are an effective antimicrobial agent against a broad spectrum of Gram-positive and negative bacterial pathogens including antibiotic-resistant strains (Percival et al. 2007; Aziz et al. 2015, 2016). The AgNPs are also effective, fast-acting fungicide against various fungal genera such as *Aspergillus*, *Candida*, and *Saccharomyces*. Silver nano-formulations were tested to determine the inhibitory effect of fungal plant pathogens, namely, *Alternaria alternata*, *Sclerotinia sclerotiorum*, *Macrophomina phaseolina*, *Rhizoctonia solani*, *Botrytis cinerea*, and

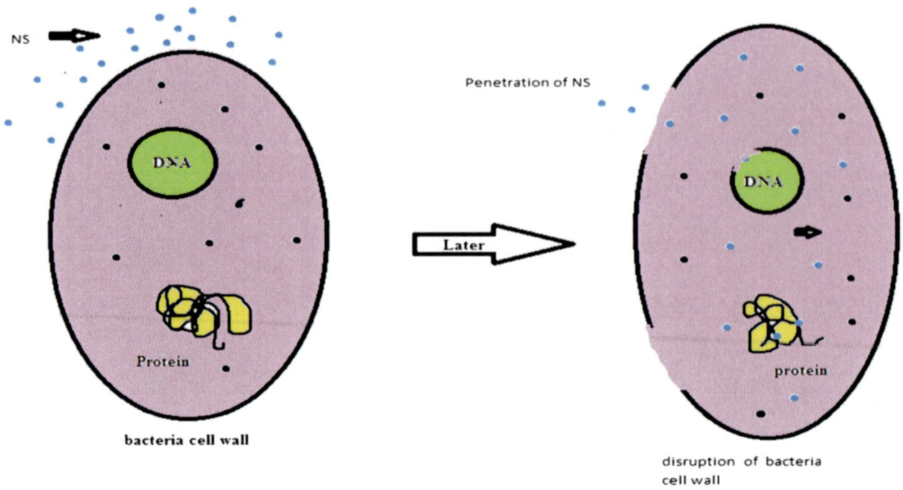

Fig. 9.6 Interaction of silver nanoparticles with bacterial cell membrane

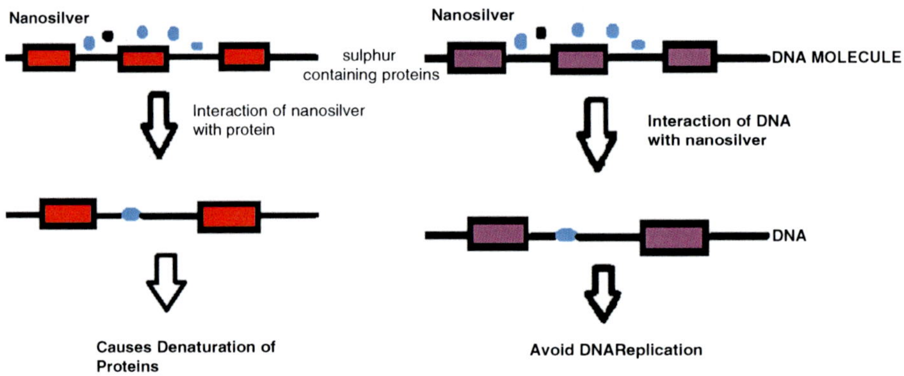

Fig. 9.7 Interaction of silver nanoparticles with proteins and DNA

Curvularia lunata; and 15 mg concentration of NPs showed excellent inhibitory activity against all the tested pathogens (Krishnaraj et al. 2012).

The possible mechanisms of action could be due to large surface area for better contact with bacteria. These silver nanoparticles get attached to the cell membrane and easily penetrate inside the bacteria as shown in Fig. 9.6. Silver nanoparticles release Ag^+ ion inside the microbial cell which may create free radicals and induce oxidative stress, thus further enhancing their bactericidal activity. Furthermore, silver nanoparticles interact with sulfur-containing proteins on microbial cell membrane causing disruption of protein and DNA (Fig. 9.7) (Liau et al. 1997; Matsumura et al. 2003; Prasad and Swamy 2013).

9.8 Miscellaneous Applications

Silver nanoparticles are used in various ointments preparations such as antiseptic soap, skin gels for healing burns, cream for dermatophytes. These are used in electronic devices like LCDs, high-intensity LEDs, and touch screens of various gadgets. AgNPs have a large number of applications in catalysis (Kamat 2002), plasmonics (Maier et al. 2001), optoelectronics (Mirkin et al. 1996), biological sensor (Gracias et al. 2002) antimicrobial activities (Shaverdi et al. 2007), DNA sequencing (Cao et al. 2001), and surface-enhanced Raman scattering (Matejka et al. 1992; Siddhanta et al. 2017). Nanoparticles are used into packaging material for foodstuff, treating contaminated water for absorption of pesticides, water purification, and air disinfection (Prasad et al. 2014, 2016, 2017b).

9.9 Conclusion

Green synthesis of silver nanoparticles using bacteria, fungi, yeast, and plant is nonhazardous and an attractive area in the field of nanotechnology. For biogenic synthesis of nanoparticles, the choice of the biological route is not a trivial decision as the product (e.g., size and shape) depends on the substrate precursor concentration, pH, temperature of incubation, ratio of biological extract and metal precursor and type of biological material. Different analytical techniques such as UV-vis spectrophotometer, XRD, FTIR, SEM, and TEM are used to characterize silver nanoparticle. Due to unique size and physical, chemical, and antimicrobial properties, AgNPs attract researchers to use them in various applications such as DNA sequencing, sensors for detection of metabolites, and cancer detection and treatment. Further improvements are required to build nanoparticles in order to apply them in the next generation of drug delivery system. However, aggregation of AgNPs could be toxic to the cells, and it is imperative to assess the toxicity effect of nanosilver in vivo before drawing any inference.

References

Ahmad N, Sharma S, Singh VN, Shamsi SF, Fatma A, Mehta BR (2011) Biosynthesis of silver nanoparticles from *Desmodium triflorum*, a novel approach towards weed utilization. Biotechnol Res Int 1:1–8

Aziz N, Faraz M, Pandey R, Sakir M, Fatma T, Varma A, Barman I, Prasad R (2015) Facile algae-derived route to biogenic silver nanoparticles: synthesis, antibacterial and photocatalytic properties. Langmuir 31:11605–11612. https://doi.org/10.1021/acs.langmuir.5b03081

Aziz N, Pandey R, Barman I, Prasad R (2016) Leveraging the attributes of *Mucor hiemalis*-derived silver nanoparticles for a synergistic broad-spectrum antimicrobial platform. Front Microbiol 7:1984. https://doi.org/10.3389/fmicb.2016.01984

Balaprasad A, Chinmay D, Absar C, Murali S (2005) Biosynthesis of gold and silver nanoparticles using *emblica officinalis* fruit extract, their phase transfer and transmetallation in an organic solution. J Nanosci Nanotech 5(10):1665–1671

Baram-Pinto D, Shukla S, Gedanken A, Sarid R (2010) Inhibition of HSV-1 attachment, entry, and cell-to-cell spread by functionalized multivalent gold nanoparticles. Small 6:1044–1050

Bhainsa KC, D'Souza SF (2006) Extracellular synthesis using the fungus *Aspergillus fumigates*. Colloids Surf B Biointerfaces 47:152–157

Cao YM, Jin R, Mirkin CA (2001) DNA-modified core–shell Ag/Au nanoparticles. J Am Chem Soc 123:7961–7962

Chandran SP, Chaudhary M, Pasricha R, Ahmad A, Sastry M (2006) Nanotriangles and silver nanoparticles using *Aloe vera* plant extract. Biotech Progr 22:577–583

Chen YH, Yeh CS (2002) Laser ablation method, use of surfactants to form the dispersed Ag nanoparticles. Colloids Surf A Physicochem Eng Asp 197:133–139

Chen N, Zheng Y, Yin J (2013) Inhibitory effects of silver nanoparticles against adenovirus type 3 in vitro. J Virol Methods 193:470–477

Coccini T, Gornati R, Rossi F (2014) Gene expression changes in rat liver and testes after lung instillation of a low dose of silver nanoparticles. J Nanomed Nanotechnol 5:227–232

Coulter JA, Hyland WB, Nicol J, Currell FJ (2013) Radio sensitizing nanoparticles as novel cancer therapeutics: pipe dream or realistic prospect. Clin Oncol 25:593–603

Deva AN, Alice ME, Santhosh C, Grace AN, Vimala R (2016) Optimization of process parameters for the microbial synthesis of silver nanoparticles using 3-level box- Behnken design. Eco Engg 87:168–174

Duran N, Marcato PD, Alves OL, Souza GIH, De Esposito E (2005) Mechanistic aspects of biosynthesis of silver nanoparticles by several *Fusarium oxysporum* strains. J Nanobiotech 3:1–8

Dwivedi AD, Gopal K (2010) Biosynthesis of silver and gold nanoparticles using *Chenopodium album* leaf extract. Coll Surf A 369:27–33

El-rafie MH, Shaheen TI, Mohamed AA, Hebeish A (2012) Biosynthesis and applications of silver nanoparticles onto cotton fabrics. Carbohyd Poly 90(2):915–920

Faraday M (1857) Experimental relations of gold (and other metals) to light. Phil Trans R Soc London 147:145–181

Gaikwad S, Ingle A, Gade A (2013) Antiviral activity of mycosynthesized silver nanoparticles against herpes simplex virus and human parainfluenza virus type 3. Int J Nanomedicine 8:4303–4314

Gracias DH, Tien J, Breen T, Hsu C, Whitesides GM (2002) Forming electrical networks in three dimensions by self-assembly. Sci 289:1170–1172

Gunther B, Kumpmann A (1992) Ultrafine oxide powders prepared by inert gas evaporation. Nanostr Mater 1:27–30

Gurav A, Kodas T, Wang L, Kauppinen E, Joutsensaari J (1994) Synthesis of fullerene-rhodium nanocomposites of nano-meter size fullerene particles via aerosol routes. Chem Phys Letter 218:304–308

Haefeli C, Franklin C, Hardy K (1984) Plasmid-determined silver resistance in *Pseudomonas stutzeri* isolated from a silver mine. J Bacteriol 158:389–392

He L, Liu Y, Mustapha A, Lin M (2010) Antifungal activity of zinc oxide nanoparticles against *Botrytis cinerea* and *Penicillium expansum*. Microbiol Res 166:207–215

Jain D, Kachhwaha S, Jain R, Srivastava G, Kothari SL (2010) Novel microbial route to synthesize silver nanoparticles using spore crystal mixture of *Bacillus thuringiensis*. Indian J Exp Bio 48(11):1152–1156

Jannathul FM, Lalitha P (2015) Apoptotic efficacy of biogenic silver nanoparticles on human breast cancer MCF-7 cell lines. Prog Biomater 4:113–121

Jeyaraj M, Rajesh M, Arun R (2013) An investigation on the cytotoxicity and caspase-mediated apoptotic effect of biologically synthesized silver nanoparticles using *Podophyllum hexandrum* on human cervical carcinoma cells. Colloid Surf B 102:708–717

Jha AK, Prasad K, Prasad K, Kulkarni AR (2009) Plant system: nature's nanofactory. Colloids Surf B Biointerfaces 73:219–223

Kabashin AV, Meunier M (2003) Synthesis of colloidal nanoparticles during femtosecond laser ablation of gold in water. J Appl Phys 94:7941–7943

Kamat PV (2002) Photophysical, photochemical and photocatalytic aspects of metal nanoparticles. J Phy Chem B 106:7729–7744

Karthikeyan AP, Kadarkarai M, Chellasamy P, Sekar P, Jiang-Shiou H, Marcello N (2012) Biolarvicidal and pupicidal potential of silver nanoparticles synthesized using *Euphorbia hirta* against *Anopheles stephensi Liston* (Diptera: Culicidae). Parasi Res 111:997–1006

Kathiresan K, Manivannan S, Nabeal MA, Dhivya B (2009) Studies on silver nanoparticles synthesized by a marine fungus, *Penicillium fellutanum* isolated from coastal mangrove sediment. Coll Surf B 71:133–137

Krishnaraj C, Ramachandran R, Mohan K, Kalaichelvan PT (2012) Optimization for rapid synthesis of silver nanoparticles and its effect on phytopathogenic fungi. Spectrochem Acta 93:95–99

Kruis FE, Fissan H, Rellinghaus B (2000) Sintering and evaporation characteristics of gas-phase synthesis of size-selected PbS nanoparticles. Mater Sci Eng 69:329–334

Kumar AS, Abyaneh MK, Gosavi Sulabha SW, Ahmad A, Khan MI (2007) Nitrate reductase mediated synthesis of silver nanoparticles from $AgNO_3$. Biotechnol Lett 29:439–445

Lara HH, Trevino EG, Turrent LI, Singh DK (2011) Silver nanoparticles are broad-spectrum bactericidal and virucidal compounds. J Nanobiotechnol 9:30

Li S, Qui L, Shen Y, Xie A, Yu X, Zhang L, Zhang Q (2007) Green synthesis of silver nanoparticles using *Capsicum annum L.* extract. Green Chem 9:852–858

Liau S, Read D, Pugh W, Furr J, Russell A (1997) Interaction of silver nitrate with readily identifiable groups, relationship to the antibacterial action of silver ions. Lett Appl Microbio 25:279–283

Locatelli E, Naddaka M, Uboldi C (2014) Targeted delivery of silver nanoparticles and alisertib: in vitro and in vivo synergistic effect against glioblastoma. Nanomedicine 9(6):839–849

Lu L, Sun RW, Chen R, Hui CK, Ho CM, Luk JM, Lau GK, Che CM (2008) Silver nanoparticles inhibit hepatitis B virus replication. Antivir Ther 13:253–262

Mafune F, Kohno J, Takeda Y, Kondow T, Sawabe H (2000) Structure and stability of silver nanoparticles in aqueous solution produced by laser ablation. J Phys Chem 104:8333–8337

Maier SA, Brongersma ML, Kik PG, Meltzer S, Requicha AAG, Atwater HA (2001) Plasmonics- a route to nanoscale optical devices. Adv Mater 19:1501–1505

Matejka P, Vlckova B, Vohlidal J, Pancoska P, Baumruk V (1992) The role of triton X-100 as an adsorbate and a molecular spacer on the surface of silver colloid, a surface enhanced Raman scattering study. J Phys Chem 96(3):1361–1366

Matsumura Y, Yoshikata K, Kunisaki SI, Tsuchido T (2003) Mode of bactericidal action of silver zeolite and its comparison with that of silver nitrate. Appl Env Microbiol 69:4278–4281

Mirkin CA, Letsinger RL, Mucic RC, Storhof JJ (1996) A DNA-based method for rationally assembling nanoparticles into macroscopic materials. Nature 382:607–609

Mude N, Ingle A, Gade A, Rai M (2009) Synthesis of silver nanoparticles using callus extract of *Carica papaya* – a first report. J Plant Biochem Biotech 18(1):83–86

Musarrat J, Dwivedi S, Singh BR, Al-Khedhairy AA, Azam A, Naqvi A (2010) Production of antimicrobial silver nanoparticles in water extracts of the fungus *Amylomyces rouxii* strain KSU-09. Bioresour Technol 101:8772–8776

Nanda A, Lakshmipathy M, Elumalai D, Kaleena PK, Nayak BK (2015) Application of biosynthesized silver nanoparticles as a novel vector control agent. Der Pharm Lett 7(8):228–231

Nayak RR, Pradhan N, Behera D, Pradhan KM, Mishra S, Sukla LB, Mishra BK (2011) Green synthesis of silver nanoparticle by *Penicillium purpurogenum* NPMF: the process and optimization. J Nanopart Res 13:3129

Ong C, Lim JZ, Ng CT, Li JJ, Yung LY, Bay BH (2013) Silver nanoparticles in cancer: therapeutic efficacy and toxicity. Curr Med Chem 20(6):772–781

Percival SL, Bowler PG, Dolman J (2007) Antimicrobial activity of silver- containing dressings on wound microorganisms using an in vitro biofilm model. Int Wound J 4:186–191

Prasad R (2014) Synthesis of silver nanoparticles in photosynthetic plants. J Nanopart, Article ID 963961, https://doi.org/10.1155/2014/963961

Prasad R, Swamy VS (2013) Antibacterial activity of silver nanoparticles synthesized by bark extract of *Syzygium cumini*. J Nanopart https://doi.org/10.1155/2013/431218

Prasad R, Kumar V, Prasad KS (2014) Nanotechnology in sustainable agriculture: present concerns and future aspects. Afr J Biotechnol 13(6):705–713

Prasad R, Pandey R, Barman I (2016) Engineering tailored nanoparticles with microbes: quo vadis. WIREs Nanomed Nanobiotechnol 8:316–330. https://doi.org/10.1002/wnan.1363

Prasad R, Pandey R, Varma A, Barman I (2017a) Polymer based nanoparticles for drug delivery systems and cancer therapeutics. In: Kharkwal H, Janaswamy S (eds) Natural polymers for drug delivery. CAB International, UK, pp 53–70

Prasad R, Bhattacharyya A, Nguyen QD (2017b) Nanotechnology in sustainable agriculture: recent developments, challenges, and perspectives. Front Microbiol 8:1014. https://doi.org/10.3389/fmicb.2017.01014

Rogers JV, Parkinson CV, Choi YW, Speshock JL, Hussain SM (2008) A preliminary assessment of silver nanoparticles inhibition of monkeypox virus plaque formation. Nanoscale Res Lett 3:129–133

Rudge S, Peterson C, Vessely C, Koda J, Stevens S, Catterall L (2001) Adsorption and desorption of chemotherapeutic drugs from a magnetically targeted carrier (MTC). J Control Release 74:335–340

Sadowski IH, Maliszewska B, Grochowalska I, Polowczyk T, Kozlecki M (2008) Synthesis of silver nanoparticles using microorganisms. Mater Sci Poland 26:2–10

Saifuddin N, Wong CW, Nuryasumira AA (2009) Rapid biosynthesis of silver nanoparticles using culture supernatant of bacteria with microwave irradiation. Eur J Chem 6:61–70

Sarsar V, Selwal KK, Selwal MK (2014) Nanosilver: potent antimicrobial agent and its biosynthesis. Afr J Biotech 13(4):546–554

Sarsar V, Selwal MK, Selwal KK (2015) Biofabrication, characterization and antibacterial efficacy of extracellular silver nanoparticles using novel fungal strain of *Penicillium atramentosum* KM. J Saudi Chem Soc 19:682–688

Sarsar V, Selwal MK, Selwal KK (2016) Biogenic synthesis, optimisation and antibacterial efficacy of extracellular silver nanoparticles using novel fungal isolate *Aspergillus fumigatus MA*. IET Nanobiotech. https://doi.org/10.1049/iet-nbt.2015.0058 www.ietdl.org

Saxena A, Tripathi RM, Singh RP (2010) Biological synthesis of silver nanoparticles by using onion (*Allium cepa*) extract and their antibacterial activity. Dig J Nanomat Biostru 5(2):427–432

Selvi CG, Madhavan J, Santhanam A (2011) Cytotoxic effect of silver nanoparticles synthesized from Padina tetrastromatica on breast cancer cell line. Nanosci Nanotechnol 7(3):1–6

Senapati S, Mandal D, Ahmad A, Khan MI, Sastry M, Kumar R (2004) Fungus mediated synthesis of silver nanoparticles, a novel biological approach. Ind J Phys 78:101–105

Shahverdi AR, Minaeian S, Shahverdi HR, Jamalifar H, Nohi AA (2007) Rapid synthesis of silver nanoparticles using culture supernatants of Enterobacteriaceae: a novel biological approach. Process Biochem 42:919–923

Shivaji S, Madhu S, Singh S (2011) Extracellular synthesis of antibacterial silver nanoparticles using psychrophilic bacteria. Process Biochem 46:1800–1807

Siddhanta S, Paidi SK, Bushley K, Prasad R, Barman I (2017) Exploring morphological and biochemical linkages in fungal growth with label-free light sheet microscopy and Raman spectroscopy. ChemPhysChem 18(1):72–78

Singhal G, Bhavesh R, Kasariya K, Sharma AR, Singh RP (2010) Biosynthesis of silver nanoparticles using *Ocimum sanctum* (Tulsi) leaf extract and screening its antimicrobial activity. J Nanopart Res 13(7):2981–2988

Sylvestre JP, Kabashin AV, Sacher E, Meunier M, Luong JHT (2004) Stabilization and size control of gold nanoparticles during laser ablation in aqueous cyclodextrins. J Am Chem Soc 126:7176–7177

Tefry JC, Wooley DP (2012) Rapid assessment of antiviral activity and cytotoxicity of silver nanoparticles using a novel application of the tetrazolium-based colorimetric assay. J Virol Methods 183:19–24

Urbanska K, Pajak B, Orzechowski B, Sokolowska J, Grodzik M, Sawosz M, Sysa P (2015) The effect of silver nanoparticles (AgNPs) on proliferation and apoptosis of in vitro cultured glioblastoma multiforme (GBM) cells. Nanoscale Res Lett 10:98

Vahabi K, Mansoori GA, Karimi S (2011) Biosynthesis of silver nanoparticles by fungus *Trichoderma Reesei* (a route for large-scale production of AgNPs). Int Sci J 1(1):65–79

Volesky B, Holan ZR (1995) Biosorption of heavy metals. Biotechnol Prog 11:235–250

Wei X, Luo M, Li W, Yang L, Liang X, Xu L (2012) Synthesis of silver nanoparticles by solar irradiation of cell-free *Bacillus amyloliquefaciens* extracts and $AgNO_3$. Bioresour Technol 103:273–278

Xiang DX, Chen Q, Pang L, Zheng CL (2011) Inhibitory effects of silver nanoparticles on H1N1 influenza a virus in vitro. J Virol Methods 178:137–142

Chapter 10
Fungal Bionanotechnology, When Knowledge Merge into a New Discipline to Combat Antimicrobial Resistance

Juan Bueno

> *Nature doth thus kindly heal every wound. By the mediation of a thousand little mosses and fungi, the most unsightly objects become radiant of beauty. There seem to be two sides of this world, presented us at different times, as we see things in growth or dissolution, in life or death. And seen with the eye of the poet, as God sees them, all things are alive and beautiful.*
>
> Henry David Thoreau (1817–1862)

> *If a healthy soil is full of death, it is also full of life: worms, fungi, microorganisms of all kinds ... Given only the health of the soil, nothing that dies is dead for very long.*
>
> Wendell Barry (1934–)

Abstract Currently antimicrobial resistance (AMR) is defined as the ability of any microorganism (bacteria, fungi, virus, and parasites) to stop the pharmacological action of an antimicrobial agent; for that reason, the World Health Organization (WHO) has declared AMR as a global threat to public health and left luggage in one of its strategic objectives of Global Action Plan: "increase investment in new medicines, diagnostic tools, vaccines and other interventions." In this way new innovative approaches are necessary for the treatment of infectious diseases caused by multidrug-resistant (MDR) pathogens that have the ability to tackle resistance mechanisms and inhibit microbial invasion and tissue damage. In this issue, nanotechnology has emerged as a powerful tool to produce new antibiotics and design new formulations that increase anti-infective activity of current medications as well as novel devices with which to prevent the acquisition of nosocomial infections in healthcare settings. But also, a strong concern has arisen, due to the risk of toxicity that the extensive use of nanomaterials can cause in ecosystems and humans, in this way an answer to this question is the development of biosynthetic methods oriented

J. Bueno (✉)
Fundación Centro de Investigación de Bioprospección y Biotecnología de la Biodiversidad (BIOLABB), Colombia, South America
e-mail: juanbueno@biolabb.org; juangbueno@gmail.com

R. Prasad (ed.), *Fungal Nanotechnology*, Fungal Biology,
DOI 10.1007/978-3-319-68424-6_10

to obtain an eco-friendly production of materials useful in medicine, agriculture and industry. This approach involves the development of a discipline known as bionanotechnology that currently is an interesting branch of nanotechnology, whose models of synthesis most used are fungi, bacteria, and plant extracts; among them, fungi have several advantages for their ubiquity, their ease of isolation and culture in vitro, as well as their metabolic and enzymatic capacity. Fungal bionanotechnology (or myconanotechnology) provides an important antimicrobial drug discovery platform with applications in nanoparticle synthesis, antimicrobial devices, and synthetic biology. Also the implementation of proper antimicrobial screening platforms with the ability to predict clinical response and safe use in different environments is of critical importance (very important in the research work with metallic nanoparticles). The aim of this chapter is to explore comprehensively fungal bionanotechnology as an interdisciplinary approach to be applied in solutions to stop the emergence of AMR without the environmental risks of chemical synthesis of nanomaterials and wide applications as a radical model of translational science.

10.1 Introduction

Microorganisms possess the molecular machinery to become resistant to antimicrobial drugs; this gene collection known as resistome constitutes an evolutionary path of adaptation that allows the survival in different environments, which makes antimicrobial resistance a natural phenomenon aggravated by the misuse of antibiotics (Brown and Wright 2016; Pawlowski et al. 2016). In this way, AMR is one of the greatest challenges in clinical practice being a threat that grows dramatically especially in developing countries, among them India, Bangladesh, Pakistan, Nigeria, Brazil, and other countries of African and Asian continents (Syed and Bana 2016). This crisis of infectious disease therapeutics is represented by these events that are occurring simultaneously: broad transmission of MDR strains, increasing of immunocompromised hosts, emergence of new pathogenic microorganisms, and extension of resistance to other drugs such as extensive-resistant (XDR) *Mycobacterium tuberculosis* (Saylor et al. 2009). Also multidrug-resistant (MDR) pathogens which are resistant to most common antibiotics represent a huge public health problem in infectious diseases. MDR microorganisms have been classified by the Infectious Diseases Society of America as "ESKAPE" pathogens (*Enterococcus faecium*, *Staphylococcus aureus*, *Klebsiella pneumoniae*, *Acinetobacter baumannii*, *Pseudomonas aeruginosa*, and *Enterobacter* spp.). MDR infection produces twice as high mortality in affected patients (Lukowiak et al. 2016).

To make matters worse, research on new antimicrobial agents has declined in recent years. For that reason in the last 2 years, AMR has gained prominence on the agenda of G7 leaders considering it one of the great threats of our time, where it was introduced in one of the fundamental pillars of the global vision of health (Utt and Wells 2016). Also, in May 2015, the World Health Assembly (WHA) of the WHO impulsed the AMR Global Action Plan (GAP) composed of five strategic objectives, in which its objective 5 expresses "Develop the economic case for sustainable investment that takes account of the needs of all countries, and increase investment in new medicines, diagnostic tools, vaccines and other interventions" (Utt and Wells 2016). But it is very important to keep in mind that the process of obtaining a new drug is long and costly, usually taking 13 years to reach the clinical phases (Bai et al. 2016).

For that reason, the development of new, more effective, and less toxic formulations is a great need to be applied in medicine, agriculture, and industry, because of which nanoformulations have shown to be an important alternative to improve activity and pharmacokinetic (PK) and pharmacodynamic (PD) profiles in infection models of antimicrobial agents (Gautam et al. 2014; Azzopardi et al. 2015); this approach solves several problems of current anti-infectious therapies using nano-sized materials of 1–100 nm with unique physical and chemical properties (small size, high surface-to-volume ratio, and amenable for surface modification) useful as carriers of therapeutic and diagnostic agents, because they can cross with high-efficiency physiological barriers (Bletcher et al. 2011; Sharma et al. 2012). Among these materials are polymer-based nanoparticles (biodegradable), nanospheres, nanocapsules, liposomes, and dendrimers. Equally, other materials have been described as magnetic nanoparticles, metallic (gold and silver), semiconductors (TiO2, CdSe, and so on), and carbon-based structures (fullerenes, carbon nanotubes, graphene), among others (DeVries et al. 2015), as well as nanobodies which are recombinant, antigen-specific, single-domain, variable fragments of camelid heavy-chain-only antibodies that can be conjugated to drug carriers in nanoformulation (Hassanzadeh-Ghassabeh et al. 2013). In addition, this technology can be applied to the development of healthcare products, molecular diagnostics, and therapeutically products in different types of associations (Saude et al. 2013).

Following this line of reasoning, the synthesis of nanomaterials is of radical importance for industry and development of new medications, but the possibility of releasing potentially toxic substances into the environment makes it necessary to seek for and develop clean, nontoxic, and biocompatible processes in order to obtain technologies that improve "green" economic growth (Manivasagan et al. 2016).

Biosynthesis of nanomaterials is a new and promising choice to develop "green chemistry" procedures using the biological systems present in microorganisms (bacteria, fungi, algae, cyanobacteria, actinomycetes, myxobacteria) and plants that can produce inorganic materials either intra- or extracellularly (Mohanpuria et al. 2008; Boroumand Moghaddam et al. 2015). Among the microbial systems, fungi are most commonly used, due to presence of P450 enzymes, which are of difficult expression in bacteria (Rodríguez et al. 2015; Yadav et al. 2015). Also, DNA editing techniques are widely applicable to this type of organisms to create models of synthetic biology with application in microbial factories (Kavšček et al. 2015).

The potential of fungal bionanotechnology and its applications for the solution of the threat of antibiotic resistance in a transdisciplinary model will be explored during the course of this chapter; this translational science approach requires the integration of antimicrobial drug discovery, antimicrobial resistance diagnostic, and antimicrobial stewardship in an infection control program.

10.2 Nanobiotechnology

The conjugation of nanotechnology and biotechnology has multiple promising applications. The combination between engineering, molecular biology and nanobiotechnology have led to develop innovations in industry subsectors as biopharmaceuticals, drug delivery, suppliers, instrumentation, devices and diagnostics (Maine et al. 2014; Prasad et al. 2017a); using several nano-objects with important biomedical applications as nanotubes, nanochannels, nanoparticles, nanopores, nanocapacitors, and nanofibers (Morais et al. 2014). So, in this order of ideas, nanomolecules have shown great activity against MDR microorganisms as metallic nanoparticles of zinc (ZnO), copper (CuO), and iron (Fe_2O_3) with activity against *S. aureus*, *Bacillus subtilis*, *P. aeruginosa*, and *E. coli* (Kandi and Kandi 2015). Equally, nanoencapsulation of antibiotics and new drugs has several advantages, among them decrease in adverse events, increased solubility of injectable formulations for intravenous administration, obtaining of sustained release formulations which decreases the number of doses, increase of intracellular concentration of antimicrobials, and potentiation of antimicrobial activity (Abed and Couvreur 2014). Nanobiotechnology is an integrative discipline with the ability to produce innovations for different biomedical applications in new anti-infective treatments and new diagnostic methods and medical devices with a fundamental role to tackle AMR (Khan et al. 2015a, b; Prasad et al. 2017a).

10.3 Fungal Bionanotechnology in Antimicrobial Drug Discovery

10.3.1 Biosynthesis of Metallic Nanomaterials

One of the strategies with the greatest impact and projection in the fight against AMR is the use of metallic nanoparticles, in which their ability to interact on a larger microbial surface can alter cell membranes and block respiration in MDR strains; in addition nanoparticles can alter proteins containing sulfur and DNA (Allahverdiyev et al. 2011), so they are a powerful alternative for the development of new antimicrobials.

Biosynthesis of cadmium selenide (CdSe) nanoparticles using fungi was reported in 1989 in a model of *Candida albicans*, demonstrating the ability of fungi to synthesize metallic nanoparticles (Xue et al. 2016). In this way, fungi possess an important secretome adapted to their environment, constituted by secreted proteins that degrade the biomass (Meinken et al. 2014), particularly hydrolases, working as exodigesters of biopolymers (starch, cellulose, pectin, xylan, proteins, and lipids) representing a promising microbial cell factory (Girard et al. 2013; Meyer et al. 2015). Various fungal species have been used as propitious resources for nanoparticle biomanufacture, such as those belonging to the genera *Fusarium*, *Aspergillus*, *Verticillium*, and *Penicillium* (Alghuthaymi et al. 2015), with the advantage in reducing costs because fungi perform extracellular biosynthesis that eliminates additional extractive processes such as sonication (Boroumand Moghaddam et al. 2015; Raliya et al. 2016). In this order of ideas, the most known antimicrobial nanomaterial produced by fungi is silver nanoparticles (AgNPs), which are synthesized for extracellular enzymes like naphthoquinones and anthraquinones that facilitate the reduction (Prabhu and Poulose 2012). Also, there are 30 fungal species that synthesize gold nanoparticles (AuNPs), product of the reduction of Au^{3+} in the cell wall by the secretome (Kitching et al. 2015).

In addition, in the great diversity of microbial cell factories, yeasts represent a model with multiple advantages due to the possibility of having more control in laboratory conditions and get a more rapid growth with the use of few nutrients; among them, *Candida glabrata*, *Saccharomyces pombe*, and *Rhodosporidium diobovatum* have been used to obtain AgNPs, AuNPs, and sulfide nanoparticles (Ingale and Chaudhari 2013). In the particular case of AgNP biosynthesis, the use of silver-tolerant yeast strain known as MKY3 has been more suitable to produce a large number of nanoparticles (Boroumand Moghaddam et al. 2015). In other wise, in the looking for bactericidal agents for MDR strains, a promising choice are metal oxide nanoparticles, that have more powerful antimicrobial activity (Dizaj et al. 2014). In this way, fungi like *Fusarium oxysporum* have been employed in biotechnological approaches for the production of metal oxide nanoparticles which include TiO_2, Sb_2O_3, SiO_2, ZnO, $BaTiO_3$, and ZrO_2 (Li et al. 2011). These particles have excellent antimicrobial activity against Gram positive and Gram negative bacteria, due to their capacity of induce reactive oxygen species (ROS) production causing accumulation of nanoparticles in the bacterial cytoplasm which produces death and cell lysis (Raghupathi et al. 2011; Azam et al. 2012).

Also as interesting as the type of particles that can synthesize the fungi is the bioprospective search for new organisms with metabolic capacity to produce nanomaterials more efficiently. An important kind of microorganism that colonizes inter- or intracellularly host plant without causing disease is the endophytes. These fungal endophytes are an important source of active metabolites and have a secretome with high chemical diversity (Raja et al. 2015). As a result, the fungi *Guignardia mangiferae* isolated from leaves of *Citrus* sp. have been detected as an organism capable of synthesizing AgNPs with in vitro antibacterial, antifungal, and cytotoxic activities (Balakumaran et al. 2015). Also, *Aspergillus versicolor* ENT7 isolated from *Centella asiatica* has shown a major microbial factory of nanoparticles (Netala et al. 2016).

Finally, the most important projection of metallic nanoparticles is the possibility of non-covalently attaching them with other molecules and discovered antibiotics as like antimicrobial adjuvant with the end of increase their bactericidal activity, this procedure known as nanoparticle functionalization is an important strategy for reduce the emergence of resistance both existing anti-infectives and those under development (Vigderman and Zubarev 2013).

10.3.2 Drug Carriers

Nanotechnology-based drug delivery systems are an approach that looks to increase drug concentration and therapeutic effect (Hamidi et al. 2013); these delivery systems have to accomplish the following characteristics: achieve drug therapeutic concentrations in affected tissues, prevent the release of the drug before reaching therapeutic target, protect the pharmaceutical formulation from degradation and elimination, and have the ability to distribute the drug efficiently in the human body (Abed and Couvreur 2014; Prasad et al. 2017b). Nanocarriers have been used successfully with antimicrobials as gentamycin, antituberculosis drugs, and ampicillin, as well as antifungal drug amphotericin B (Saha et al. 2007).

In this development of new drug carriers, chitosan (poly-N-acetylglucosamine) is a biopolymer with perspectives as drug carrier because they are compatible, biodegradable, and mucoadhesive and with reduced toxicity. Also their mucoadhesive properties allow this material to be attached to the intestinal mucosa increasing the bioavailability of conjugated drugs (Sonia and Sharma 2011). Currently, chitosan is obtained from shrimp and crab shell, but fungal chitosan from *Rhizopus* spp. can be an interesting source, more environmentally friendly (Pochanavanich and Suntornsuk 2002). The biomedical applications of chitosan nanoparticles are very broad and include wound dressing, orthopedic tissue engineering, drug delivery carrier, and hemodialysis (Kong et al. 2010). Equally, chitosan nanoparticles had also been employed in the form of microspheres in order to release antimicrobial agents in a controlled manner in the gastric cavity, such as ciprofloxacin, amoxycillin, and metronidazole (Qi et al. 2004; Ibrahim et al. 2015).

Otherwise, the hydrophilic and polycationic characteristics of chitosan make it an effective polymer to develop antimicrobial coatings and new antibiotic formulations, as well as to stabilize metallic nanoparticles (Zhu et al. 2014).

10.3.3 Biosurfactant

Nanoemulsions, defined as oil-in-water emulsions with an average droplet diameter ranging from 50 to 1000 nm, have been investigated for their antimicrobial activity (Huh and Kwon 2011). In this order of ideas, the soybean oil nanoemulsions are the most studied, presenting high antimicrobial activity (Texeira et al. 2007). Looking for new oils to be introduced in emulsions, there are microorganisms with high lipid content known as "oleaginous." Among them, fungi accumulate 20–25% lipid, mainly triacylglycerol, which has biosurfactant activity (Ratledge 1997). Biosurfactants have strong anti-infective activities such as antifungal, antibacterial, antimycoplasmal, and antiviral activities (Singh and Cameotra 2004). On the other hand, the mechanism of antimicrobial action of biosurfactants is affecting the adhesion of microorganisms, also producing the cellular lysis by the changes in membrane structure as well as through disrupting protein conformations that can alter the membrane functions such as transport and energy production (Banat et al. 2010). In addition, biosurfactants are ecologically safe and can be applied in bioremediation processes (Pacwa-Płociniczak et al. 2011). Among the fungi that produce biosurfactants are *C. bombicola*, *C. lipolytica*, *C. ishiwadae*, *C. batistae*, *Aspergillus ustus*, *Ustilago maydis*, and *Trichosporon ashii*, sophorolipids being the major type of biosurfactant produced (Bhardwaj et al. 2013). Sophorolipids are efficient microbicides, spermicides, anti-HIV drugs, anticancer agents, and immunomodulators (Bluth et al. 2006; Shah et al. 2005, 2007; Sleiman et al. 2009). Also, present antibiofilm activity, useful for the design and development of new antimicrobial adjuvants agents with the ability of inhibit biofilm formation during anti infective therapy (DeRienzo et al. 2015), for which they have great perspectives of use in the pharmaceutical and cosmetic industry (Van Bogaert et al. 2007). The antimicrobial, antibiofilm, and anti-adhesive properties of biosurfactants can be used to obtain new medical devices, as well as for the design of new procedures in healing and wound care (Sambanthamoorthy et al. 2014).

10.3.4 Antisense Technology

An important use of nanoparticles is the possibility to enhance the efficacy of the drugs used in gene silencing (Vigderman and Zubarev 2013). A promising use of synthetic RNA silencing as peptide nucleic acids (PNAs) and phosphorothioate morpholino oligomers (PMOs) is to inhibit growth of bacteria (Good and Stach 2011). These antisense molecules are active against several bacteria, including *K. pneumoniae*, *E. coli*, *S. enterica*, *Burkholderia*, *S. aureus*, and *Streptococcus pyogenes*, being tested in animal models of infectious disease, increasing the survival, and reducing the microbial load in the affected organs (Otsuka et al. 2017).

The strategy of using "antisense" agents to inhibit resistance mechanisms at the nucleic acid level is a promising approach to restore susceptibility to antibiotics inhibiting the activity of microbial intracellular nucleases (Woodford et al. 2009; Beaman et al. 2014). For this purpose the use of nanoparticles and the search of new pathways of synthesis are necessary to make these therapies more effective and specific.

10.3.5 Synthesis of Monoclonal Antibodies and Recombinant Proteins

It has been estimated that about 30% of the new drugs approved will be antibodies or antibody derivatives, which include Fc-fusion proteins, antibody–drug conjugates (ADCs), immunocytokines (antibody–cytokine fusion), and antibody–enzyme fusions (Liu 2015). In this way, looking for new antimicrobial drugs that prevent the disturbance of the normal functioning of the microbiome that permits the antibiotic resistance spread through horizontal gene transfer, antibodies have been considered as a choice to obtain new anti-infective therapeutic combinations (Saylor et al. 2009; DiGiandomenico and Sellman 2015). Currently, monoclonal antibodies (mAbs) that represent the most important biopharmaceutical products today are produced using mammalian cells with elevated costs. For that reason, there is a new trend to develop antibody fragments in microbial organisms, which are easy to manipulate and cultivate (Spadiut et al. 2014). In this way expression systems in the yeasts *S. cerevisiae* and *Pichia pastoris* have been designed (Vicentin et al. 2014). Other interesting topics in the synthesis of monoclonal antibodies are nanoantibodies (Nbs), which are functional antibodies from camelids with light chains in the N-terminal domain and large biotechnological applications. They can be expressed in microorganisms as yeast (*S. cerevisiae*) and have a high stability and solubility (Harmsen and De Haard 2007). Nanoantibodies can be produced in a scalable fermentation system easily without posttranslational modifications, which makes it an ideal system for designing new medicines and diagnostic techniques (Siontorou et al. 2013; Wang et al. 2016). Nbs have several advantages in comparison with conventional antibodies as small size (15 kDa), high affinity and specificity, high stability and solubility, low immunogenicity, and low manufacturing costs (Siontorou et al. 2013). Nanobodies have been successfully evaluated in inhibition of bacterial biofilm, bacterial toxin secretion, enhancing of antibiotic activity and prevention of bacterial adhesion (Steeland et al. 2016). Antibodies produced in mycotic cells are an important advance in the search for new biopharmaceutical and diagnostic techniques useful in the treatment and prevention of infectious diseases.

10.4 Fungal Bionanotechnology in Antimicrobial Resistance Diagnostic

10.4.1 Biosensors

Biosensors combine a biological system with a physicochemical detector. In this order of ideas, microorganisms, due to its low implementation cost, durability, and efficiency in different environmental conditions, are the most suitable biosensing platform for biosensor development (Lei et al. 2006). In this promising application, eukaryotic microbial biosensors using bioluminescent yeast strain of *Saccharomyces cerevisiae* have shown the ability to detect any toxic chemical that interferes with the cell metabolism resulting in a quantitative decrease in bioluminescence (Hollis et al. 2000).

Otherwise, in antifungal drug discovery, the cell wall is an essential target, making this an important screening platform for the evaluation of activity. In this way, an interesting model using yeast *S. cerevisiae* AT-1, transformed with the gene MLP1/YKL161c, has been shown to be a useful indicator of cell wall perturbations, showing to be promising for the development of new platforms for high-throughput antifungal screenings and identification of new compounds (Rodríguez-Peña et al. 2008).

Another interesting approach of fungal bionanotechnology in biosensors development is the application of the plasmonic properties of nanosilver produced by fungi, defined plasmon as quantum of plasma oscillation, these nanoparticles have electron oscillations with optical properties very useful in new devices (Sotiriou and Pratsinis 2011). For that reason, the development of biosensors that use silver plasmonic nanoparticles has a greater commercial interest because biosensors developed using this technology have less interference, especially with the fungal biosynthesis of triangular nanosilver particles as produced by *Fusarium oxysporum* and *Aspergillus fumigatus* (Boroumand Moghaddam et al. 2015). Plasmonic nanoparticles have a wide range of applications in the detection of biomolecules and microorganisms related to Alzheimer's disease, protein interactions, cancer, and infectious diseases (Wark et al. 2007; Xu et al. 2012; Samanta and Medintz 2016).

On the other hand, the use of magnetic particles (MPs) conjugated with biological molecules as antibodies is an important source of new diagnostic methods of greater precision and reproducibility, with a higher limit of detection than ELISA (Li et al. 2011). Following this line of reasoning, MPs can be produced extracellularly by fungi as *F. oxysporum* and *Verticillium* sp., for various diagnostic applications as biological detection, immunoseparation, and magnetic labeling of biologically active compounds (Bharde et al. 2006; Safarik and Safarikova 2009). For example, MPs have the ability to detect pathogenic microorganisms in clinical samples in an amount corresponding to 20–50 colony-forming units (CFU)/mL in 2 h using immunomagnetic separation (Billington et al. 2014).

Finally, the development of biosensors requires extensive integration with the development of new devices; for that reason, the inclusion of nano-biomaterials in Bio-MicroElectroMechanical Systems (BioMEMS) has become very important in recent years. BioMEMS are devices constructed by microfabrication, with the ability to perform identification, immobilization, growth, separation, purification, and manipulation of cells, with interesting biomedical applications in detection and diagnostics. These devices can potentiate the synthesized nano-biomaterials and improve their properties (Bhattacharya et al. 2007).

10.5 Fungal Bionanotechnology in New Antimicrobial Nanomaterials

10.5.1 Antimicrobial Hydrophobins

Hydrophobins are bioactive surface proteins produced by filamentous fungi as a result of mycotic growth and interaction with the environment (Bayry et al. 2012). Hydrophobins have important biophysical properties such as high adhesive strength, high surface activity, and self-assembly ability. For that reason, these molecules are strong candidates for the development of antimicrobial surface coating of biomaterials such as surgical instruments and medical implants (Ribeiro et al. 2012). The applications of fungal hydrophobins in coating can prevent bacterial adhesion and biofilm formation as well as can be used in immobilization of molecules and cells (Linder et al. 2005; Wösten and Scholtmeijer 2015). Equally, microbial contamination can lead to the decrease in the quality of surgical materials; in this order of ideas, the design and implementation of new anti-adhesive materials that avoid microbial adhesion and infection is of great interest; in that way hydrophobins can be assembled in hydrophilic–hydrophobic interfaces into amphipathic membranes increasing antifouling activity of several surfaces (Dumitrescu et al. 2013; Krasowska and Sigler 2014).

In conclusion, the impact of these coated materials using fungal hydrophobins in monolayers is of great potential for healthcare units and purification systems, as well as to prevent the effects of degradation induced by biofilms (Rieder et al. 2011; Khalesi et al. 2015).

10.5.2 Nanofiber

A nanofiber (NF) is considered as a fiber with an approximate size of 100–1000 nm in diameter. These fibers have been obtained from synthetic polymers, such as poly(ethylene oxide) (PEO), poly(vinyl alcohol) (PVA), poly(L-lactide) (PLA), poly(glycolic acid) (PGA), polyvinylpyrrolidone (PVP), and mineral (hydroxyapatite). Nanofibers have been a topic of interest in recent years due to its large surface area, versatility, and biocompatibility. In this order of ideas, biopolymeric nanofibers fabricated from chitosan and chitin which is a biopolymer existing in the cell walls of fungi have wide uses in biomedicine (Ding et al. 2014). Equally, chitin and chitosan have the advantages to be biocompatible, biodegradable, with low toxicity, and antimicrobial activity (Cheung et al. 2015). In the search for new sources of chitin and chitosan, mycelial cell walls are the most promising for this obtainment; in this way the mycelia of various fungi including *Absidia coerulea*, *Absidia glauca*, *A. niger*, *Gongronella butleri*, *Mucor rouxii*, and *Rhizopus oryzae* have been suggested as chitosan sources (Muñoz et al. 2015). Also, the cell walls of mushrooms also consist of chitin NFs in complex with glucans and are present in species of mushrooms as *Pleurotus eryngii*, *Agaricus bisporus*, *Lentinula edodes*, *Grifola frondosa*, and *Hypsizygus marmoreus* (Ifuku 2014). In NF production, electrospinning process has great relevance because it is a technique that can produce polymer nanofibers containing chitin or chitosan, with potential applications in areas such as drug release, dental, tissue engineering, wound healing, protective clothing, cosmetics, biosensors, and medical implants (Jayakumar et al. 2010; Chaudhari et al. 2016).

10.5.3 Graphene

Graphene is defined as two-dimensional sheet of hexagonally arranged carbon atoms packed into a honeycomb lattice (Allen et al. 2009). This material with unique properties has various applications in biosensing, diagnosis, and antimicrobial drug discovery (Gurunathan et al. 2014). Graphene has recently attracted attention as an anti-infective agent due to its ability to induce oxidative stress by ROS and lipid peroxidation in microbial cells (Nanda et al. 2016; Zhang et al. 2016).

In this order of ideas, biosynthesis of graphene, looking to improve the reduction of graphene oxide, has attracted the attention of researchers and industrialists. In this way, mushroom extracts from *Ganoderma* spp. have shown the ability to reduce graphene oxide with low toxicity and absence of toxic waste, due to abundance of polysaccharides (Gurunathan et al. 2014; Muthoosamy et al. 2015).

Graphene oxide nanosheets are an excellent material for the development of antimicrobial surfaces with low cost and scalability, in combination with other nanotechnological approaches that improve antibiofilm and anti-adhesion activity (de Faria et al. 2014; Perreault et al. 2015).

10.6 Conclusions and Perspectives

Fungal bionanotechnology is an integral field of development and knowledge for the obtaining of products and services of wide impact in the health of the populations. In this enterprise, it is important to be clear that the applications of science must be focused on solving problems with a view to achieving real innovations. In this way, there is an urgent need for new antimicrobial drugs with excellent chemical quality, high activity, and low toxicity (Moloney 2016). An important alternative is the development of nanoantibiotics that improve the pharmacokinetic/pharmacodynamic (PK/PD) index looking to increase antibiotic efficacy (Azzopardi et al. 2015). In addition, the design of nanostructure carriers will increase the transport of bioactive molecules to infected organs and tissues achieving greater therapeutic potential which results in a higher quality of life of the patients (Saude et al. 2013; Graef et al. 2016). Also, medical devices designed with these materials will avoid contamination produced by biofilms and subsequent infection (Borse et al. 2016).

Therefore, these new approaches that impact global public health must be produced on an industrial scale to reduce risks to populations; in addition, this process of industrialization should be innovative in the integration of disciplines such as what must occur with biotechnology and nanotechnology for the obtention of nanomaterials (Dorcheh and Vahabi 2016).

Equally it is necessary that these products which combine antimicrobial drugs, devices, and biological molecules are regulated. Also, the unknown risks of nanodrugs and nanoformulations must be evaluated over time (Paradise et al. 2009). But it is also important to emphasize that with these new drugs, antimicrobial stewardship programs should also be designed, because this bad medical practice is associated with the emergence of resistance (LaPlante et al. 2016).

On the other hand, we are facing a great bioprospective search of new microorganisms with applications in bionanotechnology, so the great potential of organisms to obtain development and innovations is quite high, with a huge genetic sequence diversity estimated in 7.27×10^{-3} only in yeasts (for humans is 1×10^{-3}) which shows us the enormous potential of industrial applications that we can access (Steensels et al. 2014). Without counting the possibilities of the transformation of strains mediated by genetic engineering (Meyer 2008; Castro et al. 2014), the ethical and political implications must be assessed monitoring opportunities and risks (Grunwald et al. 2004).

In addition, new forms of biosynthesis must be taken into account as the use of microorganism in biofilm state for the synthesis of nanomaterials mostly electrochemically active biofilms (EABs) (Iravani 2014; Ng et al. 2015).

Finally, we must consider clearly not to repeat the mistakes of the past and focus much of the effort to determine whether nanotechnology can induce antimicrobial resistance, because the possibility of inducing resistance has been reported related to AgNPs, and this can have major consequences without the proper pharmacovigilance strategies (Graves et al. 2015).

References

Abed N, Couvreur P (2014) Nanocarriers for antibiotics: a promising solution to treat intracellular bacterial infections. Int J Antimicrob Ag 43:485–496

Alghuthaymi MA, Almoammar H, Rai M, Said-Galiev E, Abd-Elsalam KA (2015) Myconanoparticles: synthesis and their role in phytopathogens management. Biotechnol Biotechnol Equip 29:221–236

Allahverdiyev AM, Kon KV, Abamor ES, Bagirova M, Rafailovich M (2011) Coping with antibiotic resistance: combining nanoparticles with antibiotics and other antimicrobial agents. Expert Rev Anti-Infect Ther 9:1035–1052

Allen MJ, Tung VC, Kaner RB (2009) Honeycomb carbon: a review of graphene. Chem Rev 110:132–145

Azam A, Ahmed AS, Oves M, Khan MS, Habib SS, Memic A (2012) Antimicrobial activity of metal oxide nanoparticles against Gram-positive and Gram-negative bacteria: a comparative study. Int J Nanomedicine 7:6003–6009

Azzopardi EA, Ferguson EL, Thomas DW (2015) Development and validation of an *in vitro* pharmacokinetic/pharmacodynamic model to test the antibacterial efficacy of antibiotic polymer conjugates. Antimicrob Agents Chemother 59:1837–1843

Bai H, Lv F, Liu L, Wang S (2016) Supramolecular antibiotic switches: a potential strategy for combating drug resistance. Chem Eur J 22:11114–11121

Balakumaran MD, Ramachandran R, Kalaichelvan PT (2015) Exploitation of endophytic fungus, *Guignardia mangiferae* for extracellular synthesis of silver nanoparticles and their *in vitro* biological activities. Microbiol Res 178:9–17

Banat IM, Franzetti A, Gandolfi I, Bestetti G, Martinotti MG, Fracchia L, Smyth TJ, Marchant R (2010) Microbial biosurfactants production, applications and future potential. Appl Microbiol Biotechnol 87:427–444

Bayry J, Aimanianda V, Guijarro JI, Sunde M, Latge JP (2012) Hydrophobins-unique fungal proteins. PLoS Pathog 8:e1002700

Beaman GM, Dennison SR, Phoenix DA (2014) Antimicrobial therapy based on antisense agents. In: Phoenix DA, Harris F, Dennison SR (eds) Novel antimicrobial agents and strategies. Wiley-VCH Verlag GmbH & Co. KGa A, Weinheim, pp 357–386

Bharde A, Rautaray D, Bansal V, Ahmad A, Sarkar I, Yusuf SM, Sanyal M, Sastry M (2006) Extracellular biosynthesis of magnetite using fungi. Small 2:135–141

Bhardwaj G, Cameotra SS, Chopra HK (2013) Biosurfactants from fungi: a review. J Pet Environ Biotechnol 4:1–6

Bhattacharya S, Jang J, Yang L, Akin D, Bashir R (2007) BioMEMS and nanotechnology based approaches for rapid detection of biological entities. J Rapid Methods Autom Microbiol 15:1–32

Billington C, Hudson JA, D'Sa E (2014) Prevention of bacterial foodborne disease using nanobiotechnology. Nanotechnol Sci Appl 7:73–83

Blecher K, Nasir A, Friedman A (2011) The growing role of nanotechnology in combating infectious disease. Virulence 2:395–401

Bluth MH, Kandil E, Mueller CM, Shah V, Lin YY, Zhang H, Dresner L, Lempert L, Nowakowski M, Gross R, Schulze R, Zenilman ME (2006) Sophorolipids block lethal effects of septic shock in rats in a cecal ligation and puncture model of experimental sepsis. Crit Care Med 34:188–195

Boroumand Moghaddam A, Namvar F, Moniri M, Md Tahir P, Azizi S, Mohamad R (2015) Nanoparticles biosynthesized by fungi and yeast: a review of their preparation, properties, and medical applications. Molecules 20:16540–16565

Borse V, Pawar V, Shetty G, Mullaji A, Srivastava R (2016) Nanobiotechnology perspectives on prevention and treatment of ortho-paedic implant associated infection. Curr Drug Deliv 13:175–185

Brown ED, Wright GD (2016) Antibacterial drug discovery in the resistance era. Nature 529:336–343

Castro L, Blázquez ML, González FG, Ballester A (2014) Mechanism and applications of metal nanoparticles prepared by bio-mediated process. Rev Adv Sci Eng 3:199–216

Chaudhari AA, Vig K, Baganizi DR, Sahu R, Dixit S, Dennis V, Singh SR, Pillai SR (2016) Future prospects for scaffolding methods and biomaterials in skin tissue engineering: a review. Int J Mol Sci 17(12):E1974

Cheung RCF, Ng TB, Wong JH, Chan WY (2015) Chitosan: an update on potential biomedical and pharmaceutical applications. Mar Drugs 13:5156–5186

De Rienzo MAD, Banat IM, Dolman B, Winterburn J, Martin PJ (2015) Sophorolipid biosurfactants: possible uses as antibacterial and antibiofilm agent. New Biotechnol 32:720–726

De Vries R, Andrade CA, Bakuzis AF, Mandal SM, Franco OL (2015) Next-generation nanoantibacterial tools developed from peptides. Nanomedicine 10:1643–1661

DiGiandomenico A, Sellman BR (2015) Antibacterial monoclonal antibodies: the next generation? Curr Opin Microbiol 27:78–85

Ding F, Deng H, Du Y, Shi X, Wang Q (2014) Emerging chitin and chitosan nanofibrous materials for biomedical applications. Nanoscale 6:9477–9493

Dizaj SM, Lotfipour F, Barzegar-Jalali M, Zarrintan MH, Adibkia K (2014) Antimicrobial activity of the metals and metal oxide nanoparticles. Mater Sci Eng C-Bios 44:278–284

Dorcheh SK, Vahabi K (2016) Biosynthesis of nanoparticles by fungi: large-scale production. In: Merillon JM, Rmawat KG (eds) Fungal metabolites. Springer International Publishing, Switzerland, pp 1–20

Dumitrescu I, Iordache OG, Mocioiu AM, Nicula G (2013) Antimicrobial functionalization of textile materials with hydrophobins and Ag/ZnO composite nanopowders. Ind Textila 64:303–311

de Faria AF, Martinez DST, Meira SMM, de Moraes ACM, Brandelli A, Souza Filho AG, Alves OL (2014) Anti-adhesion and antibacterial activity of silver nanoparticles supported on graphene oxide sheets. Colloids Surf B Biointerfaces 113:115–124

Gautam N, Puligujja P, Balkundi S, Thakare R, Liu XM, Fox HS, McMillan J, Gendelman HE, Alnouti Y (2014) Pharmacokinetics, biodistribution, and toxicity of folic acid-coated antiretroviral nanoformulations. Antimicrob Agents Chemother 58:7510–7519

Girard V, Dieryckx C, Job C, Job D (2013) Secretomes: the fungal strike force. Proteomics 13:597–608

Good L, Stach JE (2011) Synthetic RNA silencing in bacteria – antimicrobial discovery and resistance breaking. Front Microbiol 2:185

Graef F, Gordon S, Lehr CM (2016) Anti-infectives in drug delivery-overcoming the Gram-negative bacterial cell envelope. Curr Top Microbiol Immunol 398:475–496

Graves JL Jr, Tajkarimi M, Cunningham Q, Campbell A, Nonga H, Harrison SH, Barrick JE (2015) Rapid evolution of silver nanoparticle resistance in *Escherichia coli*. Front Genet 6:42

Grunwald A (2004) The case of nanobiotechnology. EMBO Rep 5:S32–S36

Gurunathan S, Han J, Park JH, Kim JH (2014) An in vitro evaluation of graphene oxide reduced by *Ganoderma* spp. in human breast cancer cells (MDA-MB-231). Int J Nanomedicine 9:1783–1797

Hamidi M, Azadi A, Rafiei P, Ashrafi H (2013) A pharmacokinetic overview of nanotechnology-based drug delivery systems: an ADME-oriented approach. Crit Rev Ther Drug Carrier Syst 30:435–467

Harmsen MM, De Haard HJ (2007) Properties, production, and applications of camelid single-domain antibody fragments. Appl Microbiol Biotechnol 77:13–22

Hassanzadeh-Ghassabeh G, Devoogdt N, De Pauw P, Vincke C, Muyldermans S (2013) Nanobodies and their potential applications. Nanomedicine 8:1013–1026

Hollis RP, Killham K, Glover LA (2000) Design and application of a biosensor for monitoring toxicity of compounds to eukaryotes. Appl Environ Microbiol 66:1676–1679

Huh AJ, Kwon YJ (2011) "Nanoantibiotics": a new paradigm for treating infectious diseases using nanomaterials in the antibiotics resistant era. J Control Release 156:128–145

Ibrahim HM, El-Bisi MK, Taha GM, El-Alfy EA (2015) Chitosan nanoparticles loaded antibiotics as drug delivery biomaterial. J Appl Pharm Sci 5:085–090

Ifuku S (2014) Chitin and chitosan nanofibers: preparation and chemical modifications. Molecules 19:18367–18380

Ingale AG, Chaudhari AN (2013) Biogenic synthesis of nanoparticles and potential applications: an eco-friendly approach. J Nanomed Nanotechol 4:1–7

Iravani S (2014) Bacteria in nanoparticle synthesis: current status and future prospects. Int Sch Res Not 2014:359316

Jayakumar R, Prabaharan M, Nair SV, Tamura H (2010) Novel chitin and chitosan nanofibers in biomedical applications. Biotechnol Adv 28:142–150

Kandi V, Kandi S (2015) Antimicrobial properties of nanomolecules: potential candidates as antibiotics in the era of multi-drug resistance. Epidemiol Health 37:e2015020

Kavšček M, Stražar M, Curk T, Natter K, Petrovič U (2015) Yeast as a cell factory: current state and perspectives. Microb Cell Factories 14:94

Khalesi M, Gebruers K, Derdelinckx G (2015) Recent advances in fungal hydrophobin towards using in industry. Protein J 34:243–255

Khan M, Al-Marri AH, Khan M, Shaik MR, Mohri N, Adil SF, Kuniyil M, Alkhathlan HZ, Al-Warthan A, Tremel W, Tahir MN, Siddiqui MR (2015a) Green approach for the effective reduction of graphene oxide using *Salvadora persica* l. root (miswak) extract. Nanoscale Res Lett 10:281

Khan I, Khan M, Umar MN, Oh DH (2015b) Nanobiotechnology and its applications in drug delivery system: a review. IET Nanobiotechnol 9:396–400

Kitching M, Ramani M, Marsili E (2015) Fungal biosynthesis of gold nanoparticles: mechanism and scale up. Microb Biotechnol 8:904–917

Kong M, Chen XG, Xing K, Park HJ (2010) Antimicrobial properties of chitosan and mode of action: a state of the art review. Int J Food Microbiol 144:51–63

Krasowska A, Sigler K (2014) How microorganisms use hydrophobicity and what does this mean for human needs? Front Cell Infect Microbiol 4:112

LaPlante K, Cunha C, Morrill H, Rice L, Mylonakis E (2016) Antimicrobial stewardship: principles and practice. CABI, UK

Lei Y, Chen W, Mulchandani A (2006) Microbial biosensors. Anal Chim Acta 568:200–210

Li X, Xu H, Chen ZS, Chen G (2011) Biosynthesis of nanoparticles by microorganisms and their applications. J Nanomater 2011:270974

Linder MB, Szilvay GR, Nakari-Setälä T, Penttilä ME (2005) Hydrophobins: the protein-amphiphiles of filamentous fungi. FEMS Microbiol Rev 29:877–896

Liu L (2015) Antibody glycosylation and its impact on the pharmacokinetics and pharmacodynamics of monoclonal antibodies and fc-fusion proteins. J Pharm Sci 104:1866–1884

Lukowiak A, Kedziora A, Strek W (2016) Antimicrobial graphene family materials: progress, advances, hopes and fears. Adv Colloid Interf Sci 236:101–112

Maine E, Thomas VJ, Bliemel M, Murira A, Utterback J (2014) The emergence of the nanobiotechnology industry. Nat Nanotechnol 9:2–5

Manivasagan P, Venkatesan J, Sivakumar K, Kim SK (2016) Actinobacteria mediated synthesis of nanoparticles and their biological properties: a review. Crit Rev Microbiol 42:209–221

Meinken J, Asch DK, Neizer-Ashun KA, Chang GH, Cooper JRCR, Min XJ (2014) FunSecKB2: a fungal protein subcellular location knowledgebase. Comput Mol Biol 4:1–17

Meyer V (2008) Genetic engineering of filamentous fungi – progress, obstacles and future trends. Biotechnol Adv 26:177–185

Meyer V, Fiedler M, Nitsche B, King R (2015) The cell factory aspergillus enters the big data era: opportunities and challenges for optimising product formation. Adv Biochem Eng Biotechnol 149:91–132

Mohanpuria P, Rana NK, Yadav SK (2008) Biosynthesis of nanoparticles: technological concepts and future applications. J Nanopart Res 10:507–517

Moloney MG (2016) Natural products as a source for novel antibiotics. Trends Pharmacol Sci 37:689–701

Morais MGD, Martins VG, Steffens D, Pranke P, da Costa JAV (2014) Biological applications of nanobiotechnology. J Nanosci Nanotechnol 14:1007–1017

Muñoz G, Valencia C, Valderruten N, Ruiz-Durántez E, Zuluaga F (2015) Extraction of chitosan from *Aspergillus niger* mycelium and synthesis of hydrogels for controlled release of betahistine. React Funct Polym 91:1–10

Muthoosamy K, Bai RG, Abubakar IB, Sudheer SM, Lim HN, Loh HS, Huang NM, Chia CH, Manickam S (2015) Exceedingly biocompatible and thin-layered reduced graphene oxide nanosheets using an eco-friendly mushroom extract strategy. Int J Nanomedicine 10:1505–1519

Nanda SS, Yi DK, Kim K (2016) Study of antibacterial mechanism of graphene oxide using Raman spectroscopy. Sci Rep 6:28443

Netala VR, Kotakadi VS, Bobbu P, Gaddam SA, Tartte V (2016) Endophytic fungal isolate mediated biosynthesis of silver nanoparticles and their free radical scavenging activity and anti microbial studies. 3. Biotech 6:1–9

Ng CK, Mohanty A, Cao B (2015) Biofilms in bio-nanotechnology. In: Singh OV (ed) Bio-nanoparticles: biosynthesis and sustainable biotechnological implications. Wiley, Hoboken, pp 83–100

Otsuka T, Brauer AL, Kirkham C, Sully EK, Pettigrew MM, Kong Y, Geller BL, Murphy TF (2017) Antimicrobial activity of antisense peptide-peptide nucleic acid conjugates against nontypeable *Haemophilus influenzae* in planktonic and biofilm forms. J Antimicrob Chemother 72:137–144

Pacwa-Płociniczak M, Płaza GA, Piotrowska-Seget Z, Cameotra SS (2011) Environmental applications of biosurfactants: recent advances. Int J Mol Sci 12:633–654

Paradise J, Tisdale AW, Hall RF, Kokkoli E (2009) Evaluating oversight of human drugs and medical devices: a case study of the FDA and implications for nanobiotechnology. J Law Med Ethics 37:598–624

Pawlowski AC, Wang W, Koteva K, Barton HA, McArthur AG, Wright GD (2016) A diverse intrinsic antibiotic resistome from a cave bacterium. Nat Commun 7:13803

Perreault F, de Faria AF, Nejati S, Elimelech M (2015) Antimicrobial properties of graphene oxide nanosheets: why size matters. ACS Nano 9:7226–7236

Pochanavanich P, Suntornsuk W (2002) Fungal chitosan production and its characterization. Lett Appl Microbiol 35:17–21

Prabhu S, Poulose EK (2012) Silver nanoparticles: mechanism of antimicrobial action, synthesis, medical applications, and toxicity effects. Int Nano Lett 2:32

Prasad R, Bhattacharyya A, Nguyen QD (2017a) Nanotechnology in sustainable agriculture: recent developments, challenges, and perspectives. Front Microbiol 8:1014. https://doi.org/10.3389/fmicb.2017.01014

Prasad R, Pandey R, Varma A, Barman I (2017b) Polymer based nanoparticles for drug delivery systems and cancer therapeutics. In: Kharkwal H, Janaswamy S (eds) Natural polymers for drug delivery. CAB International, UK, pp 53–70

Qi L, Xu Z, Jiang X, Hu C, Zou X (2004) Preparation and antibacterial activity of chitosan nanoparticles. Carbohydr Res 339:2693–2700

Raghupathi KR, Koodali RT, Manna AC (2011) Size-dependent bacterial growth inhibition and mechanism of antibacterial activity of zinc oxide nanoparticles. Langmuir 27:4020–4028

Raja HA, Kaur A, El-Elimat T, Figueroa M, Kumar R, Deep G, Agarwal R, Faeth SH, Cech NB, Oberlies NH (2015) Phylogenetic and chemical diversity of fungal endophytes isolated from *Silybum marianum* (L) Gaertn. (milk thistle). Mycology 6:8–27

Raliya R, Saharan V, Saran R, Choudhary K, Tarafdar JC, Biswas P (2016) Intervention of fungi in Nano-particle technology and applications. In: Prasad R (ed) Advances and applications through fungal nanobiotechnology. Springer International Publishing, Cham, pp 241–251

Ratledge C (1997) Microbial Lipids. In: Rehm HJ, Reed G (eds) Biotechnology: products of secondary metabolism, 2nd edn. Wiley-VCH Verlag GmbH, Weinheim, pp 135–186

Ribeiro M, Monteiro FJ, Ferraz MP (2012) Infection of orthopedic implants with emphasis on bacterial adhesion process and techniques used in studying bacterial-material interactions. Biomatter 2:176–194

Rieder A, Ladnorg T, Wöll C, Obst U, Fischer R, Schwartz T (2011) The impact of recombinant fusion-hydrophobin coated surfaces on *E. coli* and natural mixed culture biofilm formation. Biofouling 27:1073–1085

Rodriguez A, Kildegaard KR, Li M, Borodina I, Nielsen J (2015) Establishment of a yeast platform strain for production of p-coumaric acid through metabolic engineering of aromatic amino acid biosynthesis. Metab Eng 31:181–188

Rodriguez-Peña JM, Diez-Muñiz S, Nombela C, Arroyo J (2008) A yeast strain biosensor to detect cell wall-perturbing agents. J Biotechnol 133:311–317

Safarik I, Safarikova M (2009) Magnetic nano-and microparticles in biotechnology. Chem Pap 63:497–505

Saha B, Bhattacharya J, Mukherjee A, Ghosh A, Santra C, Dasgupta AK, Karmakar P (2007) In vitro structural and functional evaluation of gold nanoparticles conjugated antibiotics. Nanoscale Res Lett 2:614–622

Samanta A, Medintz IL (2016) Nanoparticles and DNA- a powerful and growing functional combination in bionanotechnology. Nanoscale 8:9037–9095

Sambanthamoorthy K, Feng X, Patel R, Patel S, Paranavitana C (2014) Antimicrobial and antibiofilm potential of biosurfactants isolated from lactobacilli against multi-drug-resistant pathogens. BMC Microbiol 14:197

Saúde AC, Cherobim MD, Amaral AC, Dias SC, Franco OL (2013) Nanoformulated antibiotics: the next step for pathogenic bacteria control. Curr Med Chem 20:1232–1240

Saylor C, Dadachova E, Casadevall A (2009) Monoclonal antibody-based therapies for microbial diseases. Vaccine 27:G38–G46

Shah V, Doncel GF, Seyoum T, Eaton KM, Zalenskaya I, Hagver R, Azim A, Gross R (2005) Sophorolipids, microbial glycolipids with anti-human immunodeficiency virus and sperm-immobilizing activities. Antimicrob Agents Chemother 49:4093–4100

Shah V, Badia D, Ratsep P (2007) Sophorolipids having enhanced antibacterial activity. Antimicrob Agents Chemother 51:397–400

Sharma A, Kumar Arya D, Dua M, Chhatwal GS, Johri AK (2012) Nano-technology for targeted drug delivery to combat antibiotic resistance. Expert Opin Drug Deliv 9:1325–1332

Singh P, Cameotra SS (2004) Potential applications of microbial surfactants in biomedical sciences. Trends Biotechnol 22:142–146

Siontorou CG (2013) Nanobodies as novel agents for disease diagnosis and therapy. Int J Nanomedicine 8:4215–4227

Sleiman JN, Kohlhoff SA, Roblin PM, Wallner S, Gross R, Hammerschlag MR, Zenilman ME, Bluth MH (2009) Sophorolipids as antibacterial agents. Ann Clin Lab Sci 39:60–63

Sonia TA, Sharma CP (2011) Chitosan and its derivatives for drug delivery perspective. In: Jayakumar R, Prabaharan M, RAA M (eds) Chitosan for biomaterials I. Springer-Verlag, Berlin, pp 23–53

Sotiriou GA, Pratsinis SE (2011) Engineering nanosilver as an antibacterial, biosensor and bioimaging material. Curr Opin Chem Eng 1:3–10

Spadiut O, Capone S, Krainer F, Glieder A, Herwig C (2014) Microbials for the production of monoclonal antibodies and antibody fragments. Trends Biotechnol 32:54–60

Steeland S, Vandenbroucke RE, Libert C (2016) Nanobodies as therapeutics: big opportunities for small antibodies. Drug Discov Today 21:1076–1113

Steensels J, Snoek T, Meersman E, Picca Nicolino M, Voordeckers K, Verstrepen KJ (2014) Improving industrial yeast strains: exploiting natural and artificial diversity. FEMS Microbiol Rev 38:947–995

Syed MA, Bana NF (2016) Developing countries need action plans to combat the challenge of antimicrobial resistance. Arch Clin Microbiol 7:12

Teixeira PC, Leite GM, Domingues RJ, Silva J, Gibbs PA, Ferreira JP (2007) Antimicrobial effects of a microemulsion and a nanoemulsion on enteric and other pathogens and biofilms. Int J Food Microbiol 118:15–19

Utt E, Wells C (2016) The global response to the threat of antimicrobial resistance and the important role of vaccines. Pharm Policy Law 18:179–197

Van Bogaert IN, Saerens K, De Muynck C, Develter D, Soetaert W, Vandamme EJ (2007). Microbial production and application of sophorolipids. Appl Microbiol Biotechnol 76: 23–34

Vicentin EC, Françoso KS, Rocha MV, Iourtov D, Dos Santos FL, Kubrusly FS, Sakauchi MA, Raw I, Nosten F, Rénia L, Rodrigues MM, Russell B, Soares IS (2014) Invasion-inhibitory antibodies elicited by immunization with *Plasmodium* vivax apical membrane antigen-1 expressed in *Pichia pastoris* yeast. Infect Immun 82:1296–1307

Vigderman L, Zubarev ER (2013) Therapeutic platforms based on gold nanoparticles and their covalent conjugates with drug molecules. Adv Drug Deliv Rev 65:663–676

Wang Y, Fan Z, Shao L, Kong X, Hou X, Tian D, Sun Y, Xiao Y, Yu L (2016) Nanobody-derived nanobiotechnology tool kits for diverse biomedical and biotechnology applications. Int J Nanomedicine 11:3287–3303

Wark AW, Lee HJ, Qavi AJ, Corn RM (2007) Nanoparticle-enhanced diffraction gratings for ultrasensitive surface plasmon biosensing. Anal Chem 79:6697–7701

Woodford N, Wareham DW, UK Antibacterial Antisense Study Group (2009) Tackling antibiotic resistance: a dose of common antisense? J Antimicrob Chemother 63:225–229

Wösten HA, Scholtmeijer K (2015) Applications of hydrophobins: current state and perspectives. Appl Microbiol Biotechnol 99:1587–1597

Xu X, Chen Y, Wei H, Xia B, Liu F, Li N (2012) Counting bacteria using functionalized gold nanoparticles as the light-scattering reporter. Anal Chem 84:9721–9728

Xue B, He D, Gao S, Wang D, Yokoyama K, Wang L (2016) Biosynthesis of silver nanoparticles by the fungus *Arthroderma fulvum* and its antifungal activity against genera of *Candida*, *Aspergillus* and *Fusarium*. Int J Nanomedicine 11:1899–1906

Yadav A, Kon K, Kratosova G, Duran N, Ingle AP, Rai M (2015) Fungi as an efficient mycosystem for the synthesis of metal nanoparticles: progress and key aspects of research. Biotechnol Lett 37:2099–2120

Zhang N, Hou J, Chen S, Xiong C, Liu H, Jin Y, Wang J, He Q, Zhao R, Nie Z (2016) Rapidly probing antibacterial activity of graphene oxide by mass spectrometry-based metabolite fingerprinting. Sci Rep 6:28045

Zhu X, Radovic-Moreno AF, Wu J, Langer R, Shi J (2014) Nanomedicine in the management of microbial infection – overview and perspectives. Nano Today 9:478–498

Chapter 11
Fungal Nanotechnology and Biomedicine

Niraj Kumari, Anal K. Jha, and K. Prasad

Abstract The chapter details different members of fungi-mediated biosynthetic processes for inorganic (metallic and oxide) nanoparticles and their applications in biomedicine. The biosynthetic mechanism has been discussed. The synthesis of nanoparticles broadly banks upon the variation of key parameters like concentration, temperature, pH, and other medium conditions. Applications of fungal nanosynthesis are discussed like antimicrobial, cosmetics, drug delivery, diagnostics, etc. In this chapter, different procedures for biosynthesis of inorganic (metal and oxide) nanomaterials negotiated by different cohorts of fungi and their use in biomedicine have been delineated. The biosynthetic mechanism has been discussed. The cues of synthesis broadly bank upon variation of key factors such as temperature, solute concentration, and pH, along with other medium requirements. Applications of fungus-negotiated preparedness of nanoparticles such as antimicrobial, drug delivery, cosmetics, diagnostics, etc. are also discussed.

11.1 Introduction

Nanotechnology is one among the most assuring and new areas of research in modern science. It comprises the synthesis, manipulation, and utilization of materials that range in size of less than a micron approximating to individual atoms. The concept

N. Kumari
Aryabhatta Centre for Nanoscience and Nanotechnology, Aryabhatta Knowledge University, Patna 800001, India

A.K. Jha
University Department of Chemistry, T.M. Bhagalpur University, Bhagalpur 812007, India

Aryabhatta Centre for Nanoscience and Nanotechnology, Aryabhatta Knowledge University, Patna 800001, India

K. Prasad (✉)
Aryabhatta Centre for Nanoscience and Nanotechnology, Aryabhatta Knowledge University, Patna 800001, India

University Department of Physics, T.M. Bhagalpur University, Bhagalpur 812007, India
e-mail: k.prasad65@gmail.com

R. Prasad (ed.), *Fungal Nanotechnology*, Fungal Biology,
DOI 10.1007/978-3-319-68424-6_11

of nanotechnology was given by Richard Feynman through his famous lecture entitled "There is a plenty of rooms at the bottom" at the American institute of Technology. The term nanotechnology was coined by Prof. Norio Taniguchi of Tokyo University of Science. The word "nano" originates from the Greek word *nanos* meaning "dwarf," and when it is used as a prefix, it implies 10^{-9}. Nanotechnology is defined as small structures/materials having the dimensions lesser than 100 nm. For a few example, the hydrogen atom is ~0.1 nm, a strand of DNA is 2.5 nm, a virus may be ~100 nm, a typical bacterium is ~1000–3000 nm, and a red blood corpuscle is ~7000 nm in diameter, while an average human hair is ~10,000 nm wide. Nanoscience and technology is the branch of science and technology which deals on (1) the development of synthetic methods and surface analytical tools for building structures and materials, (2) to understand the change in chemical and physical nature due to miniaturization, and (3) the use of such properties in the progress of novel and functional materials and devices. Nanoparticles are delineated as particulate dispersions of solid particles with at least one dimension at a size range of 10–1000 nm. The most significant notability of nanoparticles is their surface area to volume facet ratio, avowing them to collaborate with other particles easier. It is well entrenched that due to the quantum effect (spatial confinements), large surface area, large surface energy, and reduced nanofabrication, there is an alteration in the properties of conventional materials at nano-level, and the performance of surfaces starts to lead the performance of bulk materials (Prasad et al. 2017). By amenably manipulating the size, shape, and composition of the nanoscale materials, it is quite possible to tune their physical, mechanical, optical, thermal, electrical, magnetic, chemical, photo-electrochemical, catalytic, and biological properties (Suman et al. 2010; Daniel and Astruc 2004; Schmid 1992; Krolikowska et al. 2003; Valtchev and Tosheva 2013). Many synthetic procedures have emerged with time for the preparation of metals and/or oxide nanoparticles. There are two main ways for the synthesis of metals and/or oxide nanoparticles: a bottom-up approach (self-assembly) and a top-down approach. The bottom-up approach includes synthesis of a material either atom by atom, molecule by molecule, or by its self-arrangements, which is mostly based on Gibbs free energy, so that synthesized nanomaterials are in a state closure to a thermodynamic equilibrium state. The top-down approach mainly uses conventional methods to the synthesis of nanoparticles. Nanoparticles can be synthesized by using a variety of methods, i.e., physical, chemical, biological, and amalgamated techniques. A conventional method of synthesis includes physical and chemical methods. Some of the physical methods which are being employed for the synthesis of nanoparticles are ball milling, mechanical grinding, radiolysis, ultrasonication, spray pyrolysis, photo irradiation, electrospinning, laser ablation, etc., while the chemical methods mainly include chemical reduction, coprecipitation, sol-gel method, salvothermal, micellar medium, spray pyrolysis, template-based methods, etc. (Fig. 11.1). Each of these processes is either capital or labor, energy intensive, and time-consuming, employs toxic chemicals, and also craves high temperature for reaction (Pingali et al. 2005; Flores et al. 2013; Kalishwaralal et al. 2008; Shin et al. 2004; Wang et al. 2007). Therefore, it is required to evolve the synthetic protocols that should be reliable,

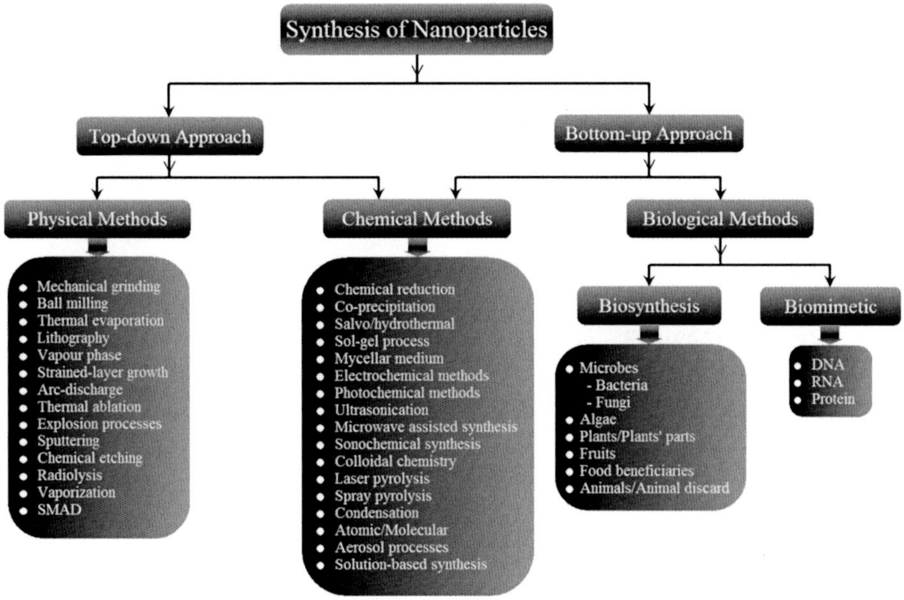

Fig. 11.1 Different ways for the synthesis of nanoparticles

nontoxic, clean, high-yield, low-cost, less time-exhausting, and eco-friendly which occurs at ambient conditions for the synthesis of metal/oxide nanoparticles of controlled size, shape, and monodispersity, which is possible through biological methods. So these methods for the synthesis of metal/oxide nanoparticles involve microbial flora like bacteria, fungi, viruses, and algae. The use of fungi in producing nanoparticles has received significant interest as they offer certain advantages over the use of other microbes for the synthesis of metal/oxide nanoparticles. The ease of scaling up, downstream processing, economic feasibility, and ubieties of mycelia offering an increased surface area are important advantages to consider, which secrete significantly higher amounts of proteins than bacteria; this would amplify the nanoparticle synthesis productivity (Chandran et al. 2006; Jia et al. 2009; Prasad 2014; Song et al. 2010; Xie et al. 2007). During the last two decades, the biosynthesis of noble metal/oxide nanoparticles (silver, gold, iron, platinum, zinc, zinc oxide, iron oxide palladium, etc.) has been noticed significantly due to the thriving need to develop eco-friendly technologies in material synthesis. In recent years, for the synthesis of various types of nanomaterials, nanobiotechnology has emanated as an upcoming field. The growth of ecologically benign, "green" synthesis cues is in consonance with the recent RoHS and WEEE legislation stipulated by the EU. Living cells are the superlative known examples of machines which operate at the nanoscale and enforce a number of works pasturing from generation of energy to extraction of targeted materials with very high efficiency. Microorganisms are natural producers of nanoscale materials. The cell factories have proved to be encouraging tools for the upcoming technologies and biomedical applications. Fungi have remarkable advantage in terms of metabolic flux as well as cellular level organization. Yeast being a member of the kingdom fungi has been taken into regular use as media supplement in different culture procedures, and this organism itself has been a very good source of different enzymes and vitamins. They are handy, nontoxic, and quite amenable in terms of culturing. In the recent past, yeast and other fungal members have been considered for the synthesis of metal, oxide, and chalcogenide nanoparticles (Agnihotri et al. 2009; Ahmad et al. 2002, 2003; Balaji et al. 2009; Bansal et al. 2005; Basavaraja et al. 2008; Bhainsa and D'Souza 2006; Bhambure et al. 2009; Dameron et al. 1989; Durán et al. 2005; Fayaz et al. (2009a, b); Gharieb et al. 1999; Jha and Prasad 2010a, b; Jha et al. 2008a, b, 2009a, c, 2010; Kalishwaralal et al. 2008; Kathiresan et al. 2009; Kowshik et al. 2002, 2003; Kumar et al. 2008; Mukherjee et al. 2001a, b, 2002; Philip 2009a, b; Prasad et al. 2007, 2010, 2014, 2016; Reese and Winge 1988; Sanghi and Verma 2009; Sastry et al. 2003; Senapati et al. 2005; Shaligram et al. 2009; Shankar et al. 2003; Narayanan and Sakthivel 2010b; Ray et al. 2011; Haq et al. 2015; Aziz et al. 2015; Beveridge and Murray 1980; Cunningham and Lundie 1993; Deplanche and Macaskie 2008; Deplanche et al. 2008; Du et al. 2007; Fortin and Beveridge 2000; Labrenz et al. 2000; Mokhtari et al. 2009; Nair and Pradeep 2002; Prasad and Jha 2009; Saifuddin et al. 2009; Southam and Beveridge 1994; Ghodake et al. 2013; Liu et al. 2013, 2014; Tanvir et al. 2012). This chapter will focus on how material science and biology can work together to create a "green" way of synthesizing metal/oxide nanoparticles for a wide ambit of its application in biomedicine.

11.2 Myconanotechnology

Fungus nanotechnology (myconanotechnology) is an emerging field, in which fungi can be effectively used for the formation/fabrication of nanoparticles with desirable shape and size Breierová et al. (2002, 2004), Butt and Ecker (1987), Čertik et al. (2005), Chae et al. (1994), Choi et al. (1998), Dameron and Winge (1990), Gadda and Fitzpatrick (1998), Gan (1991), GonzálezChávez et al. (2004), González et al. (2005), Pantidos and Horsfall (2014), Shanti and Karl (2006), Song et al. (2003). The nanotransformation involving synthesis of metal, oxide, or chalcognide nanoparticles apparently is the cumulative response of the biological system which is being taken into use and its immediate chemical ambience along with basic signal transduction, metabolic fluxes, and metabolite content of the organism being engaged for the purpose. Fungi are perceived as eukaryotic, heterotrophic, and absorptive organisms having chitinous cell walls, which reproduce mostly asexually or sexually by producing spores and multiply by budding or even by cell elongation of the hyphal tip. They are characterized by a distinctive, multinucleate vegetative (somatic) thallus called mycelium and survive by obtaining nourishment systemically or through rhizoids. They have got naturally bestowed property to depolymerize the complex molecules like proteins, carbohydrates, and lipids as source of carbon and energy, and due to this, they are ecologically considered among primary

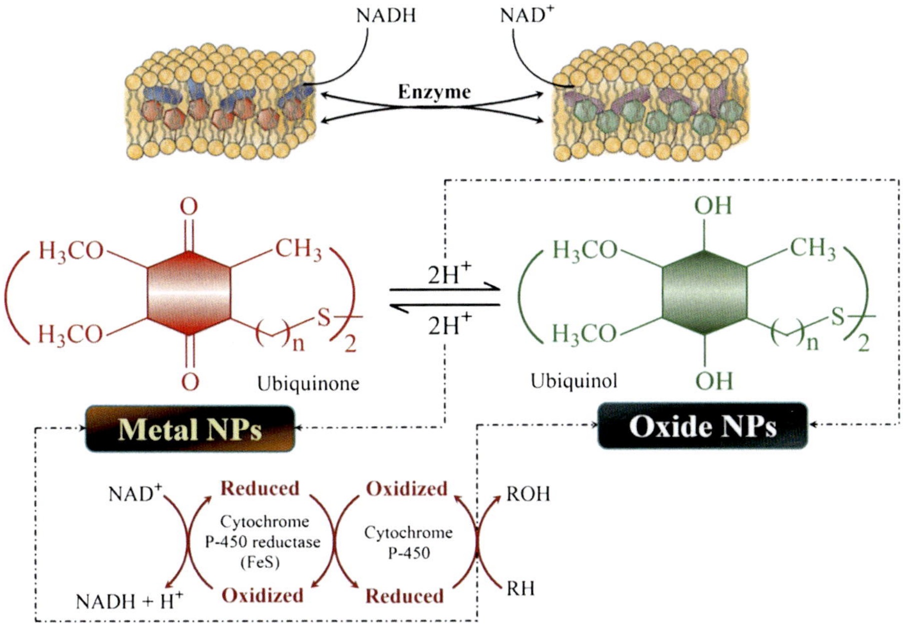

Fig. 11.2 Biosynthesis of metal/oxide nanoparticles

decomposers. Due to microbial dimension and eukaryotic cell structure, they have developed enrapturing modes of adaptability in order to continue to live right from the stale decaying food to living human/animal body. It produces secondary metabolites like alkaloids that are identical in structure to plant metabolites and also contains all respiratory enzymes that are essential for the life of an organism. Here the membrane coerced (and cytosolic) oxidoreductases and quinines have played an important role in nanotransformation. An integer of simple hydroxy/methoxy derivatives of benzoquinones and toluquinones are produced by fungi in response to a metallic stress. The presence of these metabolites triggers a redox reaction due to tautomerization (keto to enol) leading to synthesis of nanoparticles. The oxidoreductases are pH responsive and work in a substitute manner. During lower value of pH, oxidase gets triggered whereas a higher pH value activates the reductase. The speculated mechanism for the synthesis of metallic and oxide nanoparticles is illustrated in Fig. 11.2. It is very well affirmed that the reduction potential of the culture medium seems to decide the fate of metal ions exposed to the fungus. A raised amount of reduction potential means at higher value of pH integration of metal NPs takes place, and an oxide NP is resulted under reverse situations. The synthesis of fungus-mediated metal/oxide nanoparticles and its conformation by different instruments are represented in the form of a flowchart in Fig. 11.3. It is known that pH along with other parameters like concentration of reducing agent, kinetics, mixing ratio, incubation time, and solution chemistry are deciding the moirai of metal ions to bio-reductant of fungus. Then, metal/oxide nanoparticles are formed that are

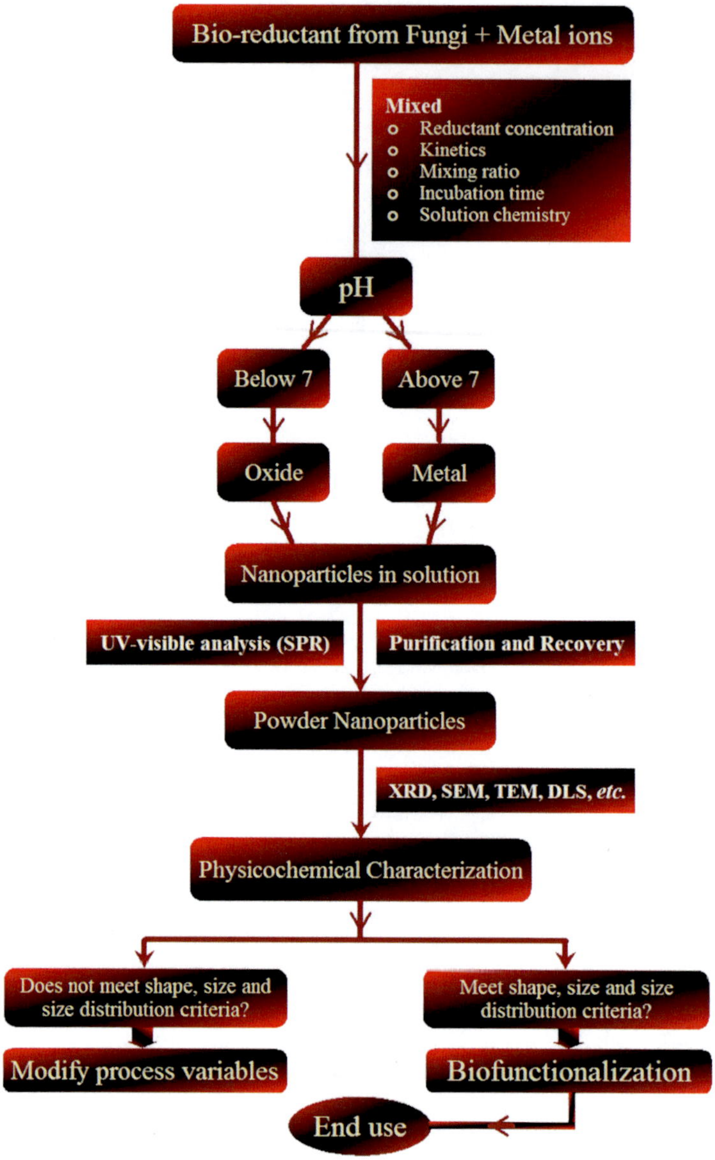

Fig. 11.3 Postulated flowchart for the conformation of fungal biosynthesis of nanoparticles

conformed by UV spectroscopy in solution and used in further study (Jha and Prasad 2010a, c; Halliwell 2006; Huang et al. 2007; JaroszWilkołazka et al. 2006; Liu et al. 2014; Haq et al. 2015; Mehra and Winge 1991; Mehra et al. 1988; Marchiol 2012; Nair and Pradeep 2002; Paraszkiewicz and Długoński 2009; Paraszkiewicz et al. 2007, 2010; Philip 2009a, b; Reese and Winge 1988; Saifuddin et al. 2009; Sanghi and Verma 2009; Schafer and Buettner 2001; Schmid 1992; Shin et al. 2004; Song et al. 2010; Southam and Beveridge 1994; Stadtman and Levine 2003;

Stadtman et al. 2003; Suman et al. 2010; Thakkar et al. 2010; Tamás et al. 2005; Tanvir et al. 2012; Ulla et al. 2000; Vesentini et al. 2006; Wang et al. 2007; Wood et al. 2003; Wright et al. 1996; Wysocki and Tamás 2010; Xie et al. 2007; Yompakdee et al. 1996; Avery 2001; Limon Pacheco and Gonsebatt 2009; Aziz et al. 2015; Beveridge and Murray 1980; Cunningham and Lundie 1993; Deplanche and Macaskie 2008; Deplanche et al. 2008; Du et al. 2007; Fortin and Beveridge 2000; Jha et al. 2008a, 2009a, 2010; Labrenz et al. 2000; Mokhtari et al. 2009; Nair and Pradeep 2002; Prasad and Jha 2009).

11.3 Methodology

Metallic/oxide/chalcogenide nanoparticles were primed by using fungus like *Saccharomyces cerevisiae* (baker's yeast), *Fusarium oxysporum*, *Aspergillus niger*, *Aspergillus fumigates*, *Penicillium brevicompactum*, *Volvariella volvacea*, *Agaricus bisporus*, *Tricholoma crassum*, etc. (Table 11.1). Fungal cells (*F. oxysporum*, *S. cerevisiae*) were allowed to incubate as suspension culture in the presence of suitable carbon and nitrogen source for 36 h. This was treated as source culture. A finite small portion of it (25 ml) was filtered and four times diluted by mixing 30% EtOH-bearing nutrients. This prepped culture was again allowed to incubate for another 24 h until it attains a light straw color/golden yellow color. Now, 20 ml of 0.025(M) metal ion solution was mixed to the culture solution, and the culture was allowed to grow. The culture solution was heated on steam bath up to 60 °C for 10–20 min until the appearance of deposition at the bottom of the flask, indicative of the inception of transformation. The value of pH of the culture solution is suitably adjusted at this stage depending upon targeted task synthesis of a metal or an oxide or chalcogenide. It is allowed to cool further in the laboratory ambience. After 2–3 days, the culture solution is noticed to have distinctly marked deposits. It is filtered through a Whatman filter paper and dried for subsequent characterization. The formations of nanoparticles were checked by UV-visible spectroscopy.

11.4 Application of Fungal Nanotechnology

Fungus-negotiated fabrication of metal/oxide/chalcogenide nanoparticles comes under fungus nanotechnology. Nanoparticles involve usage of nanotechnology for the welfare of human health and well-being. The use of nanotechnology in various sectors of biomedicine has refashioned the field of medical science where nanoparticles of dimensions ranging between 1 and 100 nm are designed and used for diagnostics, as therapeutics, and as biomedical tools for research. Nanoparticles have been created with chemically flexible surfaces in order to attach a divergency of ligands which turn into biological tools like biosensors, molecular scale fluorescent tags, imaging agents, and targeted molecular delivery vehicles, and they can easily be distinguished between diseased and healthy tissue. This can provide therapy at a

Table 11.1 List of fungi used for biosynthesis of different nanoparticles

	Nanoparticles	Location	Methods	References
Yeast				
Schizosaccharomyces pombe	CdS	Intracellular	Reduction	Kowshik et al. (2002)
Candida glabrata	CdS	Intracellular	Reduction	Dameron et al. (1989)
Torulopsis sp.	PbS	Intracellular	Reduction	Kowshik et al. (2002)
Candida albicans	Ag	Extracellular	Reduction	Li et al. (2011)
MKY3	Ag	Extracellular	Reduction	Kathiresan et al. (2009)
Pichta capsulate	Ag	Extracellular	Reduction	Marchiol (2012)
Rhodosporidium diobovatum	Ag, Pb	Extracellular	Reduction	Gericke and Pinches (2006)
Yarrowia lipolytica	Ag	Extracellular	Reduction	Kathiresan et al. (2009)
Saccharomyces cerevisiae	Sb_2O_3	Intracellular		Jha et al. (2009b)
Fungus				
Rhizopus stolonifer	Au	Extracellular	Reduction	Narayanan and Sakthivel (2010b)
Aureobasidium pullulans	Au	Intracellular	Reduction	Bankura et al. (2012)
Aspergillus terreus	Ag, Au-Ag	Extracellular	Reduction	Li et al. (1999)
Thraustochytrium sp.	Ag, Zn	Extracellular	Reduction	Du et al. (2011)
Hypocrea lixii	Cu	Extracellular	Reduction	Deplanche et al. (2010)
Cylindrocladium floridanum	Au	Extracellular	Enzyme mediated	Zhang et al. (2012)
Neurospora oryzae	Ag	Intracellular/extracellular	Reduction	Saha et al. (2010)
Pediococcus pentosaceus	Ag	Extracellular	Biosorption and reduction	Shahverdi et al. (2007)
Fusarium oxysporum	CdS, PbS, ZnS, MoS	Extracellular	Enzyme mediated	Prasad and Jha (2010)
	TiO_2, $BaTiO_3$, ZrO_2	Extracellular	Reduction	Kashyap et al. (2012)
				Bansal et al. (2006)
	Ag, Au. Ni, Pt	Intracellular/extracellular	Reduction	Karbasian et al. (2008), Syed and Ahmad (2012)

Aspergillus fumigatus	Ag	Extracellular	Reduction	Bhainsa and D'Souza (2006)
Neurospora crassa	Pt	Intracellular and extracellular	Reduction	Prasad et al. (2015)
Verticillium sp.	Au	Intracellular	Reduction	Mukherjee et al. (2001a)
T. reesei	Ag	Extracellular	Reduction	Bharde et al. (2006)
Coriolus versicolor	Ag	Extracellular and intracellular	Reduction	Aziz et al. (2015)
	Ag, Au, Fe	Extracellular	Reduction	Phanjom and Ahmed (2015), Binupriya et al. (2010), Tarafdar and Raliya (2013)
Cochliobolus lunatus	Cu, Al	Extracellular	Reduction	Deplanche and Macaskie (2008)
Pestalotia sp.	Ag	Intracellular	Reduction	Balaji et al. (2009)
Phanerochaete chrysosporium	Ag	Extracellular	Reduction	Sanghi and Verma (2009)
Phoma sorghina	Ag	Extracellular	Reduction	Clemens et al. (1999)
Aspergillus clavatus	Ag	Extracellular	Reduction	Vigneshwaran et al. (2007)
Sclerotium rolfsii	Au	Extracellular	Reduction	Kalishwaralal et al. (2010)
Aspergillus clavatus	Ag, Au	Extracellular	Reduction	Verma et al. (2010, (2011)
Trichoderma viride	CdS, Ag	Extracellular	Reduction	Ahmad et al. (2002)
Aspergillus flavus	Ag-Au, Ag	Extracellular	Reduction	Chen et al. (2003)
				Jha et al. (2008a, 2009a, 2010)
Penicillium brevicompactum	Ag, Au	Extracellular	Reduction	Fayaz et al. (2009a, b)
Fusarium culmorum	Ag	Extracellular	Reduction	Bawaskar et al. (2010)
Cryphonectria sp.	Ag	Intracellular	Reduction	Bharde et al. (2006)

(continued)

Table 11.1 (continued)

	Nanoparticles	Location	Methods	References
Cochliobolus lunatus	Ag	Intracellular	Reduction	Mehra and Winge (1991)
				Mehra et al. (1988)
Rhizopus oryzae	Au	Cell surface	Reduction	Das et al. (2012)
Pediococcus pentosaceus	Pt, Ag	Extracellular	Biosorption and reduction	Shahverdi et al. (2007)
Verticillium luteoalbum	Au, Ag, Se, Te, Pt, Pd	Extracellular	Reduction	Alghuthaymi et al. (2015)
Rhizopus nigricans	Ag	Extracellular	Reduction	Mohammadian et al. (2007)
Trichoderma harzianum	Ag	Extracellular	Reduction	Gherbawy et al. (2013)
Pleurotus sajor caju	Au, Ag	Extracellular	Reduction	Khandel and Shahi (2016)
Phanerochaete chrysosporium	Ag	Extracellular	Reduction	Sanghi and Verma (2009)
Pestalotia sp.	Ag	Extracellular	Reduction	Raheman et al. (2011)
Fusarium semitectum	Au, Au-Ag	Extracellular	Reduction	Sawle et al. (2008)
Coriolus versicolor	Ag, Au-Ag	Extracellular	Enzyme mediated	Khandel and Shahi (2016)
Bipolaris nodulosa	Au, Ag	Intracellular	Reduction	Uddin et al. (2008)
Cladosporium cladosporioides	Ag	Extracellular	Reduction	Balaji et al. (2009)
Helminthosporium solani	Pt, Zn, Cu	Extracellular	Reduction	Chen et al. (2003)
Aspergillus ochraceus	Ag	Extracellular	Reduction	Khandel and Shahi (2016)
Tricholoma crassum	Au	Extracellular	Reduction	Sawle et al. (2008)
Alternaria alternata	Ag, Cd	Extracellular	Reduction	Gajbhiye et al. (2009)
Trichoderma asperellum	Ag	Extracellular	Reduction	Mukherjee et al. (2008)
Phoma glomerata	Pb, Ag	Extracellular	Reduction	Birla et al. (2009)
Phoma infestans	Ag	Extracellular	Reduction	Birla et al. (2009)
Agaricus bisporus	Ag	Extracellular	Reduction	Ma et al. (2008)

Penicillium brevicompactum WA2315	Ag	Extracellular	Reduction	Shaligram et al. (2009)
Fusarium solani	Ag	Extracellular	Reduction	Ingle et al. (2009)
Penicillium fellutanum	Ag	Extracellular	Reduction	Kathiresan et al. (2009)
Colletotrichum sp.	Ag	Extracellular	Reduction	Philip et al. (2009b)
Yarrowiali lipolytica	Au	Extracellular	Reduction	Agnihotri et al. (2009)
Volvariella volvacea	Ag	Extracellular	Reduction	Mittal et al. (2013)
Penicillium citrinum	Ag	Extracellular	Reduction	Shaligram et al. (2009)
Mucor hiemalis	Ag	Extracellular	Reduction	Aziz et al. (2016)

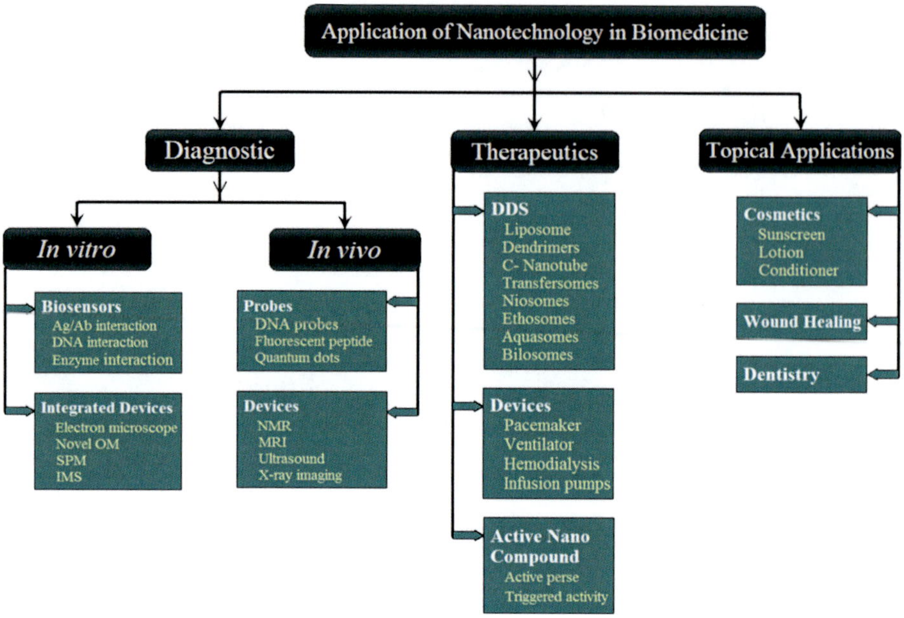

Fig. 11.4 Application of nanotechnology in the field of medical sciences

molecular level with the help of tools and technique, thus treating the disease and assisting in the study of the pathogenesis. Diagnostic methods with greater degree of sensitivity aid in early detection of the disease and provide better prognosis. Designing of drugs with greater degree of cell specificity improves efficacy and minimizes adverse effects (Barkalina et al. 2014; Surendiran et al. 2009; Nikalje 2015; Cormode et al. 2009; Ninganagouda et al. 2014; Rai et al. 2013; Fakruddin et al. 2012; Daniel and Astruc 2004; Du et al. 2007; Fayaz et al. 2009a, b; Fortin and Beveridge 2000; Krolikowska et al. 2003; Mokhtari et al. 2009; Valtchev and Tosheva 2013; Flores et al. 2013; Ghodake et al. 2013; Krolikowska et al. 2003; Liu et al. 2013; Prabhu and Poulose 2012; Ray et al. 2011; Tripathi et al. 2007; Zhang et al. 2012). Moreover, the biological procedures also occur at the nanoscale, and by dint of their amenability to biological functionalization, the nanoparticles are finding important applications in the field of medicine (Fig. 11.4).

11.4.1 Antimicrobial Nanotechnology

Nanoparticle is a new field of science and technology and interacts with biological molecules at nanoscale; it broadens the field of inquest and development. Interactions of nanoparticles with biomolecules can be understood both in the intracellular and extracellular medium. Nanoparticles are gaining greater importance by a dint of

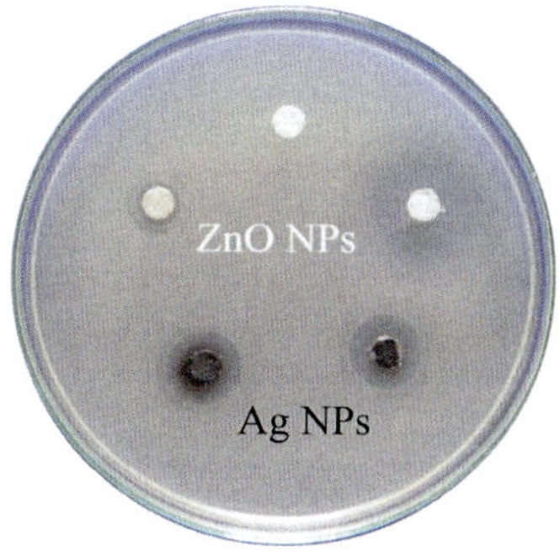

Fig. 11.5 Antimicrobial activity of silver and zinc oxide nanoparticles on *S. aureus*

their larger surface area and high antimicrobial activity at the nanoscale which enables them to have a better contact with the microorganisms as compared to their massive counterparts. The nanosize changes the physiochemical properties of the metals; hence, different nanoparticles are being used as the potent antimicrobial agents. It is an effective and fast-acting antimicrobial agent against a broad spectrum of common bacteria and fungi that have been utilized in various processes in the field of medical science. One of those is silver nanoparticle (Ag NPs) which showed effective antimicrobial (antibacterial and antifungal) activity (Aziz et al. 2015, 2016). It is found to exhibit a very strong antibacterial activity against both Gram-positive and Gram-negative bacteria which interact with building blocks of the bacterial membrane and damage the cells. The antibacterial activity of Ag NPs was studied against *S. aureus* using agar well diffusion method. After the incubation time, zones of inhibition (clear zones) were observed against the test organism with Ag NPs as well as ZnO NPs (Fig. 11.5).

It is generally believed that the high closeness of silver toward sulfur and phosphorus is the crux element of the antimicrobial effect. This study suggests that the mode of action of Ag NPs is that it attached to the sulfur-containing proteins on the bacterial cell membrane, and it can also merge with phosphorus moieties in DNA resulting in the inactivation of DNA replication, leading to increased permeability of the membrane and inhibition of enzyme functions and finally causing cell death (Prasad and Swamy 2013). It shows that Ag-ions released by active surfaces of Ag NPs induce generation of intracellular reactive oxygen species (ROS) in bacterial cells and uncouple respiratory electron transport from oxidative phosphorylation to protons and phosphate. The antimicrobial activities of these nanoparticles depend on chemisorbed Ag^+, formed on the surface of Ag NPs by dint of extreme sensitivity to oxygen. Silver nanoparticles have been studied to prepare different composites to be used as disinfecting filters, coating materials for several devices and dependable mean for antibiotic delivery which were explained the basis of inhibitory effect of Ag NPs on a variety of bacteria. These have also been used to treat burns, wounds, and infections (Singh et al. 2008; Sinha et al. 2009; El-Rafie et al. 2010; Fatima et al. 2015; Verma et al. 2010; Durán et al. 2007; Aziz et al. 2015; Birla et al. 2009; Bankura et al. 2012; Flores et al. 2013; Fortin and Beveridge 2000; Ghodake et al. 2013; Krolikowska et al. 2003; Liu et al. 2013; Prabhu and Poulose 2012; Ray et al. 2011; Tripathi et al. 2007; Zhang et al. 2012; Barkalina et al. 2014; Surendiran et al. 2009;

Nikalje 2015; Cormode et al. 2009; Ninganagouda et al. 2014; Rai et al. 2013; Fakruddin et al. 2012).

11.4.2 Cosmetics Nanotechnology

Cosmetics are products used to augment the beauty, cleanness, appearance or odor, and attractive features of the human body. Cosmetic products contain biologically agile additives that claim to have medicinal or drug-like benefits, which are manufactured under pharmaceutical called cosmetic pharmaceuticals or cosmeceuticals. It is topically applied and influences the biological function of the skin. The origin of cosmetics was attributed to Egyptians, circa 4000 BC. The latest technology advancement is incorporated into innovative formulation of highly differentiated multifunctional products which focused on treatment and aesthetics that contains active ingredients. This technology emphasizes the uses of delivery systems in cosmetics which analyze new approaches for obtaining sophisticated cosmetic products and assay the most common methods for enhancing the skin penetration properties. The main end in view of topical therapy is to modulate the limiting function of the skin and to administer an active ingredient to skin layers or compartment that minimizes systemic involvement. The latest technology offers smart approaches for novel cosmetic delivery systems by coupling the drug to a shipper particle such as nanoparticles, liposome, niosomes, microemulsions, microspheres, etc. which modulate the release and absorption characteristics of the drug. The goal of nanobiotechnology is the development of nanoscale's biologically active components and analytical instruments for the investigation of cells at molecular levels (Arora et al. 2012; Corley et al. 2009; Morganti 2010; Raj et al. 2012; Fortin and Beveridge 2000; Kawadkar et al. 2011; Krolikowska et al. 2003). Nanotechnology is the engineering of materials on the nanoscale which have many shapes and sizes, and they can be soft or hard, soluble, or insoluble for different applications. Some of these nanoparticles used in an expanding number of commercial products include metals (Ag, Au), metal oxides (TiO_2, ZnO, Fe_2O_3, SiO_2), lipophilic nanoparticles, carbon nanotubes, fullerenes, and quantum dots (QDs). Lipophilic nanoparticles have been designed for transcutaneous drug delivery in the form of liposomes and solid lipid nanoparticles. Nanoemulsions and liposomal nanoparticles do not muddle with the integrity of the skin lipid bilayers and are not washed out while cleansing the skin; these formulations are believed to have a great future in the cosmetic products. TiO_2 and ZnO are widely used in cosmeceuticals such as sunscreen, lotion, and conditioner which reduce UV radiation that caused skin cancer because nanoparticles have large surface area and reflect/scatter UVA and UVB rays. Nanoparticles must be able to penetrate the skin barrier, deliver their contents, and clear from the body without adverse side effects. Nanoparticle skin penetration is affected by anatomical locations, epidermal thickness, and hair follicle density of species. Nanoparticle accumulation in hair follicles and stratum corneum penetration through barrier-flawed skin are common trends which occur in many species. QDs are commonly used in cosmetic product as hair removal agents which are detected in ex vivo human skin, and they also penetrate the mouse skin treated with ultraviolet B radiation. The effect of UVB radiation enhances nanoparticle (TiO_2 and ZnO) stratum corneum penetration that was approved in vivo study of nanoparticles applied to pigs in sunscreen. The nanoparticles must be able to breach the stratum corneum barrier and enter cells through receptor-mediated processes. Therefore, many techniques including tape stripping, gene gun, electroporation, microneedles, and ultrasound have been developed to muddle the stratum corneum to help in nanoparticle delivery (Lindemann et al. 2003; Polat et al. 2011; Kim et al. 2012). The therapeutic applications in the skin have focused in three main areas such as (1) antimicrobials and wound healing, (2) skin cancer imaging and targeted therapeutics, and (3) immunomodulation and vaccine delivery. The penetration of ZnO and TiO_2 nanoparticles constitutes minimal health concern. ZnO is soluble in acidic environments, so the stratum corneum of the skin induces dissociation and penetration of ionic Zn which is an essential mineral and therefore does not poses toxicity effects. TiO_2 nanoparticles are highly insoluble and susceptible to agglomeration that hinders their penetration.

11.4.3 Diagnostic Nanotechnology

Current methods for diagnosing most diseases are dependent upon the apparent symptoms before medical professionals can see the patient suffers from a specific illness, but the treatment has a decreased chance of being effective. Therefore, earlier a disease is diagnosed; better the chance for a cure. Biomedical laboratory diagnosis plays a key role in today's healthcare. Nanotechnology is spreading the options commonly available, which will culminate in greater subtlety and far better efficacy and economy. Nanobiotechnology is that branch of nanotechnology which deals with biological and biochemical applications/uses. It extends the limits of diagnostics to the nanoscale and generates nanodevices for diagnosing genetic diseases to cancer such as screening tests for genetic abnormalities; mutation detection for cystic fibrosis, neurodegenerative, and cardiovascular disease; and diagnostic imaging during pregnancy and infection. The diagnostic applications of nanotechnology show great promise to meet the accurate demands of the clinical laboratory for sensitivity, cost-effectiveness, and multiplexing capabilities. Nanoparticles can be used for both quantitative and qualitative in vitro and in vivo detection of disease cells. The various applications of nanomaterials in diagnosis have led to generation of nanodiagnostic tools/devices. Nanoparticles have emerged the characteristic properties that they possess at the nanoscale with the feasible immobilization of targeted ligands (proteins, peptides, amino acids) on the surface for efficient diagnostics. Therefore, they become ideal candidates for sensitive detection and highly efficient contrast agents for imaging, and it helps to create biomarkers with well-distinguished colors. One of the first applications of nanoparticles will be improved fluorescent markers for diagnostic and screening purposes. Conventional fluorescent markers require a complex color. Gold nanoparticles, quantum dots, and magnetic nanoparticles are fluorescent materials used to detect the early stage of cancer and other diseases. These nanoparticles are attached to biomarkers, which may vary in their pattern of adherence and research attempts are ongoing to decipher the state/status of biomarkers for different types of cancers. The use of nanosystems for reading the state of biomarkers becomes important because of the high surface area offered by the nanoparticle surface. Some nanotechnology-based tests are being developed which appear to be extremely simple, specific, and fast in detection of viruses, *E.coli*, DNA proteins, antibodies, etc. Nanotechnology could significantly improve diagnostic capabilities. Nanoparticles will increase the efficiency and accuracy of diagnosis from samples of body fluids. Illinois (USA)-based company Northbrook has developed a technique that allows doctors to optically detect the genetic compositions of biological specimens by individual target probes using spherical gold nanoparticles. When target molecules labelled with quantum dots (QDs) show fluorescence that will facilitate direct investigation of intracellular signalling complex by optical techniques, i.e., confocal fluorescence microscopy. (Kawadkar et al. 2011; Maojo et al. 2010; Cormode et al. 2009; Fortin and Beveridge 2000; Kawadkar et al. 2011; Krolikowska et al. 2003; Barkalina et al. 2014; Surendiran et al. 2009; Nikalje 2015; Ninganagouda et al. 2014; Rai et al. 2013; Fakruddin et al. 2012; Daniel and Astruc 2004; Du et al. 2007; Fayaz et al. 2009a, b; Krolikowska et al. 2003; Mokhtari et al. 2009; Valtchev and Tosheva 2013).

11.4.4 Therapeutics Nanotechnology

Nanobiotechnology has been applied in almost every area of medical sciences for human healthcare such as stem cell-based therapies to important therapeutic areas like infections, cardiovascular diseases, neurological disorders, and cancer. A great transformation is taking place in biotechnology and medical field due to nanotechnology. Initial test of various drug delivery systems, cancer, or tumor therapies or detection has been successful by using nanotechnology. Nanoparticles are very small, easy to inject, and targeted toward a specific portion in a body. In recent years, gold nanorods which have strong scattering and absorption property in the infrared are used successfully to detect and destroy cancer cells without affecting the healthy cells. It is quite important as it would avoid killing of healthy cells unlike in chemotherapy. Nanotechnology can aptly provide new formulations of drugs with less side effects and amenable routes for drug delivery. Nanoparticles are found in the form of drug carriers because of large surface area of materials and small size due to which they are easily transported into cells and nuclei and specificity to the target can be achieved as desired. This is achieved by different ways like nanospheres, nanocapsules, nanopores, liposome, micelles, vesicles, dendrimers, etc. Drugs can be encapsulated in nanocapsules and addressed toward craved parts of a body. Drug can then be fast or slowly delivered, as desired, by opening the capsule using some external stimulus like magnetic field or infrared light or under some physiological conditions. The action of the nanoparticle drug system can be seen in the forms, i.e., of skin administration, oral administration, nasal administration, and ocular administration. Development of new drug delivery systems based on this method is being tried for conditions like cancer, diabetes, bacterial infection, fungal infections, and viral infections and in gene therapy. The main advantages of this method of treatment are targeting of the drug and enhanced safety (Vaidyanathan et al. 2009; Dean et al. 1997; Fortin and Beveridge 2000; Kawadkar et al. 2011; Krolikowska et al. 2003; Singh et al. 2008; Sinha et al. 2009; El-Rafie et al. 2010; Fatima et al. 2015; Verma et al. 2010; Durán et al. 2007; Aziz et al. 2015; Birla et al. 2009; Bankura et al. 2012; Flores et al. 2013; Ghodake et al. 2013).

11.4.5 Liposomes

The synergy between the brain and blood cells between fungal pathogens and infection sites confides on complex reaction between cells and surface characteristics. Nanofabrication unscrambles the complexity of these interactions by modifying surface characteristics with nanoscale resolutions, which can lead to hybrid biological systems. This hybrid material can be used to screen drugs, as sensors, or as medical devices and implants. An Irish drug company, Elan, developed a polymer coating capable of changing the surface of drugs that have poor water solubility. Liposomes are original models of nanoscaled drug delivery devices and being composed of a lipid bilayer. They can be used as targeted drug delivery systems, and these can be loaded with drugs either in the aqueous compartment or in the lipid membrane shown in Fig. 11.6. Water-soluble drugs are loaded in aqueous compartment, and lipid-soluble drugs are incorporated in the liposomal membrane. It can be targeted to specific cell/tissue/organ by passive as well as active methods. Active targeted drugs are immunoliposomes and ligand-directed liposomes which are conjugated with an antibody directed toward the tumor antigen. For example, these liposomes are being tried with epirubicin, doxorubicin, and hamycin drugs for treatment of diseases. The targeted liposomal preparations of drugs are found to have a better efficacy than nontargeted liposomal drugs (Du et al. 2011; Fortin and Beveridge 2000; Kawadkar et al. 2011; Krolikowska et al. 2003; Kumar et al. 2008).

Nanobiotechnology ensures treatment of ailments which cannot be accomplished through normal pharmaceuticals. By convention, the traditional pharmaceutical industry concentrates mainly on universally confirmed around 500 ailments only, while this technology ensures manipulation of targets even on solid substrates through tethering and thereby controlling the chemical reactions in a much quicker process involving few materials. This leap of development will not only drastically reduce the cost of novel discovery of potential drugs rather would also ensure a plethora of choices among population of highly specific drugs.

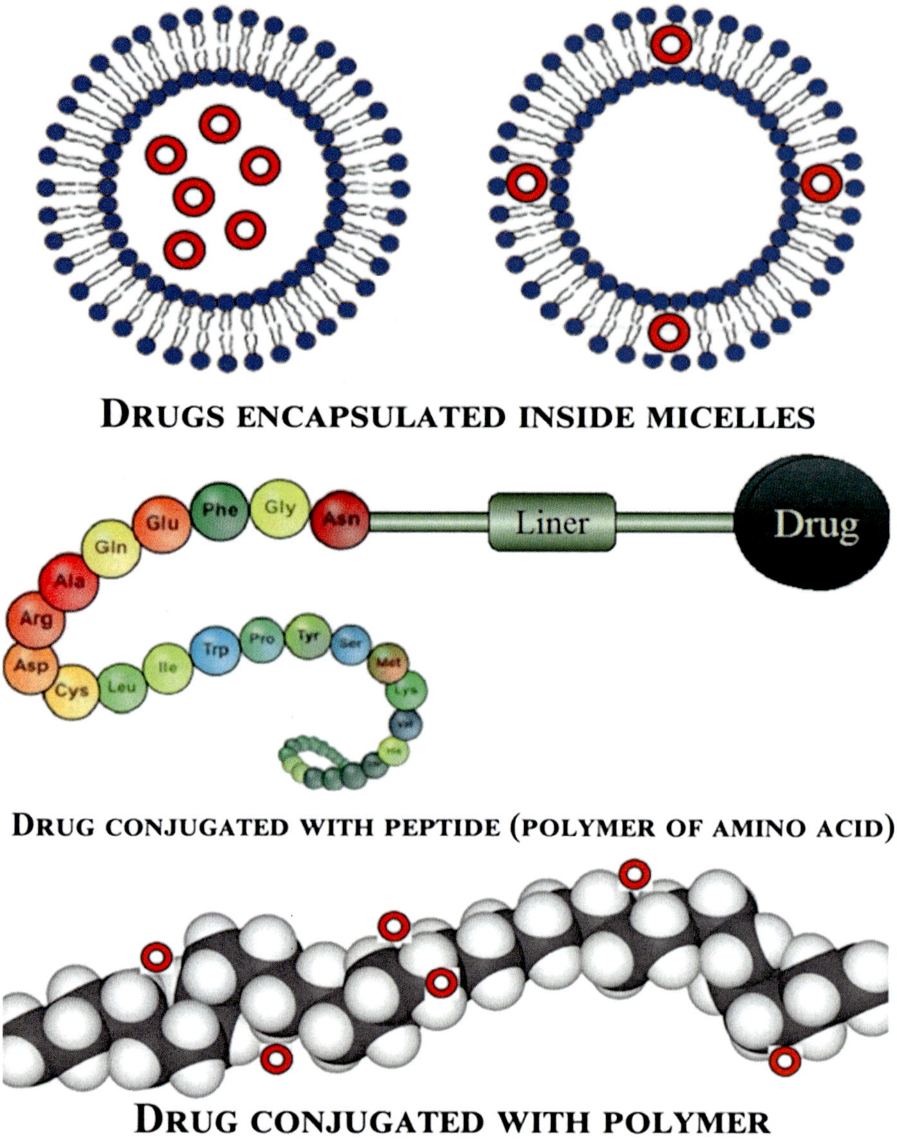

Fig. 11.6 Targeted drugs for drug delivery system

11.5 Conclusion

With changing pace of time, the demand for nanomaterials in diagnostics, prognostics, therapeutics, and cosmetics is bound to rise, and this can very well be met by going for green options like benign fungi-based synthesis. These naturally bestowed decomposers are indeed wonderfully amenable time-surviving nanofabrication tools for fabricating regular as well as superspecialized nanomaterials for drug delivery and diagnostics with the scope for scaling up.

References

Agnihotri M, Joshi S, Kumar AR, Zinjarde S, Kulkarni S (2009) Biosynthesis of gold nanoparticles by the tropical marine yeast *Yarrowia lipolytica* NCIM 3589. Mater Lett 63:1231–1234

Ahmad A, Mukherjee P, Mandal D, Senapati S, Khan MI, Kumar R, Sastry M (2002) Enzyme mediated extracellular synthesis of CdS nanoparticles by the fungus, *Fusarium oxysporum*. J Am Chem Soc 124:12108–12109

Ahmad A, Mukherjee P, Senapati S, Mandal D, Khan MI, Kumar R, Sastry M (2003) Extracellular biosynthesis of silver nanoparticles using the fungus *Fusarium oxysporum*. Colloids Surf B: Biointerfaces 28:313–318

Alghuthaymi MA, Almoammar H, Rai M, Said-Galiev E, Abd-Elsalam KA (2015) Myconanoparticles: synthesis and their role in phytopathogens management. Biotechnol Biotechnological Equip 29:221–236

Arora N, Agarwal S, Murthy RSR (2012) Latest technology advances in cosmeceuticals. Int J Pharm Sci Drug Res 4:168–182

Avery SV (2001) Metal toxicity in yeasts and the role of oxidative stress. Adv Appl Microbiol 49:111–142

Aziz N, Faraz M, Pandey R, Sakir M, Fatma T, Varma A, Barman I, Prasad R (2015) Facile algae derived route to biogenic silver nanoparticles: synthesis, antibacterial and photocatalytic properties. Langmuir 31:11605–11612

Aziz N, Pandey R, Barman I, Prasad R (2016) Leveraging the attributes of *Mucor hiemalis*-derived silver nanoparticles for a synergistic broad-spectrum antimicrobial platform. Front Microbiol 7:1984. https://doi.org/10.3389/fmicb.2016.01984

Balaji DS, Basavaraja S, Deshpande R, Mahesh DB, Prabhakar BK, Venkataraman A (2009) Extracellular biosynthesis of functionalized silver nanoparticles by strains of *Cladosporium cladosporioides* fungus. Colloid Surf B: Biointerfaces 68:88–92

Bankura P, Maity D, Mollick MM, Mondal D, Bhowmick B, Bain MK, Chakraborty A, Sarkar J, Acharya K, Chattopadhyay D (2012) Synthesis, characterization and antimicrobial activity of dextran stabilized silver nanoparticles in aqueous medium. Carbohydr Polym 1:1159–1165

Bansal V, Rautaray D, Bharde A, Ahire K, Sanyal A, Ahmad A, Sastry M (2005) Fungus mediated biosynthesis of silica and titania particles. J Mater Chem 15:2583–2589

Bansal V, Poddar P, Ahmad A, Sastry M (2006) Room temperature biosynthesis of ferroelectric barium titanate nanoparticles. J Am Chem Soc 128:11958–11963

Barkalina N, Charalambous C, Jones C, Coward K (2014) Nanotechnology in reproductive medicine: emerging applications of nanomaterials. Nanomedicine: NBM XX:1–18

Basavaraja S, Balaji SD, Lagashetty A, Rajasab AH, Venkataraman A (2008) Extracellular biosynthesis of silver nanoparticles using the fungus *Fusarium semitectum*. Mater Res Bull 43:1164–1170

Bawaskar M, Gaikwad S, Ingle A, Rathod D, Gade A, Duran N, Marcato P, Rai M (2010) A new report on mycosynthesis of silver nanoparticles by *Fusarium culmorum*. Curr Nanosci 1:376–380

Beveridge TJ, Murray RGE (1980) Sites of metal deposition in the cell wall of *Bacillus subtilis*. J Bacteriol 141:876–887

Bhainsa KC, D'Souza SF (2006) Extracellular biosynthesis of silver nanoparticles using the fungus *Aspergillus fumigatus*. Colloid Surf B: Biointerfaces 47:160–164

Bhambure R, Bule M, Shaligram NS, Kamat M, Singhal R (2009) Extracellular biosynthesis of gold nanoparticles using *Aspergillus niger*–its characterization and stability. Chem Eng Technol 32:1036–1041

Bharde A, Rautray D, Bansal V, Ahmad A, Sarkar I, Yusuf SM, Sanyal M, Sastry M (2006) Extracellular biosynthesis of magnetite using fungi. Small 2:135–114

Binupriya AR, Sathishkumar M, Vijayaraghavan K, Yun S-I (2010) Bioreduction of trivalent aurum to nano-crystalline gold particles by active and inactive cells and cell-free extract of Aspergillus oryzae var. viridis. J Hazard Mater 177:539–545

Birla SS, Tiwari VV, Gade AK, Ingle AP, Yadav AP, Rai MK (2009) Fabrication of silver nanoparticles by *Phoma glomerata* and its combined effect against *Escherichia coli, Pseudomonas aeruginosa* and *Staphylococcus aureus*. Lett Appl Microbiol 1:2173–2179

Breierová E, Vajczikova I, Sasinkova V, Stratilova E, Fisera M, Gregor T, Sajbidor J (2002) Biosorption of cadmium ions by different yeast species. Z Naturforsch C 57:634–639

Breierová E, Gregor T, Juršíková P, Stratilová E, Fišera M (2004) The role of pullulan and pectin in the uptake of Cd and Ni ions by *Aureobasidium pullulans*. Ann Microbiol 54:247–255

Butt TR, Ecker DJ (1987) Yeast metallothionein and applications in biotechnology. Microbiol Rev 51:351–364

Čertik M, Breierová E, Juršíková P (2005) Effect of cadmium on lipid composition of *Aureobasidium pullulans* grown with added extracellular polysaccharides. Int Biodeterior Biodegrad 55:195–202

Chae HZ, Chung SJ, Rhee SG (1994) Thioredoxin dependent peroxide reductase from yeast. J Biol Chem 269:27670–27678

Chandran SP, Chaudhary M, Pasricha R, Ahmad A, Sastry M (2006) Synthesis of gold nanotriangles and silver nanoparticles using *Aloe vera* plant extract. Biotechnol Prog 22:577–583

Chen JC, Lin ZH, Ma XX (2003) Evidence of the production of silver nanoparticles via pretreatment of *Phoma sp.*3.2883 with silver nitrate. Lett Appl Microbiol 37:105–108

Choi JH, Lou W, Vancura A (1998) A novel membrane bound glutathione S transferase functions in the stationary phase of the yeast *Saccharomyces cerevisiae*. J Biol Chem 273:29915–29922

Clemens S, Kim EJ, Neumann D, Schroeder JI (1999) Tolrance to toxic metals by a gene family of phytochelatin synthase from plants and yeast. EMBO J 18:3325–3333

Corley EA, Scheufele DA, Hu Q (2009) Of risks and regulations: how leading U.S. nanoscientists form policy stances about nanotechnology. J Nanopart Res 11:1573–1585

Cormode DP, Skajaa T, Fayad ZA, Mulder WJM (2009) Nanotechnology in medical imaging: probe design and applications. Arterioscler Thromb Vasc Biol 29:992–1000

Cunningham DP, Lundie LL (1993) Precipitation of cadmium by *Clostridium thermoaceticum*. Appl Environ Microbiol 59:7–14

Dameron CT, Winge DR (1990) Peptide mediated formation of quantum semiconductors. Trends Biotechnol 8:3–6

Dameron CT, Reese RN, Mehra RK, Kortan AR, Carroll PJ, Steigerwald ML, Brus LE, Winge DR (1989) Biosynthesis of cadmium sulphide quantum semiconductor crystallites. Nature 338:596–597

Daniel MC, Astruc D (2004) Gold nanoparticles: assembly, supramolecular chemistry, quantum size related properties, and applications toward biology, catalysis, and nanotechnology. Chem Rev 104:293–346

Das SK, Dickinson C, Lafir F, Brougham DF, Marsili E (2012) Synthesis, characterization and catalytic activity of gold nanoparticles biosynthesized with Rhizopus oryzae protein extract. Green Chem 14:1322–1334

Dean RT, Fu S, Stocker R, Davies MJ (1997) Biochemistry and pathology of radical mediated protein oxidation. Biochem J 324:1–18

Deplanche K, Macaskie LE (2008) Biorecovery of gold by *Escherichia coli* and *Desulfovibrio desulfuricans*. Biotechnol Bioeng 99:1055–1064

Deplanche K, Woods RD, Mikheenko IP, Sockett RE, Macaskie LE (2008) Manufacture of stable palladium and gold nanoparticles on native and genetically engineered flagella scaffolds. Biotechnol Bioeng 101:873–880

Deplanche K, Caldelari I, Mikheenko IP, Sargent F, Macaskie LE (2010) Involvement of hydrogenases in the formation of highly catalytic Pd(0) nanoparticles by bioreduction of Pd(II) using *Escherichia coli* mutant strains. Microbiology 1:2630–2640

Du L, Jiang H, Liu X, Wang E (2007) Biosynthesis of gold nanoparticles assisted by *Escherichia coli* DH5α and its application on direct electrochemistry of haemoglobin. Electrochem Commun 9:1165–1170

Du Z, Du XJ, Mao CQ, Wang J (2011) Tailor-made dual pH-sensitive polymer–doxorubicin nanoparticles for efficient anticancer drug delivery. J Am Chem Soc 17:17560–17563

Durán N, Marcato PD, Alves OL, De Souza GIH, Esposito E (2005) Mechanistic aspects of biosynthesis of silver nanoparticles by several *Fusarium oxysporum* strains. J Nanobiotechnol 3:8. https://doi.org/10.1186/1477315538

Durán N, Marcato PD, De Souza GIH, Alves OL, Esposito E (2007) Antibacterial effect of silver nanoparticles produced by fungal process on textile fabrics and their effluent treatment. J Biomed Nanotechnol 3:203–208

El-Rafie MH, Mohamed AA, Shaheen TI, Hebeish A (2010) Antimicrobial effect of silver nanoparticles produced by fungal process on cotton fabrics. Carbohydr Polym 80:779–782

Fakruddin M, Hossain Z, Afroz H (2012) Prospects and applications of nanobiotechnology: a medical perspective. J Nanobiotechnol 10:1–8

Fatima F, Bajpai P, Pathak N, Singh S, Priya S, Verma SR (2015) Antimicrobial and immunomodulatory efficacy of extracellularly synthesized silver and gold nanoparticles by a novel phosphate solubilizing fungus *Bipolaris tetramera*. BMC Microbiol 15:52–61

Fayaz AM, Balaji K, Kalaichelvan KT, Venkatesan R (2009a) Fungal based synthesis of silver nanoparticles-an effect of temperature on the size of particles. Colloids Surf B: Biointerfaces 74:123–126

Fayaz AM, Balaji K, Girilal M, Kalaichelvan PT, Venkatesan R (2009b) Mycobased synthesis of silver nanoparticles and their incorporation into sodium alginate films for vegetable and fruit preservation. J Agric Food Chem 57:6246–6252

Flores CY, Miñán AG, Grillo CA, Salvarezza RC, Vericat C, Schilardi PL (2013) Citrate capped silver nanoparticles showing good bactericidal effect against both planktonic and sessile bacteria and a low cytotoxicity to osteoblastic cells. ACS Appl Mater Interfaces 5:3149–3159

Fortin D, Beveridge TJ (2000) From biology to biotechnology and medical applications. In: Aeuerien E (ed) Biomineralization. Wiley VCH, Weinheim

Gadda G, Fitzpatrick PF (1998) Biochemical and physical characterization of the active FAD containing form of nitroalkane oxidase from *Fusarium oxysporum*. Biochemistry 37:6154–6164

Gajbhiye B, Kesharwani JG, Ingle AP, Gade AK, Rai MK (2009) Fungus mediated synthesis of silver nanoparticles and its activity against pathogenic fungi in combination of fluconazole. Nanomedicine: NBM 5:282–286

Gan Z (1991) Yeast thioredoxin genes. J Biol Chem 266:1692–1696

Gericke M, Pinches A (2006) Biological synthesis of metal nanoparticles. Hydrometallurgy 83:132–140

Gharieb MM, Kierans M, Gadd GM (1999) Transformation and tolerance of tellurite by filamentous fungi: accumulation, reduction, and volatilization. Mycol Res 103:299–305

Gherbawy YA, Shalaby IM, El-sadek MSA, Elhariry HM, Banaja AA (2013) The anti-fasciolasis properties of silver nanoparticles produced by Trichoderma harzianum and their improvement of the anti-fasciolasis drug triclabendazole. Int J Mol Sci 14:21887–21898

Ghodake G, Lim SR, Lee DS (2013) Casein hydrolytic peptides mediated green synthesis of antibacterial silver nanoparticles. Colloids Surf B: Biointerfaces 108:147–151

González Chávez M, Carrillo Gonzalez R, Wright SF, Nichols KA (2004) The role of glomalin, a protein produced by arbuscular mycorrhizal fungi, in sequestering potentially toxic elements. Environ Pollut 130:317–323

González GM, Azcón AC, Mooney M, Valderas A, MacDiarmid CW, Eide DJ, Ferrol N (2005) Characterization of a *Glomus intraradices* gene encoding a putative Zn transporter of the cation diffusion facilitator family. Fungal Genet Biol 42:130–140

Halliwell B (2006) Reactive species and antioxidants-redox biology is a fundamental theme of aerobic life. Plant Physiol 141:312–322

Haq M, Rathod V, Singh D, Singh AK, Ninganagouda S, Hiremath J (2015) Dried mushroom *Agaricus bisporus* mediated synthesis of silver nanoparticles from Bandipora District (Jammu and Kashmir) and their efficacy against methicillin resistant *Staphylococcus aureus* (MRSA) strains. Nanosci Nanotechnol Int J 5:1–8

Huang J, Li Q, Sun D, Lu Y, Su Y, Yang X, Wang H, Wang Y, Shao W, He N, Hong J, Chen C (2007) Biosynthesis of silver and gold nanoparticles by novel sundried *Cinnamomum camphora* leaf. Nanotechnology 18:105104–105115

Ingle A, Rai M, Gade A, Bawaskar M (2009) Fusarium solani: a novel biological agent for the extracellular synthesis of silver nanoparticles. J Nanopart Res 11:2079–2085

Jarosz Wilkołazka A, Graz M, Braha B, Menge S, Schlosser D, Krauss GJ (2006) Species specific Cd stress response in the white rot basidiomycetes *Abortiporus biennis* and *Cerrena unicolor*. Biometals 19:39–49

Jha AK, Prasad K (2010a) Biosynthesis of metal and oxide nanoparticles using *Lactobacilli* from yoghurt and probiotic spore tablets. Biotechnol J 5:285–291

Jha AK, Prasad K (2010b) Synthesis of $BaTiO_3$ nanoparticles: a new sustainable green approach. Integr Ferroelectr 117:49–54

Jha AK, Prasad K (2010c) Understanding biosynthesis of metallic/oxide nanoparticles: a biochemical perspective. In: Kumar SA, Thiagarajan S, Wang SF (eds) Biocompatible nanomaterials synthesis, characterization and applications. NOVA Sci. Publication, USA

Jha AK, Prasad K, Kulkarni AR (2008a) Synthesis of nickel nanoparticles: bioreduction method. Nanosci Nanotechnol Ind J 2:26–29

Jha AK, Prasad K, Kulkarni AR (2008b) Yeast mediated synthesis of silver nanoparticles. Int J Nanosci Nanotechnol 4:17–21

Jha AK, Prasad K, Kulkarni AR (2009a) Synthesis of TiO_2 nanoparticles using microorganisms. Colloids Surf B: Biointerfaces 71:226–229

Jha AK, Prasad K, Prasad K (2009b) A green low cost biosynthesis of Sb_2O_3 nanoparticles. Biochem Eng J 43:303–306

Jha AK, Prasad K, Prasad K (2009c) Biosynthesis of Sb_2O_3 nanoparticles: a low cost green approach. Biotechnol J 4:1582–1585

Jha AK, Prasad K, Kulkarni AR (2010) Synthesis of Gd_2O_3 nanoparticles using *Lactobacillus* sp.: a novel green approach. Int J Green Nanotechnol: Phys Chem 2:P31–P38

Jia L, Zhang Q, Li Q, Song H (2009) The biosynthesis of palladium nanoparticles by antioxidants in *Gardenia jasminoides* Ellis: long lifetime nanocatalysts for nitrotoluene hydrogenation. Nanotechnology 20:385601

Kalishwaralal K, Deepak V, Ramkumarpandian S, Nellaiah H, Sangiliyandi G (2008) Extracellular biosynthesis of silver nanoparticles by the culture supernatant of *Bacillus licheniformis*. Mater Lett 62:4411–4413

Kalishwaralal K, Deepak V, Ramkumarpandian S, Nellaiah H, Sangiliyandi G (2010) Biosynthesis of silver and gold nanoparticles using *Brevibacterium casei*. Colloids Surf B: Biointerfaces 77:257–262

Karbasian M, Atyabi SM, Siadat SD, Momen SB, Norouzian D (2008) Optimizing nanosilver formation by *Fusarium oxysporum* PTCC 5115 employing response surface methodology. Am J Agric Biol Sci 3:433–437

Kashyap PL, Kumar S, Srivastava SL, Sharma AK (2012) Myconanotechnology in agriculture: a perspective. World J Microbiol Biotechnol 29:191–207

Kathiresan K, Manivannan S, Nabeel MA, Dhivya B (2009) Studies on silver nanoparticles synthesized by a marine fungus, *Penicillium fellutanum* isolated from coastal mangrove sediment. Colloids Surf B: Biointerfaces 71:133–137

Kawadkar J, Chauhan MK, Maharana M (2011) Nanobiotechnology: application of nanotechnology in diagnosis, drug discovery and drug development. Asian J Pharm Clin Res 4:23–29

Khandel P, Shahi SK (2016) Microbes mediated synthesis of metal nanoparticles: current status and future prospects. Int J Nanomater Biostruct 6:1–24

Kim YC, Jarrahian C, Zehrung D (2012) Delivery systems for intradermal vaccination. Curr Top Microbiol Immunol 351:77–112

Kowshik M, Deshmukh N, Vogel W, Urban J, Kulkarni SK, Paknikar KM (2002) Microbial synthesis of semiconductor CdS nanoparticles, their characterization, and their use in the fabrication of an ideal diode. Biotechnol Bioeng 78:583–588

Kowshik M, Ashtaputre S, Kharrazi S, Vogel W, Urban J, Kulkarni SK, Paknikar KM (2003) Extracellular synthesis of silver nanoparticles by a silver tolerant yeast strain MKY3. Nanotechnology 14:95

Krolikowska A, Kudelski A, Michota A, Bukowska J (2003) SERS studies on the structure of thioglycolic acid monolayers on silver and gold. Surf Sci 532:227–232

Kumar SA, Peter YA, Nadeau JL (2008) Facile biosynthesis, separation and conjugation of gold nanoparticles to doxorubicin. Nanotechnology 19:495101

Labrenz M, Druschel GK, ThomsenEbert T, Gilbert B, Welch SA, Kemner KM, Logan GA, Summons RE, Stasio GD, Bond PL, Lai B, Kelly SD, Banfield JF (2000) Formation of sphalerite (ZnS) deposits in natural biofilms of sulphate reducing bacteria. Science 290:1744–1747

Li Y, Duan X, Qian Y, Liao H (1999) Nanocrystalline silver particles: synthesis, agglomeration, and sputtering induced by electron beam. J Colloid Interf Sci 209:347–349

Li X, Xu H, Chen Z-S, Chen G (2011) Biosynthesis of nanoparticles by microorganisms and their applications. J Nanomater 2011:1–16

Limon Pacheco J, Gonsebatt ME (2009) The role of antioxidants and antioxidant related enzymes in protective responses to environmentally induced oxidative stress. Mutat Res 674:137–147

Lindemann U, Wilken K, Weigmann HJ (2003) Quantification of the horny layer using tape stripping and microscopic techniques. J Biomed Opt 8:601–607

Liu L, Yang J, Xie J, Luo Z, Jiang J, Yang Y (2013) The potent antimicrobial properties of cell penetrating peptide conjugated silver nanoparticles with excellent selectivity for gram positive bacteria over erythrocytes. Nanoscale 5:3834–3840

Liu L, Liu T, Tade M, Wang S, Lib X, Liu S (2014) Less is more, greener microbial synthesis of silver nanoparticles. Enzym Microb Technol 67:53–58

Ma M, Zhang Y, Yu W, Shen HY, Zhang HQ, Gu N (2008) Preparation and characterization of magnetite nanoparticles coated by amino silane. Colloids Surf A: Physicochem Eng Asp 212:219–226

Maojo V, Sanchez FM, Kulikowski C, Paton AR, Fritts M (2010) Nanoinformatics and DNA-based computing: catalyzing nanomedicine. Int Pediatr Res Found, Inc 67:481–489

Marchiol L (2012) Synthesis of metal nanoparticles in living plants. Ital J Agron 3:37–41

Mehra RK, Winge DR (1991) Metal ion resistance in fungi: molecular mechanisms and their regulated expression. J Cell Biochem 45:30–40

Mehra RK, Tarbet EB, Gray WR, Winge DR (1988) Metalspecific synthesis of two metallothioneins and gammaglutamyl peptides in Candida glabrata. Proc Natl Acad Sci USA 85:8815–8819

Mittal AK, Chisti Y, Banerjee U (2013) Synthesis of metallic nanoparticles using plant extracts. Biotechnol Adv 31:346–356

Mohammadian M, Shojaosadati SA, Habibi MR (2007) *Fusarium oxysporum* mediates photogeneration of silver nanoparticles. Sci Iran 1:323–326

Mokhtari N, Daneshpajouh S, Seyedbagheri S, Atashdehghan R, Abdi K, Sarkar S, Minaian S, Shahverdi HR, Shahverdi AR (2009) Biological synthesis of very small silver nanoparticles by culture supernatant of *Klebsiella pneumonia*: the effects of visible light irradiation and the liquid mixing process. Mater Res Bull 44:1415–1421

Morganti P (2010) Use and potential of nanotechnology in cosmetic dermatology. Clin Cosmet Investig Dermatol 3:5–13

Mukherjee P, Ahmad A, Mandal D, Senapati S, Sainkar SR, Khan MI, Ramani R, Parishcha R, Ajaykumar PV, Alam M, Sastry M, Kumar R (2001a) Bioreduction of AuCl ions by the fungus, *Verticillium* sp. and surface trapping of the gold nanoparticles formed. Angew Chem Int Ed Eng 40:3585–3588

Mukherjee P, Ahmad A, Mandal D, Senapati S, Sainkar SR, Khan MI, Parishcha R, Ajaykumar PV, Alam M, Kumar R, Sastry M (2001b) Fungus mediated synthesis of silver nanoparticles and their immobilization in the mycelial matrix: a novel biological approach to nanoparticle synthesis. Nano Lett 1:515–519

Mukherjee P, Senapati S, Mandal D, Ahmad A, Khan MI, Kumar R, Sastry M (2002) Extracellular synthesis of gold nanoparticles by the fungus *Fusarium oxysporum*. Chembiochem 3:461–463
Mukherjee P, Roy M, Mandal BP, Dey GK, Mukherjee PK, Ghatak J, Tyagi AK, Kale SP (2008) Green synthesis of highly stabilized nanocrystalline silver particles by a non-pathogenic and agriculturally important fungus T. asperellum. Nanotechnology 19:075103–075107
Nair B, Pradeep T (2002) Coalescence of nanoclusters and formation of submicron crystallites assisted by *Lactobacillus* strains. Cryst Growth Des 2:293–298
Narayanan KB, Sakthivel N (2010a) Biological synthesis of metal nanoparticles by microbes. Adv Col Interf Sci 156:1–13
Narayanan KB, Sakthivel N (2010b) Facile green synthesis of gold nanostructures by NADPH-dependent enzyme from the extract of *Sclerotium rolfsii*. Colloid Surf A 380:156–161
Nikalje AP (2015) Nanotechnology and its applications in medicine. Med Chem 5:081–089
Ninganagouda S, Rathod V, Singh D (2014) Characterization and biosynthesis of silver nanoparticles using a fungus *Aspergillus niger*. Int Lett Nat Sci 10:49–57
Pantidos N, Horsfall LE (2014) Biological synthesis of metallic nanoparticles by bacteria, fungi and plants. J Nanomed Nanotechnol 5:233. https://doi.org/10.4172/2157-7439.1000233
Paraszkiewicz K, Długoński J (2009) Effect of nickel, copper, and zinc on emulsifier production and saturation of cellular fatty acids in the filamentous fungus *Curvularia lunata*. Int Biodeterior Biodegrad 63:100–105
Paraszkiewicz K, Frycie A, Słaba M, Długoński J (2007) Enhancement of emulsifier production by *Curvularia lunata* in cadmium, zinc and lead presence. Biometals 20:797–805
Paraszkiewicz K, Bernat P, Naliwajski M, Długoński J (2010) Lipid peroxidation in the fungus *Curvularia lunata* exposed to nickel. Arch Microbiol 192:135–141
Phanjom P, Ahmed G (2015) Biosynthesis of Silver Nanoparticles by Aspergillus oryzae (MTCC No. 1846) and Its Characterizations. Nanosci Nanotechnol 5:14–21
Philip D (2009a) Honey mediated green synthesis of gold nanoparticles. Spectrochim Acta Part A 73:650–653
Philip D (2009b) Biosynthesis of Au, Ag and AuAg nanoparticles using edible mushroom extract. Spectrochim Acta Part A 73:374–381
Pingali KC, Rockstraw DA, Deng S (2005) Silver nanoparticles from ultrasonic spray pyrolysis of aqueous silver nitrate. Aerosol Sci Technol 39:1010–1014
Polat BE, Hart D, Langer R (2011) Ultrasound-mediated transdermal drug delivery: mechanisms, scope, and emerging trends. J Control Release 152:330–348
Prabhu S, Poulose EK (2012) Silver nanoparticles: mechanism of antimicrobial action, synthesis, medical applications, and toxicity effects. Int Nano Lett 2:32–41
Prasad R (2014) Synthesis of silver nanoparticles in photosynthetic plants. J Nanopart 11:963955–963961
Prasad K, Jha AK (2009) ZnO nanoparticles: synthesis and adsorption study. Nat Sci 1:129–135
Prasad K, Jha AK (2010) Biosynthesis of CdS nanoparticles: an improved green and rapid procedure. J Colloid Interf Sci 342:68–72
Prasad R, Swamy VS (2013) Antibacterial activity of silver nanoparticles synthesized by bark extract of *Syzygium cumini*. J Nanoparticles https://doi.org/10.1155/2013/431218
Prasad K, Jha AK, Kulkarni AR (2007) *Lactobacillus* assisted synthesis of titanium nanoparticles. Nanoscale Res Lett 2:248–250
Prasad K, Jha AK, Prasad K, Kulkarni AR (2010) Can microbes mediate nanotransformation? Ind J Phys 84:1355–1360
Prasad R, Kumar V, Prasad KS (2014) Nanotechnology in sustainable agriculture: present concerns and future aspects. Afr J Biotechnol 13:705–713
Prasad R, Pandey R, Barman I (2016) Engineering tailored nanoparticles with microbes: *quo vadis*. WIREs Nanomedicine Nanobiotechnol 8:316–330
Prasad R, Bhattacharyya A, Nguyen QD (2017) Nanotechnology in sustainable agriculture: recent developments, challenges, and perspectives. Front Microbiol 8:1014. https://doi.org/10.3389/fmicb.2017.01014

Raheman F, Deshmukh S, Ingle A, Gade A, Rai M (2011) Silver nanoparticles: novel antimicrobial agent synthesized from an endophytic fungus *Pestalotia sp.* isolated from leaves of *Syzygium cumini* (L). Nano Biomed Eng 3:174–178

Rai M, Ingle AP, Gupta IR, Birla SS, Yadav AP, Abd-Elsalam KA (2013) Potential role of biological systems in formation of nanoparticles: mechanism of synthesis and biomedical applications. Curr Nanosci 9:576–587

Raj S, Jose S, Sumod US, Sabitha M (2012) Nanotechnology in cosmetics: opportunities and challenges. J Pharm Bioallied Sci 4:186–193

Ray S, Sarkar S, Kundu S (2011) Extracellular biosynthesis of silver nanoparticles using the mycorrhizal mushroom *Tricholoma crassum* (BERK.) SACC: its antimicrobial activity against pathogenic bacteria and fungus, including multidrug resistant plant and human bacteria. Digest J Nanomater Biostruct 6:1289–1299

Reese RN, Winge DR (1988) Sulfide stabilization of the cadmium gamma glutamyl peptide complex of *Schizosaccharomyces pombe*. J Biol Chem 263:12832–12835

Saha S, Pal A, Kundu S, Basu S, Pal T (2010) Photochemical green synthesis of calcium-alginate stabilized Ag and Au nanoparticles and their catalytic application to 4-nitrophenol reduction. Langmuir 26:2885–2893

Saifuddin N, Wong CW, Yasumira AAN (2009) Rapid biosynthesis of silver nanoparticles using culture supernatant of bacteria with microwave irradiation. E J Chem 6:61–70

Sanghi R, Verma P (2009) Biomimetic synthesis and characterisation of protein capped silver nanoparticles. Bioresour Technol 100:501–504

Sastry M, Ahmad A, Khan MI, Kumar R (2003) Biosynthesis of metal nanoparticles using fungi and actinomycete. Curr Sci India 85:162–170

Sawle D, Salimath B, Deshpande R, Bedre MD, Prabhakar BK, Venkataraman A (2008) Biosynthesis and stabilization of Au and Au-Ag alloy nanoparticles by fungus, *Fusarium semitectum*. Sci Technol Adv Mater 9:1–6

Schafer FQ, Buettner GR (2001) Redox environment of the cell as viewed through the redox state of the glutathione disulfide/glutathione couple. Free Radic Biol Med 30:1191–1212

Schmid G (1992) Large clusters and colloids, metals in the embryonic state. Chem Rev 92:1709–1727

Senapati S, Ahmad A, Khan MI, Sastry M, Kumar R (2005) Extracellular biosynthesis of bimetallic Au Ag alloy nanoparticles. Small 1:517–520

Shahverdi AR, Fakhimi A, Shahverdi HR, Minaian S (2007) Synthesis and effect of silver nanoparticles on the antibacterial activity of different antibiotics against *Staphylococcus aureus* and *Escherichia coli*. Nanomedicine 3:168–171

Shaligram NS, Bule M, Bhambure R, Singhal RS, Singh SK, Szakacs G, Pandey A (2009) Biosynthesis of silver nanoparticles using aqueous extract from the compactin producing fungal strain. Process Biochem 44:939–943

Shankar SS, Ahmad A, Pasricha R, Sastry M (2003) Bioreduction of chloroaurate ions by geranium leaves and its endophytic fungus yields gold nanoparticles of different shapes. J Mater Chem 13:1822–1826

Shanti SS, Karl JD (2006) The significance of amino acids and amino acid derived molecules in plant responses and adaptation to heavy metal stress. J Exp Bot 57:711–726

Shin HS, Yang HJ, Kim SB, Lee SS (2004) Mechanism of growth of colloidal silver nanoparticles stabilized by polyvinyl pyrrolidone in γ irradiated silver nitrate solution. J Colloid Interf Sci 274:89–94

Singh M, Singh S, Prasad S, Gambhir IS (2008) Nanotechnology in medicine and antibacterial effect of silver nanoparticles. Digest J Nanomater Biostruct 3:115–122

Sinha S, Pan I, Chanda P, Sen SK (2009) Nanoparticles fabrication using ambient biological resources. J Appl Biosci 19:1113–1130

Song WY, Sohn EJ, Martinoia E, Lee YJ, Yang YY, Jasinski M, Forestier C, Hwang I, Lee Y (2003) Engineering tolerance and accumulation of lead and cadmium in transgenic plants. Nat Biotechnol 21:914–919

Song JY, Kwon EY, Kim BS (2010) Biological synthesis of platinum nanoparticles using *Diopyros kaki* leaf extract. Bioprocess Biosyst Eng 33:159–164

Southam G, Beveridge TJ (1994) The in vitro formation of placer gold by bacteria. Geochim Cosmochim Acta 58:4527–4530

Stadtman ER, Levine RL (2003) Free radical mediated oxidation of free amino acids and amino acid residues in proteins. Amino Acids 25:207–218

Stadtman ER, Moskovitz J, Levine RL (2003) Oxidation of methionine residues of proteins: biological consequences. Antioxid Redox Signal 5:577–582

Suman V, Prasad R, Jain VK, Varma A (2010) Role of nanomaterials in symbiotic fungus growth enhancement. Curr Sci 99:1189–1191

Surendiran A, Sandhiya S, Pradhan SC, Adithan C (2009) Novel applications of nanotechnology in medicine. Indian J Med Res 130:689–701

Syed A, Ahmad A (2012) Extracellular biosynthesis of platinum nanoparticles using the fungus Fusarium oxysporum. Colloids Surf B: Biointerfaces 97:27–31

Tamás MJ, Labarre J, Toledano MB, Wysocki R (2005) Mechanisms of toxic metal tolerance in yeast. In: Tamás MJ, Martinoia E (eds) Molecular biology of metal homeostasis and detoxification: from microbes to man. Springer, Heidelberg

Tanvir S, Oudet F, Pulvin S, Anderson WA (2012) Coenzyme based synthesis of silver nanocrystals. Enzym Microb Technol 51:231–236

Tarafdar JC, Raliya R (2013) Rapid, low-cost, and ecofriendly approach for iron nanoparticle synthesis using Aspergillus oryzae TFR9. J Nanopart 2013:4,141274. https://doi.org/10.1155/2013/141274

Thakkar KN, Mhatre SS, Parikh RY (2010) Biological synthesis of metallic nanoparticles. Nanomedicine Nanotechnol Biol Med 6:257–262

Tripathi AK, Harsh NSK, Gupta N (2007) Fungal treatment of industrial effluents: a mini review. Life Sci J 4:78–81

Uddin T, Adyanthaya S, Syed A, Selvaraj K, Ahmad A, Poddar P (2008) Structure and microbial synthesis of sub-10 nm Bi2O3 nanocrystals. J Nanosci Nanotechnol 1:3909–3913

Ulla AJ, Patrick AWV, Ulla SL, Roger DF (2000) Organic acids produced by mycorrhizal *Pinus sylvestris* exposed to elevated aluminium and heavy metal concentrations. New Phytol 146:557–567

Vaidyanathan R, Kalimuthu K, Shubaash G, Gurunathan S (2009) Nanosilver-the burgeoning therapeutic molecule and its green synthesis. Biotechnol Adv 27:924–937

Valtchev V, Tosheva L (2013) Porous nanosized particles: preparation, properties, and applications. Chem Rev 113:6734–6760

Verma VC, Kharwar RN, Gange AC (2010) Biosynthesis of antimicrobial silver nanoparticles by the endophytic fungus *Aspergillus clavatus*. Nanomedicine 5:33–40

Verma VC, Singh SK, Solanki R, Prakash S (2011) Biofabrication of anisotropic gold nanotriangles using extract of endophytic Aspergillus clavatus as a dual functional reductant and stabilizer. Nanoscale Res Lett 6:16–17

Vesentini D, Dickinson DJ, Murphy RJ (2006) Fungicides affect the production of fungal extracellular mucilaginous material (ECMM) and the peripheral growth unit (PGU) in two wood rotting basidiomycetes. Mycol Res 110:1207–1213

Vigneshwaran N, Ashtaputre NM, Varadarajan PV, Nachane RP, Paralikar KM, Balasubramanya RH (2007) Biological synthesis of silver nanoparticles using the fungus *Aspergillus flavus*. Mater Lett 61:1413–1418

Wang X, Zuo J, Keil P, Grundmeier G (2007) Comparing the growth of PVD silver nanoparticles on ultra-thin fluorocarbon plasma polymer films and self-assembled fluoroalkyl silane monolayers. Nanotechnology 18:265303

Wood ZA, Schroder E, Harris JR, Poole LB (2003) Structure, mechanism and regulation of peroxiredoxins. Trends Biochem Sci 28:32–40

Wright SF, FrankeSnyder M, Morton JB, Upadhyaya A (1996) Time course study and partial characterization of a protein on hyphae of arbuscular mycorrhizal fungi during active colonization of roots. Plant Soil 181:193–203

Wysocki R, Tamás MJ (2010) How *Saccharomyces cerevisiae* copes with toxic metals and metalloids. FEMS Microbiol Rev 34:925–951

Xie H, Lee JY, Wang DIC, Ting YP (2007) Silver nanoplates: from biological to biomimetic synthesis. ACS Nano 1:429–439

Yompakdee C, Bunya M, Shikata K, Ogawa N, Harashima S, Oshima Y (1996) A putative new membrane protein, Pho86p, in the inorganic phosphate uptake system of *Saccharomyces cerevisiae*. Gene 171:41–47

Zhang H, Smith JA, Oyanedel-Craver V (2012) The effect of natural water conditions on the antibacterial performance and stability of silver nanoparticles capped with different polymers. Water Res 46:691–699

Chapter 12
Myconanotechnology to Treat Infectious Diseases: A Perspective

Ernestina Castro-Longoria, Diana Garibo-Ruiz, and Sandy Martínez-Castro

Abstract In nanotechnology fungi have been identified as excellent candidates for the production of nanoparticles (NPs), thus presenting a cleaner alternative to produce new materials with a wide range of potential applications in biomedicine and industry. In this respect, several species have demonstrated excellent bio-reducing capacity to produce metallic NPs, presenting a number of advantages over other biological systems such as a rapid growth rate, simple nutrient requirements, and easy handling of biomass/cultures. Furthermore, they secrete proteins that are assumed to have enhanced reducing and stabilizing capacity (Prasad et al. WIREs Nanomed Nanobiotechnol 8:316–330, 2016). Recent investigations have reported the potential medical applications of fungal-mediated NPs, particularly silver nanoparticles (AgNPs) for their excellent antimicrobial activity. Fungal-mediated AgNPs (FM-AgNPs) have demonstrated successful inhibition against microorganisms that cause infectious diseases in humans, even over those considered multiresistant to conventional antimicrobial drugs. Despite the advances in this field, there is still much work to be done, particularly in finding new biomolecules to improve the biocompatibility of the produced nanomaterial. In this chapter the state of the art in FM-AgNPs against human pathogens is reviewed.

E. Castro-Longoria (✉) • D. Garibo-Ruiz
Departmento de Microbiología, Centro de Investigación Científica y de Educación Superior de Ensenada (CICESE), Ensenada, Baja California, Mexico
e-mail: ecastro@cicese.mx

S. Martínez-Castro
Residencia Médica en Cirugía General, Universidad Autónoma de Baja California, Mexicali, Baja California, Mexico

R. Prasad (ed.), *Fungal Nanotechnology*, Fungal Biology,
DOI 10.1007/978-3-319-68424-6_12

12.1 Introduction: Infectious Diseases and the Problem of Drug Resistance

Infectious diseases represent one of the most important health problems worldwide and remain as the leading cause of mortality (Williams 2002). A number of pathogen microorganisms can cause severe infections and even death if not treated efficiently. For instance, patients with diabetes mellitus are susceptible to develop foot ulcerations and present a high risk of infection, gangrene, amputation, and even death (Snyder and Hanft 2009). Although there are treatments on the market that are effective for mild to moderate lesions (Yazdanpanah et al. 2015), the probability of treatment failure increases when methicillin-resistant microorganisms, such as *S. aureus* (MRSA), are present (Vardakas et al. 2008); it is estimated that between 50% and 70% of lower limb amputations are due to diabetic foot ulcers (Leone et al. 2012). Unfortunately, over the years the increasing emergence of drug-resistance strains represent a serious threat to human health resulting in longer hospital stays, increased doses of antibiotics, the use of more toxic drugs, and increase in the rate of morbidity and mortality (Cohen 1992). Antimicrobial resistance has been provoked mostly by the indiscriminate use of systemic antibiotics worldwide (Safdar and Maki 2002). In the case of bacteria, they rapidly acquire resistance through mutation of chromosomal genes during therapy with several antibiotics including those of the quinolone and rifamycin classes (Cirz et al. 2005). Likewise, the risk for opportunistic fungal infections (candidiasis, aspergillosis, cryptococcosis, and zygomycosis) also increases with the indiscriminate use of broad-spectrum antibiotics (Diekema et al. 2002; Perea and Patterson 2002). It is precisely in severe or persistent lesions that nanotechnology offers an alternative treatment, since the effectiveness of silver nanoparticles (AgNPs) against bacteria, fungi, and viruses has been demonstrated (Kim et al. 2009; Rai et al. 2009, 2012; Hu and Kwon 2011; Elechiguerra et al. 2005; Lu et al. 2008). In fact, nanomaterials showing antimicrobial activity by themselves or improve the effectiveness of commercial drugs was termed "nanoantibiotics" by Huh and Kwon (2011).

Thus, the development of nanoantibiotics presents a novel platform to solve many problems associated with health. For instance, nanosilver has proven to be effective to inhibit human pathogens, even strains considered multidrug resistant (MDR) to conventional antimicrobial drugs (Ray et al. 2011; Rai et al. 2012; Pelgrift and Friedman 2013). Furthermore, there is a noteworthy approach in which the combination of commercial antibiotics with AgNPs becomes more effective instead of using only NPs or antibiotics. In the last 10 years, nanotechnology research has been intensified on the production of nanosilver using biological methods, which offers several advantages over physical or chemical methods, such as being more eco-friendly, facile, rapid, and cost-effective (Aziz et al. 2015; Prasad et al. 2016). Fungi are excellent candidates to further explore their potential as bio-reducing agents to produce nanomaterials, and many species have been utilized searching for a cleaner alternative (Castro-Longoria 2016). Moreover, FM-AgNPs have shown excellent antimicrobial activity and also present synergistic antimicrobial effect when combined with commercial antibiotics (Birla et al. 2009; Fayaz et al. 2010;

Dar et al. 2013; Aziz et al. 2016), which could potentially be a more effective treatment. The use of commercial products containing silver is increasing, especially in the field of wound care with a wide variety of silver-containing dressings including Acticoat®, Aquacel® Ag, Contreet®, Actisorb® Silver 220, Urgotul SSD®, and Avance® (Dowsett 2004). Those products are used to treat infections in surgical wounds, burns, and chronic wounds such as leg ulcers (Lansdown 2004). Therefore, FM-AgNPs could be incorporated into materials such as creams, liquid solutions, dressings, surgical gowns, surgical materials, etc., to treat or prevent infectious diseases. While the increased use of products containing silver in wound care has created concern about possible development of bacterial resistance, currently there is no solid evidence of emerging microbial resistance to silver (Percival et al. 2005). Nevertheless, before suggesting potential uses in the clinic for FM-AgNPs, fungal biocompatible molecules must be investigated since most of the current research on FM-AgNPs has been conducted using pathogenic fungi. Still, myconanotechnology is a promising field of research since thousands of fungal species remain to be investigated in order to find active molecules to be applied in the fabrication of more efficient nanoantibiotics to combat resistant microorganisms.

12.2 Fungal-Mediated Nanoparticles Are Effective Against Pathogenic Bacteria

Among the bacteria implicated in different infectious diseases, *Staphylococcus aureus*, *Pseudomonas aeruginosa*, *Escherichia coli*, *Klebsiella pneumoniae*, *Enterobacter aerogenes*, *Clostridium perfringens*, *Serratia marcescens*, *Proteus vulgaris*, and *Shigella dysenteriae* are the major causes of hospitalization (Anderson et al. 1998). Globally, infectious diseases are considered a serious public health problem due to the fact that they are transmitted by a variety of vectors, including person to person, by respiration of airborne particles, airborne contact of open wounds, and through injection/insertion of medical devices as well as contaminated food. Therefore, antibiotics are widely used to decrease the prevalence of such infectious diseases. However, the overuse and misuse of antibiotics have resulted in the antibiotic resistance crisis; in fact many isolated resistant bacteria lack options for treatment, representing a serious public health concern (Weinstein et al. 2005).

In this respect, nanotechnology offers promising biogenic NPs to treat bacterial infections (Aziz et al. 2015, 2016). Recently, myconanotechnology research has been focused to develop eco-friendly methods to attain metallic NPs with antimicrobial properties. An increasing number of studies report the antimicrobial capacity of fungal-mediated nanoparticles (FM-NPs) to inhibit many pathogenic bacteria, including multiresistant strains (Table 12.1). Thus, FM-NPs could be incorporated in a variety of products to treat infectious diseases not responding to conventional treatments.

Antibacterial efficiency of NPs is highly dependent on the chemical composition, shape, size, and monodispersity of particles. Therefore most of the current

Table 12.1 Minimal inhibitory concentration (MIC) and inhibition zone determined by agar disk diffusion method of fungal-mediated nanoparticles against pathogenic bacteria

Bio-reducing agent	NPs	Size (nm)	Pathogenic bacteria	MIC (μg/ml)	Mean size of inhibition zone (mm)	References
Alternaria alternata	Ag	~25–30	*B. subtilis, E. coli, P. aeruginosa, S. aureus*	No data	13 ± 1.1, 17 ± 2.1, 15 ± 0.3, 17 ± 0.5	Ibrahim and Hassan (2016)
	Fe	5.4–12.1	*B. subtilis, E. coli, P. aeruginosa, E. aureus*	No data	16.4 ± 0.7, 13.2 ± 0.6, 10.5 ± 0.3, 12.3 ± 0.5	Mohamed et al. (2015)
Alternaria solani	Ag	5–20	*E. coli, E. faecalis, S. pyogenes*	No data	10 ± 0.25, 10 ± 0.24, 10 ± 0.23	Devi et al. (2014)
Amylomyces rouxii KSU-09	Ag	5–27	*B. subtilis, Citrobacter sp., E. coli, P. aeruginosa, S. aureus, S. dysenteriae* type I	No data	No data	Musarrat et al. (2010)
Aspergillus clavatus	Ag	10–25	*E. coli, P. fluorescens*	No data	10, 14	Verma et al. (2010)
	Ag	20–30	MR: *S. aureus, S. epidermidis*	No data	20.5, 19	Saravanan and Nanda (2010)
A. flavus	Ag	5–30	*A. baumannii, Bacillus* sp., *E. coli, E. faecalis, K. pneumoniae, M. luteus, P. aeruginosa, S. aureus*	No data	15 ± 1, 15 ± 1.5, 15 ± 1.5, 15 ± 1.5, 14 ± 0.6, 14 ± 1, 14 ± 1.5, 16 ± 2	Naqvi et al. (2013)
	TiO_2	62–74	*B. subtilis, E. coli, K. pneumoniae, P. aeruginosa, S. aureus*	45, 40, 70, 80, 40	22, 35, 18, 27, 25	Rajakumar et al. (2012)

A. niger	Ag	1–20	*E. coli, P. aeruginosa, S. aureus*	No data	13 ± 1, 12 ± 1, 16 ± 1	Sagar and Ashok (2012)
	Ag	3–30	*Bacillus* sp., *E. coli, Staphylococcus* sp.	No data	0.8, 0.8, 09	
	Ag	5–35	K. pneumoniae, L. monocytogenes, M. *luteus, P. aeruginosa, S. aureus*	No data	*No data	Jaidev and Narasimha (2010)
	Ag	40–60	*Bacillus sp., E. coli, P. aeruginosa, S. aureus*	No data	16, 28, 14, 25	Kathiresan et al. (2010)
						Nithya and Ragunathan (2014)
A. terreus	Ag	1–20	*E. coli, P. aeruginosa, S. aureus*	No data	13 ± 1, 12 ± 1, 16 ± 1	Li et al. (2012)
	Ag, Au	10–50, 8–20	*B. subtilis, E. coli, E. faecalis, K. pneumoniae, P. aeruginosa, S. aureus*	6.25, 3.125, 3.125, 6.25, 6.25, 12.5	16.43 ± 0.21, 22.47 ± 0.06, 19.47 ± 0.06, 16.13 ± 0.06, 17.13 ± 0.15, 14.27 ± 0.12	
	Ag	5–30	*S. aureus*	No data		Balakumaran et al. (2016)
	Ag	5–50				
			E. faecalis, S. enterica, S. aureus, S. pyogenes	No data	20	
					8, 10, 8, 14	Abeer et al. (2013)
						Devi and Joshi (2012)
A. tubingensis	Ag	35 ± 10	*M. luteus, P. aeruginosa, S. aureus*	>0.71, 0.54, 0.28	No data	Rodrigues et al. (2013)
A. versicolor	Ag	3–40	*K. pneumoniae, P. aeruginosa, S. aureus, S. pneumoniae*	No data	13.6, 18.5, 12.1, 11.8	Netala et al. (2016)

(continued)

Table 12.1 (continued)

Bio-reducing agent	NPs	Size (nm)	Pathogenic bacteria	MIC (µg/ml)	Mean size of inhibition zone (mm)	References
Bionectria ochroleuca	Ag	35 ± 10	*M. luteus, P. aeruginosa, S. aureus*	>0.71, 0.54, 0.54	No data	Rodrigues et al. (2013)
Bipolaris nodulosa	Ag	10–60	*B. subtilis, B. cereus, E. coli, M. luteus, P. vulgaris, P. aeruginosa*	*No data	*No data	Saha et al. (2010)
Beauveria bassiana	Ag	20–34	*E. coli, S. aureus*	No data	13.3 ± 1.0, 12.1 ± 2.0	Prabakaran et al. (2016)
Candida albicans	Ag	300–800	*E. coli, S. aureus*	No data	17 ± 0.01, 21 ± 0.12	Rahimi et al. (2016)
Cordyceps militaris	Ag	15	*A. punctata, B. subtilis, E. coli, P. aeruginosa, S. aureus, V. alginolyticus, V. anguillarum, V. parahaemolyticus*	No data. 6.25 (for *V. Anguillarum*)	13.3 ± 0.3, 12.7 ± 0.9, 17.9 ± 0.6, 13. 9 ± 0.9, 13.5 ± 0.4, 13.6 ± 0.5, 16.4 ± 0.5, 16.9 ± 0.6	Wang et al. (2016)
Cryphonectria sp.	Ag	30–70	*E. coli, S. typhi, S. aureus*	No data	13 ± 1.54, 12 ± 0.29, 16 ± 0.69	Dar et al. (2013)
Fusarium sp.	Ag	5–50	*E. faecalis, S. enterica, S. aureus, S. pyogenes*	No data	10, 11, 10, 14	Devi and Joshi (2012)
Fusarium acuminatum	Ag	5–40	*E. coli, S. typhi, S. aureus, S. epidermidis*	No data	*No data	Ingle et al. (2008)
F. oxysporum	Ag	50	*E. coli, K. pneumoniae, P.aeruginosa, S. typhi, S. aureus*	No data	19, 17, 16, 14, 23	Gholami-Shabani et al. (2014)
	Ag	10–45	*S. typhi, S. aureus*	No data	*No data	
	Fe	20–40	*Bacillus, E. coli, Klebsiella sp., Proteus sp., Pseudomonas sp., Staphylococcus sp., Streptococcus sp., Vibrio sp.*	No data	*No data	Joshi et al. (2013) Abdeen et al. (2013)

F. oxysporum f. p. cubense JT1	Au	22	*Pseudomonas sp.*	No data	*No data	Thakker et al. (2013)
Neurospora intermedia	Ag	19–24	*E. coli*	No data	No data	Hamedi et al. (2014)
Macrophomina phaseolina	Ag	8–25	*S. typhi, S. aureus*	0.05, 0.03	*No data	Joshi et al. (2013)
Monascus purpureus	Ag	1–7	*E. coli, P. aeruginosa, S. typhimurium, S. aureus, S. epidermidis, S. pyogenes*	No data	11.2 ± 0.58, 7.0, 15.8 ± 0.25, 16.2 ± 0.4, 17.2 ± 0.19, 15.9 ± 0.25	El-Baz et al. (2016)
Nigrospora sphaerica	Ag	20–70	*E. coli, P. mirabilis, P. aeruginosa, S. typhi, S. aureus*	0.156 for *E. Coli*, 0.0024 for *S. aureus*	*No data	Muhsin and Hachim (2014)
N. oryzae	Ag	30 ~ 90	*B. cereus, B. subtilis, E. coli, P. aeruginosa, P. vulgaris, M. luteus*	No data	*No data	Saha et al. (2011)
Paecilomyces lilacinus	Ag	5–50	*E. faecalis, S. enterica, S. aureus, S. pyogenes*	No data	10, 11, 10, 14	Devi and Joshi (2012)
Penicillium (K1 and K10)	Ag	1–100	*B. cereus, E. coli, P. aeruginosa, S. aureus, S. marcescens*	No data	K1: 13, 14, 16, 15, NI K10: 12, 14, 15, 14, NI	Maliszewska and Sadowski (2009)
Penicillium sp.	Ag	25	MDR *E. coli, S. aureus*	No data	14, 12	Singh et al. (2014)2013
	Ag	20–200	*K. pneumoniae, P. mirabilis, S. dysenteriae, S. aureus*	No data	*No data	Nameirakpam et al. (2012)
Penicillium decumbens	Ag	30–60	*E. coli, P. vulgaris, S. aureus, V. cholerae*	No data	13 ± 0.33, 13 ± 0.66, 12 ± 0.18, 12 ± 0.58	Majeed et al. (2016)
P. diversum	Ag	5–45	*E. coli, Paratyphia, S. typhi, V. cholerae*	5 for *E. coli*	1.5, 1.6, 1.6, 1.9	Ganachari et al. (2012)

(continued)

Table 12.1 (continued)

Bio-reducing agent	NPs	Size (nm)	Pathogenic bacteria	MIC (µg/ml)	Mean size of inhibition zone (mm)	References
P. funiculosum	Ag	5–20	*E. coli, E. faecalis, S. pyogenes*	No data	13 ± 0.21, 12 ± 0.19, 11 ± 0.25	Devi et al. (2014)
P. janthinellum	Ag	8–14	*B. subtilis, E. coli, Enterobacter, Enterococcus, K. pneumoniae, Micrococcus, S. typhi, S. aureus, Streptococcus, V. cholerae*	No data	NI	Bharathidasan and Panneerselvam (2012)
Pestalotia sp.	Ag	10–40	*S. aureus, S. typhi*	No data	12, 10	Raheman et al. (2011)
Pleurotus sajor-caju	Ag	30.5 ± 4.0	*K. pneumoniae, S. aureus*	No data	*No data	Vigneshwaran et al. (2007)
	Ag	5–50	*E. coli, P. aeruginosa, S. aureus*	No data	12, 14, 11	Nithya and Ragunathan (2009)
Phoma glomerata	Ag	60–80	*E. coli, P. aeruginosa, S. aureus*	No data	*No data	Birla et al. (2009)
Phomopsis sp.	Ag	10–16	*B. subtilis, E. coli, Enterobacter, Enterococcus, K. pneumoniae, Micrococcus, S. typhi,, S. aureus, Streptococcus, V. cholerae*	No data	11, 14, 12, 10, 10, 11, 13, 10, 13, 12	Bharathidasan and Panneerselvam (2012)
Rhizopus sp.	Ag	10–30	MDR *E.coli*: *E. coli* 1, *E. coli* 2, *E. coli* 3, *E. coli* 4, *E. coli* 5, *E. coli* 6	No data	19, 18, 21, 18, 14, 17	Hiremath et al. (2014)
R. oryzae	Au	10	*B. subtilis, E. coli, P. aeruginosa, Salmonella sp., S. aureus*	No data	*No data	Das et al. (2009)

Rhizopus stolonifer	Ag	5–50	MDRstrains of *P. aeruginosa:* p1, p2	No data	33, 30.5	Rathod and Ranganath (2011)
Schizophyllum commune	Ag	51–93	*B. subtilis, E. coli, K. pneumoniae, P. fluorescens*	No data	2.5, 2.0, 2.5, 1.8	Arun et al. (2014)
Schizophyllum radiatum	Ag	10–40	*B. subtilis, E. coli, S. aureus, P. vulgaris*	No data	14, 14, 14, 13	Gudikandula et al. (2015)
Stachybotrys chartarum	Ag	65–108	*B. subtilis, E. coli*	No data	20.2, 17.0	Ghany and Mohamed (2013)
Trichoderma viride	Ag	5–40	*E. coli, M. luteus, S. typhi, S. aureus*	30, 65, 35, 80	*No data	Fayaz et al. (2010)
Trichosporon beigelii	Ag	50–100	*E. coli*	No data	*No data	Ghodake et al. (2011)
T. harzianum	Ag	51.10	*K. pneumoniae, S. aureus*	No data	13.86 ± 0.30, 9.83 ± 0.20	Ahluwalia et al. (2014)
Tricholoma crissum	Ag	5–50	MDR *E. coli*	No data	*No data	Ray et al. (2011)

MIC = 80 or 90
No data = Information is not mentioned in the article
*No data = The authors only provide images or graph
NI = Not inhibited

research is concentrated in the synthesis of AgNPs due to their antibacterial activity (Kim et al. 2007). Many studies have demonstrated the efficacy of FM-AgNPs to inhibit growth of the most important pathogenic bacteria that have a threat to public health; for example, *Bacillus subtilis*, *Escherichia coli*, *Pseudomonas aeruginosa*, *S. aureus*, and *S. typhi* are among the most widely used strains for antibacterial determinations (Table 12.1). However, most fungal species used to synthesize NPs are pathogenic species; therefore the application of those synthesized NPs in bacterial infections must be carefully evaluated. For instance, *B. nodulosa* which is a plant pathogen was used as reducing agent for AgNPs production, which resulted effective against *B. subtilis*, *B. cereus*, *E. coli*, *M. luteus*, *P. vulgaris*, and *P. aeruginosa* (Saha et al. 2010). Also AgNPs from *A. terreus* showed antimicrobial activity against *E. coli*, *P. aeruginosa*, and *S. aureus*, and in all cases AgNPs showed higher inhibitory effect than silver nitrate used as a control (Li et al. 2012). Furthermore, in a recent study, AgNPs also from *A. terreus* were successfully used to inhibit antibiotic-resistant bacteria including *E. faecalis* and *K. pneumoniae* (Fig. 12.1) (Balakumaran et al. 2016). Likewise, FM-AgNPs from *Rhizopus stolonifer* were effective against two *P. aeruginosa* MDR strains isolated from burnt patients (Rathod and Ranganath 2011). However, *A. terreus* is considered an opportunistic human pathogen (Person et al. 2010), and *Rhizopus* species can provoke aggressive and fulminant infections under certain clinical conditions (Lehrer et al. 1980). Thus, pathogenicity could be an important impediment to consider some fungal species as candidates to produce NPs to combat human infections.

On the other hand, *Cordyceps militaris* (Wang et al. 2016) and *Penicillium decumbens* (Majeed et al. 2016) were recently used for the first time to biosynthesize AgNPs. Synthesized AgNPs from *C. militaris* exhibited antibacterial activity against aquatic as well as clinical pathogenic bacteria. This fungal species could be an excellent candidate for NPs production to be applied in humans since it is not considered as pathogenic. In fact, it is categorized as a valuable edible and medicinal fungus, rich in a variety of biologically active substances (Das et al. 2010). *Penicillium decumbens* and other *Penicillium* species are reported to show bactericidal, fungicidal, and insecticidal capacity (Santamarina et al. 2002). Therefore, those fungal species represent good candidates to further explore their use as reducing agents, since more efficient antimicrobial NPs could be produced. In fact, fungal filtrate of *P. decumbens* showed bacterial inhibition, and synthesized AgNPs using this extract displayed remarkable effectivity against MDR pathogens *Proteus vulgaris*, *E. coli*, *S. aureus*, and *Vibrio cholerae*, isolated from urine, stool, and blood (Majeed et al. 2016).

Even though FM-AgNPs are effective for bacterial inhibition, it has been shown that in general terms the antibacterial activity against Gram-negative bacteria is higher than the obtained Gram-positive bacteria (Wang et al. 2016). The variation in efficiency has been explained in terms of the bacterial cell wall differences; Gram-positive bacteria possess a more rigid and thicker structure, leading to a difficult penetration of AgNPs compared to the thinner cell wall of Gram-negative bacteria

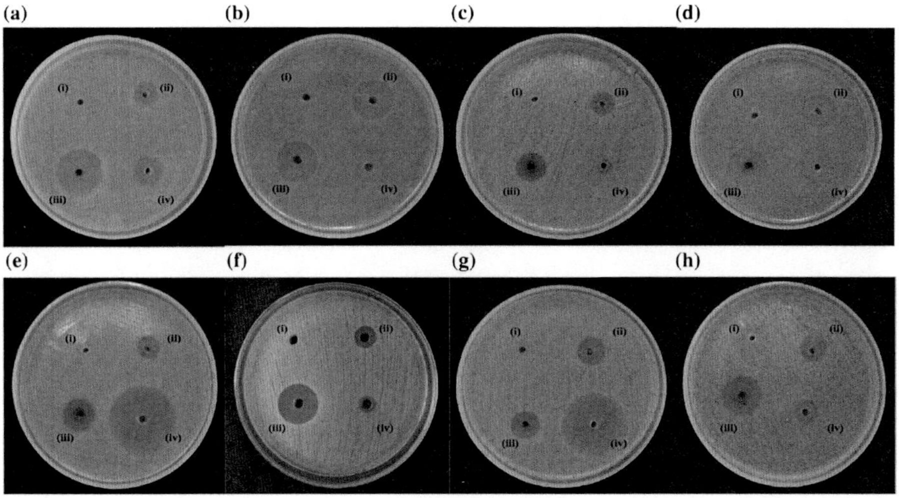

Fig. 12.1 Antimicrobial activity of *A. terreus* synthesized silver nanoparticles against human pathogenic microorganisms by well diffusion assay. (**a**) *E. coli*, (**b**) *P. aeruginosa*, (**c**) *K. pneumoniae*, (**d**) *E. faecalis*, (**e**) *B. subtilis*, (**f**) methicillin-resistant *S. aureus*, (**g**) *S. aureus*, (**h**) *C. albicans*. Each well was treated with (i) mycelial free extract (25 μL), (ii) silver nitrate (1 mM), (iii) silver nanoparticles (1 mg/mL), and (iv) streptomycin (1 mg/mL) for bacterial strains; fluconazole (1 mg/mL) in case of *C. albicans* (Reproduced from Balakumaran et al. (2016) with permission from Elsevier)

(Shrivastava et al. 2007). The mode of action of FM-AgNPs against bacteria has not been fully explained, but it has been concluded that the antibacterial efficiency of AgNPs is due to their attachment to the bacterial cell surface, which is believed to significantly disrupt respiration and permeability (Morones et al. 2005).

Besides FM-AgNPs, fungal-mediated gold NPs (FM-AuNPs) have also been tested as antimicrobial agents; however very few studies have demonstrated FM-AuNPs with antimicrobial capacity. For instance, AuNPs from *S. hygroscopicus* were tested against six different pathogens, and maximal growth inhibition was observed against *E. coli*, *S. typhimurium*, and *S. aureus*, according to the MIC profile determination (Sadhasivam et al. 2012). Remarkable antimicrobial activity was reported from nanogold-bioconjugate (NGBC) produced with the use of *R. oryzae* against pathogenic bacteria *E. coli*, *P. aeruginosa*, *B. subtilis*, *S. aureus*, *S. cerevisiae*, and the opportunistic fungus *C. albicans* (Fig. 12.2) (Das et al. 2009). Also, AuNPs from *F. oxysporum* were reported to inhibit *Pseudomonas* sp. (Thakker et al. 2013).

Other metallic FM-NPs have been also tried as nanoantibiotics, for instance, TiO_2 NPs from *A. flavus* were recently reported as a novel antimicrobial nanomaterial; they were successfully tested against *B. subtilis*, *E. coli*, *K. pneumoniae*, *P. aeruginosa*, and *S. aureus* (Rajakumar et al. 2012). Also, antibacterial activity of iron nanoparticles (FeNPs) from *A. alternata* was reported against four species of bacteria. FeNPs were described to be more effective against *B. subtilis* than *E. coli*, *S. aureus*, and *P. aeruginosa* (Mohamed et al. 2015); commercial antibiotics were more effective against those pathogens.

In general, most studies suggest the potential use of the FM-NPs as antimicrobial agent for treatment of infectious diseases. However, to ensure the biosafety of using those NPs in humans, in vivo studies are inevitable.

12.3 Fungal-Mediated Nanoparticles to Combat Pathogenic Fungi

Pathogenic fungi are a frequent cause of life-threatening infections in patients who are immunosuppressed. Even more problematic is the fact of a limited number of available antifungal drugs to treat resistant strains. Therefore, their treatment and control is becoming a major problem in the management of these diseases (Perea and Patterson 2002). Thus, AgNPs and AuNPs synthesized from fungal species seem to be alternative antifungal agents, which could be biocompatible and non-toxic and obtained by eco-friendly methods (Prasad 2016). Metallic nanoparticles of varying size using diverse fungal species as reducing agents and their antifungal activity are listed in Table 12.2. Owing to the antimicrobial properties of silver, AgNPs are the most frequently tested against pathogenic fungi. On the other hand, few studies are available on the antifungal activity of AuNPs (Thakker et al. 2013; Balakumaran et al. 2016).

In most studies using AgNPs, potent antifungal activity has been shown against *Candida* and *Fusarium* species, being the most common etiological agents of fungal infection. For instance, polydispersed spherical and hexagonal AgNPs from *A. clavatus* with a size range of 10–25 nm exhibited good antifungal activity against *C. albicans* with a fungicidal concentration (MFC) of 9.7 μg/ml (Verma et al. 2010). Also, excellent antifungal activity against pathogenic species such as *Candida* spp., *Aspergillus* spp., *Fusarium* spp., *C. neoformans*, and *Sporothrix schenckii* (MIC values of 0.125–1 μg/mL) was obtained with AgNPs from *E. nigrum* (Qian et al. 2013). FM-AgNP from *S. commune* inhibited the dermatophytic fungal pathogens *Trichophyton simii*, *Trichophyton mentagrophytes*, and *Trichophyton rubrum*; those pathogens are known to cause tinea of the scalp, nails, and skin in humans. Antifungal activity of FM-NPs was reported to be concentration dependent (Arun et al. 2014); however similar inhibition was found using the antifungal drug fluconazole as a positive control (Fig. 12.3).

Recently, AgNPs from the fungus *A. fulvum* showed antimicrobial activity against *Candida* spp., *Aspergillus* spp., and *Fusarium* spp. Furthermore, the authors presented the growth curves of *C. albicans*, *C. parapsilosis*, *C. krusei*, and *C. tropicalis* in the presence of AgNPs showing their fungistatic capacity (Xue et al. 2016). The mode of action of AgNPs against fungal pathogens is not fully described; however Vazquez-Muñoz et al. (2014) examined the ultrastructural localization of

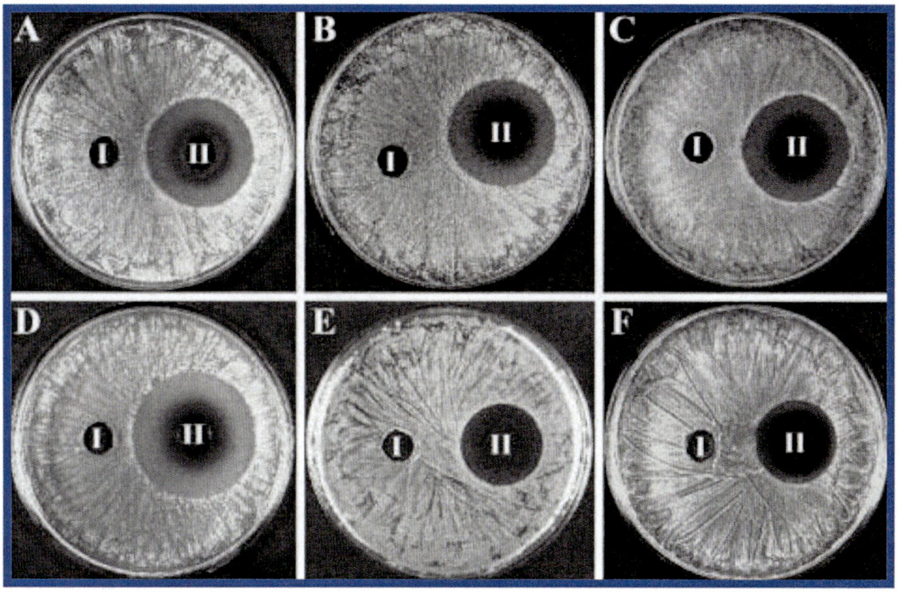

Fig. 12.2 Antimicrobial activity of nanogold-bioconjugate (NGBC) material against (**a**) *E. coli*, (**b**) *P. aeruginosa*, (**c**) *B. subtilis*, (**d**) *S. aureus*, (**e**) *S. cerevisiae*, and (**f**) *C. albicans*. Cup containing (I) dispersed solution of *R. oryzae* and (II) NGBC; concentration of bioconjugate in each plate was 50 μg/mL (Reproduced from Das et al. (2009) with permission from Elsevier)

AgNPs in *C. albicans* exposed to a commercial product and concluded that NPs aggregate outside the fungal cells and the consequent release of silver ions infiltrates through the cell wall leading to cell death. In general, FM-AgNPs have demonstrated capacity to inhibit a wide variety of fungal pathogens; however it remains to be determined which fungal species could be used to produce biocompatible nanomaterial to combat fungal infections.

12.4 Antimicrobial Drugs Enhanced by Fungal-Mediated Nanoparticles

Bacterial or fungal infections could potentially be treated with AgNPs when the pathogen is resistant to commercial drug treatments. As previously mentioned, the presence of pathogenic microorganisms resistant to multiple antibiotics is an increasing problem that could be solved by the use of AgNPs. Also, there is an important approach in which the combination of commercial antibiotics with AgNPs may become more effective instead of the separate use of only antibiotics or NPs. This approach was demonstrated by Li et al. (2005) when they studied the synergistic antibacterial effects of amoxicillin and chemically synthesized AgNPs.

Table 12.2 Minimal inhibitory concentration (MIC) and inhibition zone determined by agar disk diffusion method of fungal-mediated nanoparticles against pathogenic fungi

Bio-reducing agent	NPs	Size (nm)	Pathogenic fungi	MIC (μg/ml)	Mean size of inhibition zone (mm)	References
Alternaria alternata	Ag	20–60	*C. albicans, F. semitectum, P. glomerata, P. herbarum, Trichoderma sp.*	No data	*No data	Gajbhiye et al. (2009)
	Ag	~25–30	*A. niger*	No data	14 ± 1.6	Ibrahim and Hassan (2016)
Amylomyces rouxii KSU-09	Ag	5–27	*C. albicans, F. oxysporum*	*No data	*No data	Musarrat et al. (2010)
Arthroderma fulvum	Ag	15.5 ± 2.5	*A. flavus, A. fumigatus, A. terreus C. albicans, C. krusei, C. parapsilosis, C. tropicalis, F. moniliforme, F. solani, F. oxysporum*	2.0, 1.0, 1.0, 0.5, 0.125, 0.125, 0.250, 4.0, 2.0, 2.0	No data	Xue et al. (2016)
Aspergillus clavatus	Ag	10–25	*C. albicans*	5.83	16	Verma et al. (2010)
Aspergillus parasiticus	Ag	5–30	*C. albicans*	4.0	No data	Moazeni et al. (2014)
A. foetidus	Ag	20–40	*A. foetidus, A. flavus, A. niger, A. parasiticus, A. oryzae, F. oxysporum*	No data	1.7, 1.5, 2.0, 1.9, 1.8, 1.7	Roy et al. (2013)
A. niger	Ag	1–20	*A. flavus, A. fumigatus, C. albicans, C. krusei, C. parapsilosis, C. tropicalis*	No data	13 ± 2, 14 ± 2, 16 ± 1, 14 ± 2, 13 ± 1, 14 ± 1	Sagar and Ashok (2012)
	Ag	3–30	*A. niger*	No data	1.2	
	Ag	5–35	*A. alternata, C. albicans, F. equiseri, P. italicum*	No data	No data	Jaidev and Narasimha (2010)
						Kathiresan et al. (2010)
A. terreus	Ag	1–20	*A. flavus, A. fumigatus, C. albicans, C. krusei, C. parapsilosis, C. tropicalis*	No data	13 ± 2, 14 ± 2, 16 ± 1, 14 ± 2, 13 ± 1, 14 ± 1	Li et al. (2012)
	Ag, Au	8–20, 10–50		3.125		
			C. albicans	No data	18.13 ± 0.02	
	Ag	5–30	*A. flavus, A. fumigatus, A. niger, C. albicans*		22, 20, 19, 24	Balakumaran et al. (2016)
	Ag	5–30	*A. flavus*	No data	*No data	Abeer et al. (2013)
						Al-Othman et al. (2014)

A. tubingensis	Ag	35 ± 10	*C. albicans* (IOC4525), *C. albicans* (ATCC 36802), *C. glabrata*, *C. guilliermondii*, *C. krusei*, *C. tropicalis*	0.22, 27.30, 0.44, 0.44, 0.22, 0.44	No data	Rodrigues et al. (2013)
A. versicolor	Ag	3–40	*C. albicans*, *C.* non-*albicans*	No data	12.2, 13.6	Netala et al. (2016)
Bionectria ochroleuca	Ag	35 ± 10	*C. albicans* (IOC4525), *C. albicans* (ATCC 36802), *C. krusei*, *C. parapsilosis*	0.11, 1.75, 0.11, 0.44	No data	Rodrigues et al. (2013)
Cryphonectria sp.	Ag	30–70	*C. albicans*	No data	11 ± 1.54	Dar et al. (2013)
Epicoccum nigrum	Ag	1–22	*A. flavus*, *A. fumigatus*, *C. albicans*, *C. krusei*, *C. parapsilosis*, *C. tropicalis*, *C. neoformans*, *F. solani*, *S. schenckii*	0.5, 1.0, 0.5, 0.125, 0.125, 1.0, 0.25, 1.0, 0.25	No data	Qian et al. (2013)
Fusarium oxysporum	Ag	50	*A. alternata*, *A. fumigatus*, *C. albicans*, *C. glabrata*, *M. gypseum*, *P. citrinum*, *P. variotii*, *T. mentagrophytes*, and *T. parceramosum*	No data	22, 24, 22, 20, 17, 24, 25, 25, and 18	Gholami-Shabani et al. (2014)
Monascus purpureus	Ag	1–7	*C. albicans*, *C. glabrata*, *C. tropicalis*	No data	16.7 ± 0.25, 16 ± 0.44, 18.7 ± 0.37	El-Baz et al. (2016)
Penicillium sp.	Ag	2–200	*C. albicans*	No data	*No data	Nameirakpam et al. (2012)
Pleurotus ostreatus	Ag	4–15	*C. albicans*, *C. glabrata*, *C. krusei*, *C. parapsilosis*, *C. tropicalis*	5–7, 16, 4–11, 10, 7–28	No data	Yehia and Al-Sheikh (2014)
Rhizopus oryzae	Au	10	*C. albicans*, *S. cerevisiae*	No data	No data	Das et al. (2009)
Stemphylium vericans	Ag	25–60	*F. oxysporum*	*No data	*No data	Ravindra and Rajasab (2015)
Stachybotrys chartarum	Ag	65–108	*A. niger*, *C. albicans*	No data	0.00, 15.0	Ghany and Mohamed (2013)

MIC = 80 or 90

No data = Information is not mentioned in the article

*No data = The authors only provide images or graph

NI = Not inhibited

The authors found a remarkable increased antimicrobial efficacy when *E. coli* was exposed to a combination of amoxicillin and AgNPs. It was hypothesized that amoxicillin acts on the surface of the bacterial cells causing cell wall destruction, facilitating AgNPs internalization. Then AgNPs react with cell's DNA, resulting in more serious damage to the pathogen impeding cell recovery.

In this respect, biogenic NPs could represent a more biocompatible option rather of using chemical synthetized NPs. For instance, some studies have reported the synergistic effect of FM-AgNPs in combination with several antibiotics, increasing the drug effectivity against pathogenic microorganisms. One of the first studies reporting the synergistic effect of antibiotics and FM-AgNPs was reported by Birla et al. (2009). AgNPs were synthesized using the fungal cell filtrate from *P. glomerata* and were used in combination with ampicillin, gentamicin, kanamycin, streptomycin, and vancomycin against *E. coli*, *S. aureus*, and *P. aeruginosa*. All pathogens were more susceptible to the combination of antibiotics with AgNPs including the bacteria with MDR (Birla et al. 2009). Extracellular AgNPs from the endophytic fungus *Pestalotia* sp. also showed increased activity in combination with gentamicin and sulfamethizole against *S. aureus* and *S. typhi* (Raheman et al. 2011). Increased antibacterial activity is obtained when using various antibiotics in combination with FM-AgNPs. For instance, *E. coli*, *S. typhi*, *S. aureus*, and *M. luteus* were challenged with ampicillin, kanamycin, erythromycin, and chloramphenicol in combination with AgNPs from *T. viride*, and it was found that there is an increased antibacterial activity in all tested strains (Fayaz et al. 2010). The antibiotics imipenem, gentamicin, and ciprofloxacin combined with AgNPs from *A. flavus* also showed increased effectivity against *E. coli*, *S. aureus*, *P. aeruginosa*, *K. pneumoniae*, *Bacillus* spp., and *M. luteus* (Naqvi et al. 2013).

A remarkable increase in fold area of antibacterial activity against *E. coli*, *P. mirabilis*, *P. aeruginosa*, *S. typhi*, and *S. aureus* was reported when AgNPs from *N. sphaerica* were applied in combination with gentamicin (Muhsin and Hachim 2014). The antibacterial activity of ceftriaxone and streptomycin also increased significantly in combination with AgNPs from *S. radiatum* when used against *S. aureus*, *B. subtilis*, *E. coli*, and *P. aeruginosa* (Gudikandula et al. 2015). Recently, AgNPs from *P. decumbens* showed a broad antimicrobial activity against various pathogenic Gram-negative and Gram-positive bacteria. Furthermore, FM-AgNPs showed excellent enhancement of antimicrobial activity in combination with carbenicillin, piperacillin, cefixime, amoxicillin, ofloxacin, and sparfloxacin in a synergistic mode against *P. vulgaris*, *E. coli*, *S. aureus*, and *V. cholerae* (Majeed et al. 2016).

In most cases antibiotics showed increased activity in the presence of FM-AgNPs against tested bacterial strains; however the mode of action for the synergistic effect of AgNPs and commercial drugs has not been fully addressed. Nevertheless Fayaz et al. (2010) used several antibiotics in combination with AgNPs from *T. viride* and found the highest enhancing effect for ampicillin against all tested strains and concluded that ampicillin provokes cell wall lysis and thus AgNPs penetrate more easily into the bacterium, similar to the hypothesis presented by Li et al. (2005). The synergistic effect of ampicillin with AgNPs against bacteria is represented in Fig. 12.4.

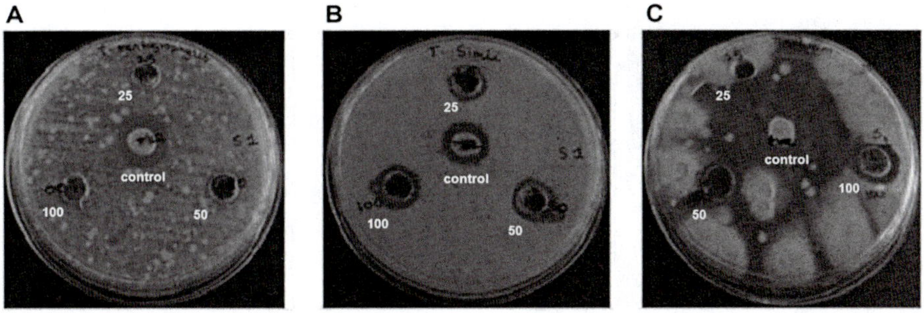

Fig. 12.3 Antifungal activity of AgNPs produced by *S. commune* against three dermatophytic fungal pathogens. (**a**) *Trichophyton mentagrophytes*, (**b**) *Trichophyton simii*, and (**c**) *Trichophyton rubrum*. Concentrations used were AgNPs at 25, 50 and 100 µg control (fluconazole) at 100 µg/well (Reproduced and modified from Arun et al. (2014) with permission from Springer)

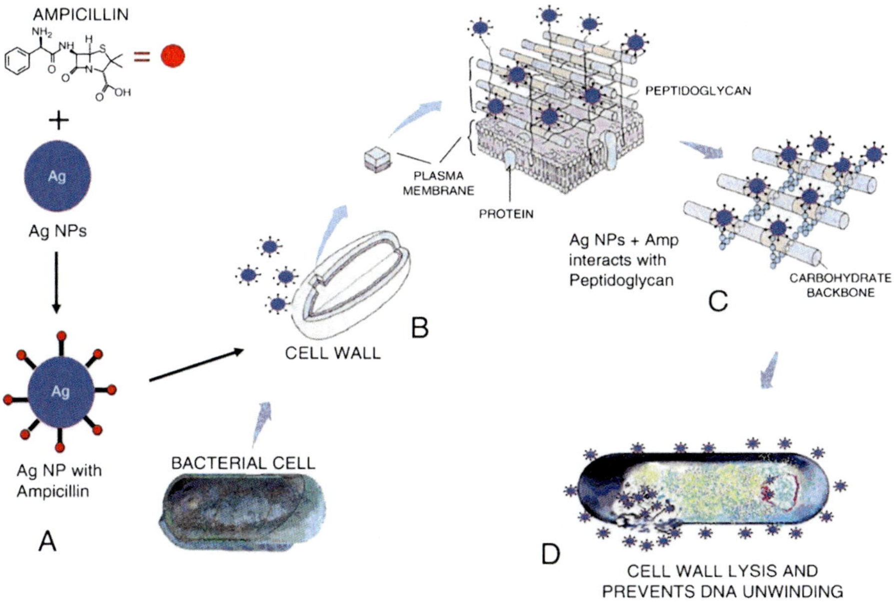

Fig. 12.4 Synergistic activity of AgNPs with ampicillin (Amp) against bacteria. (**a**) Formation of core silver nanoparticles with ampicillin. (**b**) Interaction of AgNPs-Amp complex over the cell wall of bacteria. (**c**) AgNPs-Amp complex inhibits the formation of cross-links in the peptidoglycan layer (which provides rigidity to the cell wall), leading to cell wall lysis. (**d**) AgNPs-Amp complex prevents the DNA unwinding (Reproduced from Fayaz et al. (2010) with permission from Elsevier B.V., the Netherlands)

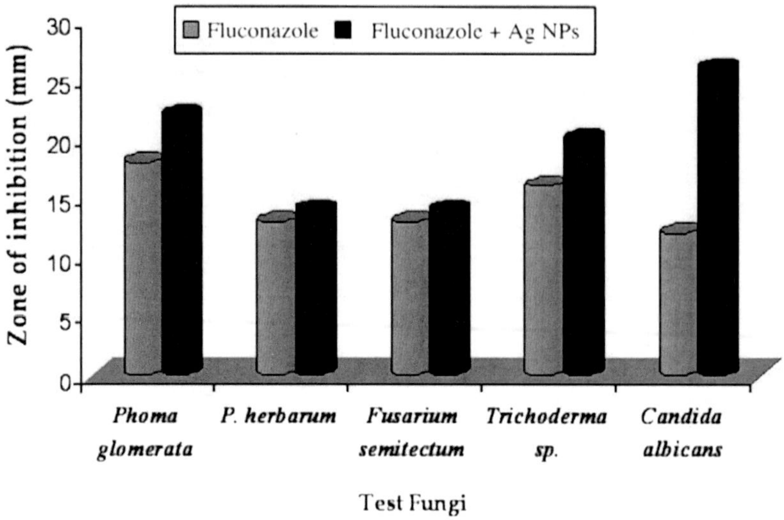

Fig. 12.5 Graphical representation of the zone of inhibition obtained for the combined effect of fluconazole and AgNPs against five fungi species (Reproduced from Gajbhiye et al. (2009) with permission from Elsevier)

The majority of studies exploring the synergistic effect of FM-AgNPs in combination with commercial antimicrobial drugs have been carried out for bacteria. Studies on fungal inhibition using antifungals combined with FM-NPs are scarce. However, some studies have reported the synergistic effect of antifungals with AgNPs, for instance, Vazquez-Muñoz et al. (2014) reported the synergistic effect of fluconazole with commercial AgNPs against *C. albicans*. Fluconazole also displays enhanced antifungal activity in combination with AgNPs from *A. alternata*; they were tested against *P. glomerata*, *P. herbarum*, *F. semitectum*, *Trichoderma* sp., and *C. albicans*. Maximum inhibition was found against *C. albicans*, followed by *P. glomerata* and *Trichoderma* sp., whereas no significant enhancement was found against *P. herbarum* and *F. semitectum* (Fig. 12.5) (Gajbhiye et al. 2009).

The synergistic result of combining commercial antibiotics with FM-NPs seems to be an excellent option to combat resistant pathogens; this approach could be advantageous because it offers the possibility to reduce the amounts of silver and commercial antibiotics to be applied in human infectious diseases. However, more research is needed to determine the synergistic effect for antifungal drugs in combination with FM-NPs.

12.5 Other Possible Applications of Fungal-Mediated Nanoparticles to Prevent Infections

The antimicrobial efficacy of AgNPs has been proven by many studies. Therefore, their incorporation into different materials has been explored and thus providing them with sterile properties. By doing this, nosocomial infections could be prevented using products coated or embedded with nanosilver, such as surgical coatings, wound dressings, contraceptive devices, and surgical instruments, and medical implants could prevent those infections; in fact some of these products are commercially available (Cheng et al. 2004; Dowsett 2004; Muangman et al. 2006; Chen et al. 2006; Lansdown 2006; Zhang et al. 2007; Cohen et al. 2007). Nevertheless, to our knowledge the use of biogenic NPs for the fabrication of commercial products is still not a reality.

Some efforts have been conducted to incorporate FM-AgNPs in cotton fabrics in order to determine the antimicrobial capacity of the modified material. Antimicrobial efficiency of cotton fabrics with FM-AgNPs from *F. oxysporum* was tested against the Gram-negative bacterium *E. coli*, Gram-positive bacterium *S. aureus*, and the fungus *C. albicans*. The authors reported good antibacterial property of the material and proposed the use of cyanobacteria to treat the effluents from the washing process of modified cotton fabric (Duran et al. 2007). In another report FM-AgNPs from *F. solani* were successfully applied to cotton fabrics, and the material showed antibacterial efficiency against *S. aureus* and *E. coli*. Modified cotton material retained excellent antibacterial property even after exposing the material to laundering for 20 cycles (El-Rafie et al. 2010). Similar results were found in cotton fabrics with FM-AgNPs from *A. alternata* with AgNPs incorporated into cotton fibers (Fig. 12.6); here the authors further evaluated the modified cotton fabric against the microbial degradation process by the soil burial test. After 14 days the material with NPs showed maximum protection against the adverse effects normally occurring by soil microflora (Ibrahim and Hassan 2016).

One important aspect for human health is water hygiene, thus looking for alternative water purification methods; gold nanoparticles from the fungus *R. oryzae* were utilized successfully to obtain potable water free from pathogens and pesticides, all in a single step (Das et al. 2009). FM-NPs have also been suggested as mosquito-controlling agents, which transmit serious human diseases; for instance, malaria is a mosquito-borne parasitic infection spread by *Anopheles stephensi*. Viral diseases of major international public health concern like dengue and chikungunya are transmitted by the mosquito *Aedes aegypti*. The use of AgNPs from *Cochliobolus lunatus*, *Isaria fumosorosea*, and *Trichoderma harzianum* was suggested as larvicide agents to control *A. aegypti*, *A. stephensi*, and *Culex quinquefasciatus* populations (Salunkhe et al. 2011; Banu and Balasubramanian 2014; Sundaravadivelan and Padmanabhan 2014). Likewise, the use of silver and gold NPs from the fungus *Chrysosporium tropicum* was proposed for mosquito control against the *A. aegypti* larvae (Soni and Prakash 2012).

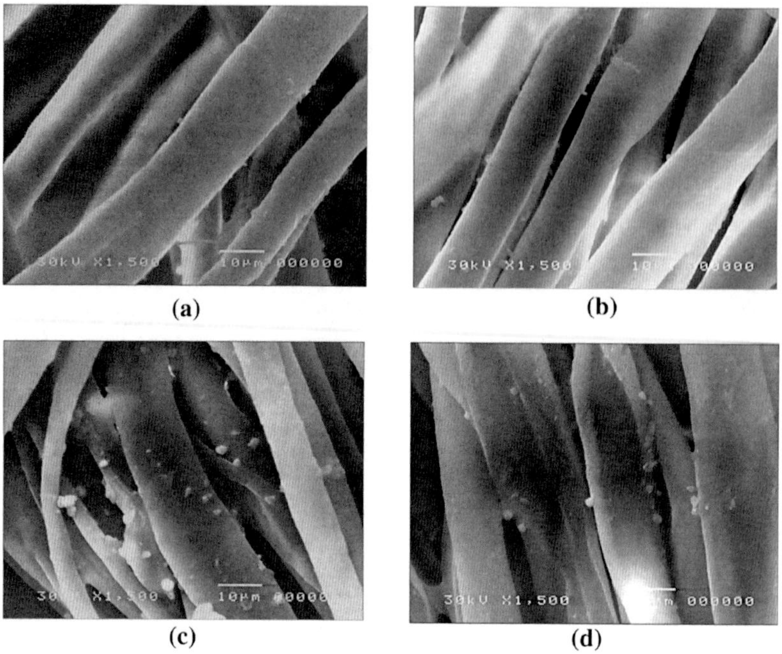

Fig. 12.6 Scanning electron microscope images of untreated cotton fabric (**a**), fabric treated with binder only (**b**), and fabric treated with the antimicrobial formulation (**c**, **d**) (Reproduced from Ibrahim and Hassan (2016) with permission from Elsevier)

On the whole FM-AgNPs could be applied in different ways to prevent/treat infectious diseases; however to suggest their use in materials intended to be in human contact, it is necessary to implement more complete studies. For instance, cytotoxic analyses of FM-NPs are needed before suggesting them for clinical trials.

12.6 Does the Size of Nanoparticles is Important for Microbicide Action?

The antimicrobial activity of FM-AgNPs has been demonstrated using different size ranges (Tables 12.1 and 12.2). Nevertheless, the efficacy of AgNPs is reported to be size and dose dependent; chemically synthesized AgNPs of average size 5, 7, 10, 15, 20, 30, 50, 63, 85, and 100 nm were tested against *E. coli*, *B. subtilis*, and *S. aureus*. Antibacterial effectivity increased with AgNPs of smaller sizes. In fact, the MIC and MBC values showed AgNPs of 5 nm with the best antibacterial activity against all the tested strains. By using the disk diffusion method (Fig. 12.7), AgNPs of the smaller sizes (5, 7, and 10 nm) also showed a clear zone of inhibition for the *E. coli* MTCC 443 strain. The authors also found different sensitivities

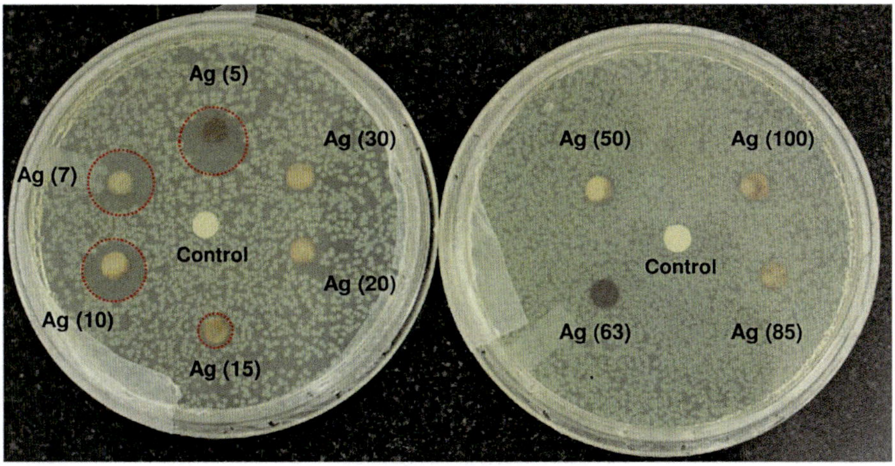

Fig. 12.7 Disk diffusion tests for different sized silver nanoparticles (chemically synthesized) against the *E. coli* MTCC 443 strain. The zone of inhibition is highlighted with a dashed circle indicating a noticeable antibacterial effect (Reproduced from Agnihotri et al. (2014) with permission from The Royal Society of Chemistry)

among the strains; *E. coli* MTCC 443 and *S. aureus* were found to be the most and least sensitive strains regardless of the size of AgNPs (Agnihotri et al. 2014). Apparently, antibacterial efficacy also is shape dependent, since truncated triangular silver nanoplates with a {111} lattice plane as the basal plane were found to display the strongest biocidal action, compared with spherical and rod-shaped nanoparticles (Pal et al. 2007).

Similar studies have not been carried out with FM-AgNPs, as shape and size control is still a challenge for biogenic NPs. Also, it is difficult to make any conclusions from the results presented in different studies because most results lack of MIC and MBC/MFC determinations. Therefore, it is important for further studies to consider the incorporation of MIC and/or MBC/MFC assays.

12.7 Final Remarks

The effectivity of FM-AgNPs against human pathogens either alone or in combination with commercial antibiotics is clear; they can inhibit a wide variety of bacterial and fungal strains. Furthermore, modified cotton material with FM-AgNPs showed excellent antibacterial properties; thus FM-AgNPs could also be incorporated into materials such as surgical coatings and wound dressings to prevent/treat infectious diseases. However, the production of more biocompatible NPs remains to be further investigated since fungal strains secrete a wide variety of compounds, and at present most synthesis protocols have used fungal filtrates of many pathogenic fungi containing all secreted molecules. Therefore a feasible alternative is to use non-pathogenic fungal strains and/or biocompatible molecules for the production of NPs; this is of prime importance in order to suggest their use in the clinic.

References

Abdeen S, Rimal Isaac RS, Geo S, Sornalekshmi S, Rose A, Praseetha PK (2013) Evaluation of antimicrobial activity of biosynthesized iron and silver nanoparticles using the fungi *Fusarium oxysporum* and *Actinomycetes* sp. on human pathogens. Nano Biomed Eng 5:39–45

Abeer RM, El-Aziz A, Al-Othman MR, Eifan SA, Mahmoud MA, Majrashi M (2013) Green synthesis of silver nanoparticles using *Aspergillus terreus* (kc462061). Dig J Nanomater Biostruct 8:1215–1225

Agnihotri S, Mukherji S, Mukherji S (2014) Size-controlled silver nanoparticles synthesized over the range 5-100 nm using the same protocol and their antibacterial efficacy. RSC Adv 4:3974–3983

Ahluwalia V, Kumar J, Sisodia R, Shakil NA, Walia S (2014) Green synthesis of silver nanoparticles by *Trichoderma harzianum* and their bio-efficacy evaluation against *Staphylococcus aureus* and *Klebsiella pneumonia*. Ind Crop Prod 55:202–206

Al-Othman MR, Abd El-Aziz ARM, Mahmoud MA, Eifan SA, El-Shikh MS, Majrashi M (2014) Application of silver nanoparticles as antifungal and antiaflatoxin b1 produced by *Aspergillus flavus*. Dig J Nanomater Biostruct 9:151–157

Anderson RM, May RM, Anderson B (1998) Infectious diseases of humans: dynamics and control. Oxford university press, Oxford

Arun G, Eyini M, Gunasekaran P (2014) Green synthesis of silver nanoparticles using the mushroom fungus *Schizophyllum* commune and its biomedical applications. Biotechnol Bioprocess Eng 19:1083–1090

Aziz N, Faraz M, Pandey R, Sakir M, Fatma T, Varma A, Barman I, Prasad R (2015) Facile algae-derived route to biogenic silver nanoparticles: synthesis, antibacterial and photocatalytic properties. Langmuir 31:11605–11612. https://doi.org/10.1021/acs.langmuir.5b03081

Aziz N, Pandey R, Barman I, Prasad R (2016) Leveraging the attributes of *Mucor hiemalis*-derived silver nanoparticles for a synergistic broad-spectrum antimicrobial platform. Front Microbiol 7:1984. https://doi.org/10.3389/fmicb.2016.01984

Balakumaran MD, Ramachandran R, Balashanmugam P, Mukeshkumar DJ, Kalaichelvan PT (2016) Mycosynthesis of silver and gold nanoparticles: optimization, characterization and antimicrobial activity against human pathogens. Microbiol Res 182:8–20

Banu AN, Balasubramanian C (2014) Optimization and synthesis of silver nanoparticles using *Isaria fumosorosea* against human vector mosquitoes. Parasitol Res 113(10):3843–3851

Birla SS, Tiwari VV, Gade AK, Ingle AP, Yadav AP, Rai MK (2009) Fabrication of silver nanoparticles by *Phoma glomerata* and its combined effect against *Escherichia coli*, *Pseudomonas aeruginosa* and *Staphylococcus aureus*. Lett Appl Microbiol 48:173–179

Bharathidasan R, Panneerselvam A (2012) Biosynthesis and characterization of silver nanoparticles using endophytic fungi *Aspergillus concius*, *Penicillium janthinellum* and *Phomosis* sp. Int J Pharm Sci Res 3:3163–3169

Castro-Longoria E (2016) Fungal biosynthesis of nanoparticles, a cleaner alternative. In: Purchase D (ed) Fungal applications in sustainable environmental biotechnology. Springer International Publishing, Cham, pp 223–238

Chen J, Han CM, Lin XW, Tang ZJ, Su SJ (2006) Effect of silver nanoparticle dressing on second degree burn wound. Zhonghua Wai Ke Za Zhi 44:50–52

Cheng D, Yang J, Zhao Y (2004) Antibacterial materials of silver nanoparticles application in medical appliances and appliances for daily use. Chin Med Equip J 4:26–32

Cirz RT, Chin JK, Andes DR, de Crécy-Lagard V, Craig WA, Romesberg FE (2005) Inhibition of mutation and combating the evolution of antibiotic resistance. PLoS Biol 3:1024–1033

Cohen ML (1992) Epidemiology of drug resistance: implications for a post-antimicrobial era. Science 257:1050–1055

Cohen MS, Stern JM, Vanni AJ, Kelley RS, Baumgart E, Field D, Libertino JA, Summerhayes IC (2007) In vitro analysis of a nanocrystalline silver-coated surgical mesh. Surg Infect 8(3):397–403

Dar MA, Ingle A, Rai M (2013) Enhanced antimicrobial activity of silver nanoparticles synthesized by *Cryphonectria* sp. evaluated singly and in combination with antibiotics. Nanomedicine 9:105–110

Das SK, Das AR, Guha AK (2009) Gold nanoparticles: microbial synthesis and application in water hygiene management. Langmuir 25:8192–8199

Das SK, Masuda M, Sakurai A, Sakakibara M (2010) Medicinal uses of the mushroom *Cordyceps militaris*: current state and prospects. Fitoterapia 81:961–968

Devi LS, Joshi SR (2012) Antimicrobial and synergistic effects of silver nanoparticles synthesized using soil fungi of high altitudes of eastern Himalaya. Mycobiology 40:27–34

Devi LS, Bareh DA, Joshi SR (2014) Studies on biosynthesis of antimicrobial silver nanoparticles using endophytic fungi isolated from the ethno-medicinal plant *Gloriosa superba* L. Proc Natl Acad Sci, India, Sect B Biol Sci 84:1091–1099

Diekema DJ, Messer SA, Brueggemann AB, Coffman SL, Doern GV, Herwaldt LA, Pfaller MA (2002) Epidemiology of candidemia: 3-year results from the emerging infections and the epidemiology of Iowa organisms study. J Clin Microbiol 40(4):1298–1302

Dowsett C (2004) The use of silver-based dressings in wound care. Nurs Stand 19(7):56–60

Duran N, Marcato PD, De S, Gabriel IH, Alves OL, Esposito E (2007) Antibacterial effect of silver nanoparticles produced by fungal process on textile fabrics and their effluent treatment. J Biomed Nanotechnol 3:203–208

El-Baz AF, El-Batal AI, Abomosalam FM, Tayel AA, Shetaia YM, Yang ST (2016) Extracellular biosynthesis of anti-*Candida* Silver nanoparticles using *Monascus purpureus*. J Basic Microbiol 56:531–540

Elechiguerra JL, Burt JL, Morones JR, Camacho-Bragado A, Gao X, Lara HH, Yacaman MJ (2005) Interaction of silver nanoparticles with HIV-1. J Nanobiotechnol 3(1):6

El-Rafie MH, Mohamed AA, Shaheen TI, Hebeish A (2010) Antimicrobial effect of silver nanoparticles produced by fungal process on cotton fabrics. Carbohydr Polym 80:779–782

Fayaz AM, Balaji K, Girilal M, Yadav R, Kalaichelvan PT, Venketesan R (2010) Biogenic synthesis of silver nanoparticles and their synergistic effect with antibiotics: a study against gram-positive and gram-negative bacteria. Nanomedicine 6:103–109

Gajbhiye M, Kesharwani J, Ingle A, Gade A, Rai M (2009) Fungus-mediated synthesis of silver nanoparticles and their activity against pathogenic fungi in combination with fluconazole. Nanomedicine 5:382–386

Ganachari SV, Bhat R, Deshpande R, Venkataraman A (2012) Extracellular biosynthesis of silver nanoparticles using fungi *Penicillium diversum* and their antimicrobial activity studies. Bio Nano Sci 2:316–321

Ghany A, Mohamed T (2013) *Stachybotrys chartarum*: a novel biological agent for the extracellular synthesis of silver nanoparticles and their antimicrobial activity. Indones J Biotechnol 18:75–82

Ghodake VP, Kininge PT, Magdum SP, Dive AS, Pillai MM (2011) Biosynthesis of silver nanoparticles using *Trichosporon beigelii* NCIM 3326 and evaluation of their antimicrobial activity. J Eng Res Stud E-ISSN 976:7916

Gholami-Shabani M, Akbarzadeh A, Norouzian D, Amini A, Gholami-Shabani Z, Imani A, Chiani M, Riazi G, Shams-Ghahfarokhi M, Razzaghi-Abyaneh M (2014) Antimicrobial activity and physical characterization of silver nanoparticles green synthesized using nitrate reductase from *Fusarium oxysporum*. Appl Biochem Biotechnol 172(8):4084–4098

Gudikandula K, Metuku RP, Burra S, Laxmi SVSSS, Nidadavolu HB, Pabba SK, Alha SCM (2015) Fungus-mediated synthesis of silver nanoparticles and their activity against gram positive and gram negative bacteria in combination with antibiotics. Walailak J Sci Technol 12(7):613–622

Hamedi S, Shojaosadati SA, Shokrollahzadeh S, Hashemi-Najafabadi S (2014) Extracellular biosynthesis of silver nanoparticles using a novel and non-pathogenic fungus, *Neurospora* intermedia: controlled synthesis and antibacterial activity. World J Microbiol Biotechnol 30:693–704

Hiremath J, Rathod V, Ninganagouda S, Singh D, Prema K (2014) Antibacterial activity of silver nanoparticles from *Rhizopus spp* against gram negative *E.coli*-MDR strains. J Pure Appl Microbiol 8:555–562

Huh AJ, Kwon YJ (2011) 'Nanoantibiotics': a new paradigm for treating infectious diseases using nanomaterials in the antibiotics resistant era. J Control Release 156:128–145

Ibrahim HMM, Hassan MS (2016) Characterization and antimicrobial properties of cotton fabric loaded with green synthesized silver nanoparticles. Carbohyd Polym 151:841–850

Ingle A, Gade A, Pierrat S, Sonnichsen C, Rai M (2008) Mycosynthesis of silver nanoparticles using the fungus *Fusarium acuminatum* and its activity against some human pathogenic bacteria. Curr Nanosci 4:141–144

Jaidev LR, Narasimha G (2010) Fungal mediated biosynthesis of silver nanoparticles, characterization and antimicrobial activity. Colloids Surf B Biointerfaces 81:430–433

Joshi PA, Bonde SR, Gaikwad SC, Gade AK, Abd-Elsalam K, Rai MK (2013) Comparative studies on synthesis of silver nanoparticles by *Fusarium oxysporum* and *Macrophomina phaseolina* and it's efficacy against bacteria and *Malassezia furfur*. J Bionanosci 7:1–8

Kathiresan K, Alikunhi NM, Pathmanaban S, Nabikhan A, Kandasamy S (2010) Analysis of antimicrobial silver nanoparticles synthesized by coastal strains of *Escherichia coli* and *Aspergillus niger*. Can J Microbiol 56:1050–1059

Kim JS, Kuk E, Yu KN, Kim JH, Park SJ, Lee HJ, Kim SH, Park YK (2007) Antimicrobial effects of silver nanoparticles. Nanomed Nanotechnol 3:95–101

Kim KJ, Sung WS, Suh BK, Moon SK, Choi JS, Kim JG, Lee DG (2009) Antifungal activity and mode of action of silver nano-particles on *Candida albicans*. Biometals 22:235–242

Lansdown AB (2004) A review of the use of silver in wound care: facts and fallacies. Br J Nurs 13(6 Suppl):6–17

Lansdown AB (2006) Silver in health care: antimicrobial effects and safety in use. Curr Probl Dermatol 33:17–34

Lehrer RI, Howard DH, Sypherd PS, Edwards JE, Segal GP, Winston DJ (1980) Mucormycosis. Ann Intern Med 93:93–108

Leone S, Pascale R, Vitale M, Esposito S (2012) Epidemiology of diabetic foot. Infez Med 1:8–13

Li P, Li J, Wu C, Wu Q, Li J (2005) Synergistic antibacterial effects of β-lactam antibiotic combined with silver nanoparticles. Nanotechnology 16:1912–1917

Li G, He D, Qian Y, Guan B, Gao S, Cui Y, Yokoyama K, Wang L (2012) Fungus-mediated green synthesis of silver nanoparticles using *Aspergillus terreus*. Int J Mol Sci 13(1):466–476

Lu L, Sun RW, Chen R, Hui CK, Ho CM, Luk JM, Lau GK, Che CM (2008) Silver nanoparticles inhibit hepatitis B virus replication. Antivir Ther 13(2):253

Majeed S, bin Abdullah MS, Dash GK, Ansari MT, Nanda A (2016) Biochemical synthesis of silver nanoparticles using filamentous fungi *Penicillium decumbens* (MTCC-2494) and its efficacy against A-549 lung cancer cell line. Chin J Nat Med 14:615–620

Maliszewska I, Sadowski Z (2009) Synthesis and antibacterial activity of silver nanoparticles. J Phys Conf Ser 146:012024

Moazeni M, Shahverdi AR, Nabili M, Noorbakhsh F, Rezaie S (2014) Green synthesis of silver nanoparticles: the reasons for and against *Aspergillus parasiticus*. Nanomed J 4:267–275

Mohamed YM, Azzam AM, Amin BH, Safwat NA (2015) Mycosynthesis of iron nanoparticles by *Alternaria alternata* and its antibacterial activity. Afr J Biotechnol 14:1234–1241

Morones JR, Elechiguerra JL, Camacho A, Holt K, Kouri JB, Ramirez JT, Yacaman MJ (2005) The bactericidal effect of silver nanoparticles. Nanotechnology 16(10):2346–2353

Muangman P, Chuntrasakul C, Silthram S, Suvanchote S, Benjathanung R, Kittidacha S, Rueksomtawin S (2006) Comparison of efficacy of 1% silver sulfadiazine and Acticoat for treatment of partial-thickness burn wounds. J Med Assoc Thail 89(7):953–958

Muhsin TM, Hachim AK (2014) Mycosynthesis and characterization of silver nanoparticles and their activity against some human pathogenic bacteria. World J Microbiol Biotechnol 30:2081–2090

Musarrat J, Dwivedi S, Singh BR, Al-Khedhairy AA, Azam A, Naqvi A (2010) Production of antimicrobial silver nanoparticles in water extracts of the fungus *Amylomyces rouxii* strain KSU-09. Bioresour Technol 101:8772–8776

Nameirakpam ND, Dheeban SP, Sutha S (2012) Biomimetic synthesis of silver nanoparticles from an endophytic fungus and their antimicrobial efficacy. Int J Biomed Adv Res 3:409–415

Naqvi SZ, Kiran U, Ali MI, Jamal A, Hameed A, Ahmed S, Ali N (2013) Combined efficacy of biologically synthesized silver nanoparticles and different antibiotics against multidrug-resistant bacteria. Int J Nanomedicine 8:3187–3195

Netala VR, Kotakadi VS, Bobbu P, Gaddam SA, Tartte V (2016) Endophytic fungal isolate mediated biosynthesis of silver nanoparticles and their free radical scavenging activity and antimicrobial studies. Biotech 6:132–141

Nithya R, Ragunathan R (2009) Synthesis of silver nanoparticle using *Pleurotus sajor caju* and its antimicrobial study. Dig J Nanomater Biostruct 4:623–629

Nithya R, Ragunathan R (2014) In vitro synthesis, characterization and medical application of silver nanoparticle by using a lower fungi. Middle East J Sci Res 21:922–928

Pal S, Tak YK, Song JM (2007) Does the antibacterial activity of silver nanoparticles depend on the shape of the nanoparticle? A study of the gram-negative bacterium *Escherichia coli*. Appl Environ Microbiol 27:1712–1720

Pelgrift RY, Friedman AJ (2013) Nanotechnology as a therapeutic tool to combat microbial resistance. Adv Drug Deliv Rev 65:1803–1815

Percival SL, Bowler PG, Russell D (2005) Bacterial resistance to silver in wound care. J Hosp Infect 60(1):1–7

Perea S, Patterson TF (2002) Antifungal resistance in pathogenic fungi. Clin Infect Dis 35:1073–1080

Person AK, Chudgar SM, Norton BL, Tong BC, Stout JE (2010) *Aspergillus niger*: an unusual cause of invasive pulmonary aspergillosis. J Med Microbiol 59:834–838

Prabakaran K, Ragavendran C, Natarajan D (2016) Mycosynthesis of silver nanoparticles from *Beauveria bassiana* and its larvicidal, antibacterial, and cytotoxic effect on human cervical cancer (HeLa) cells. RSC Adv 6:44972–44986

Prasad R (2016) Advances and applications through fungal nanobiotechnology. Springer, International Publishing, Cham. (ISBN: 978-3-319-42989-2)

Prasad R, Pandey R, Barman I (2016) Engineering tailored nanoparticles with microbes: quo vadis. WIREs Nanomed Nanobiotechnol 8:316–330. https://doi.org/10.1002/wnan.1363

Qian Y, Yu H, He D, Yang H, Wang W, Wan X, Wang L (2013) Biosynthesis of silver nanoparticles by the endophytic fungus *Epicoccum nigrum* and their activity against pathogenic fungi. Bioprocess Biosyst Eng 36:1613–1619

Raheman F, Deshmukh S, Ingle A, Gade A, Rai M (2011) Silver nanoparticles: novel antimicrobial agent synthesized from an endophytic fungus *Pestalotia* sp. isolated from leaves of *Syzygium cumini* (L). Nano Biomed Eng 3:174–178

Rahimi G, Alizadeh F, Khodavandi A (2016) Mycosynthesis of silver nanoparticles from *Candida albicans* and its antibacterial activity against *Escherichia coli* and *Staphylococcus aureus*. Trop J Pharm Res 15:371–375

Rai M, Yadav A, Gade A (2009) Silver nanoparticles as a new generation of antimicrobials. Biotechnol Adv 27:76–83

Rai MK, Deshmukh SD, Ingle AP, Gade AK (2012) Silver nanoparticles: the powerful nanoweapon against multidrug-resistant bacteria. J Appl Microbiol 112:841–852

Rajakumar G, Rahuman AA, Roopan SM, Khanna VG, Elango G, Kamaraj C, Abduz Zahir A, Velayutham K (2012) Fungus-mediated biosynthesis and characterization of TiO_2 nanoparticles and their activity against pathogenic bacteria. Spectrochim Acta A Mol Biomol Spectrosc 91:23–29

Rathod V, Ranganath E (2011) Synthesis of monodispersed silver nanoparticles by *Rhizopus Stolonifer* and its antibacterial activity against MDR strains of *Pseudomonas aeruginosa* from burnt patients. Int J Environ Sci 1(7):1582

Ravindra BK, Rajasab AH (2015) Silver nanoparticles synthesis from different fungal species and their antifungal effect. Int J Pharm Pharm Sci 7:165–170

Ray S, Sarkar S, Kundu S (2011) Extracellular biosynthesis of silver nanoparticles using the mycorrhizal mushroom *Tricholoma crassum* (berk.) sacc.: its antimicrobial activity against pathogenic bacteria and fungus, including multidrug resistant plant and human bacteria. Dig J Nanomater Biostruct 6:1289–1299

Rodrigues AG, Ping LY, Marcato PD, Alves OL, Silva MCP, Ruiz RC, Melo IS, Tasic L, De Souza AO (2013) Biogenic antimicrobial silver nanoparticles produced by fungi. Appl Microbiol Biotechnol 97(2):775–782

Roy S, Mukherjee T, Chakraborty S, Das TK (2013) Biosynthesis, characterisation & antifungal activity of silver nanoparticles synthesized by the fungus *Aspergillus foetidus* MTCC8876. Dig J Nanomater Bios 8:197–205

Sadhasivam S, Shanmugam P, Veerapandian M, Subbiah R, Yun K (2012) Biogenic synthesis of multidimensional gold nanoparticles assisted by *Streptomyces hygroscopicus* and its electrochemical and antibacterial properties. Biometals 25(2):351–360

Safdar N, Maki DG (2002) The commonality of risk factors for nosocomial colonization and infection with antimicrobial-resistant *Staphylococcus aureus*, *Enterococcus*, gram-negative bacilli, *Clostridium difficile*, and *Candida*. Ann Intern Med 136:834–844

Sagar G, Ashok B (2012) Green synthesis of silver nanoparticles using *Aspergillus niger* and its efficacy against human pathogens. Eur J Exp Biol 2:1654–1658

Saha S, Sarkar J, Chattopadhyay D, Patra S, Chakraborty A, Acharya K (2010) Production of silver nanoparticles by a phytopathogenic fungus *Bipolaris nodulosa* and its antimicrobial activity. Dig J Nanomater Biostruct 5:887–895

Saha S, Chattopadhyay D, Acharya K (2011) Preparation of silver nanoparticles by bio-reduction using *Nigrospora oryzae* culture filtrate and its antimicrobial activity. Dig J Nanomater Biostruct 6:1519–1528

Salunkhe RB, Patil SV, Patil CD, Salunke BK (2011) Larvicidal potential of silver nanoparticles synthesized using fungus *Cochliobolus lunatus* against *Aedes aegypti* (Linnaeus, 1762) and *Anopheles stephensi* Liston (Diptera; Culicidae). Parasitol Res 109(3):823–831

Santamarina MP, Roselló J, Llacer R, Sanchis V (2002) Antagonistic activity of *Penicillium oxalicum* Corrie and Thom, *Penicillium decumbens* Thom and *Trichoderma harzianum* Rifai isolates against fungi, bacteria and insects in vitro. Rev Iberoam Micol 19:99–103

Saravanan M, Nanda A (2010) Extracellular synthesis of silver bionanoparticles from *Aspergillus clavatus* and its antimicrobial activity against MRSA and MRSE. Colloids Surf B Biointerfaces 77:214–218

Shrivastava S, Bera T, Roy A, Singh G, Ramachandrarao P, Dash D (2007) Characterization of enhanced antibacterial effects of novel silver nanoparticles. Nanotechnology 18:225103

Singh D, Rathod V, Ninganagouda S, Hiremath J, Singh AK, Mathew J (2014) Optimization and characterization of silver nanoparticle by endophytic fungi *Penicillium* sp isolated from *Curcuma longa* (turmeric) and application studies against MDR *E. coli* and *S. aureus*. Bioinorg Chem Appl 2014:1–8

Snyder RJ, Hanft JR (2009) Diabetic foot ulcers--effects on QOL, costs, and mortality and the role of standard wound care and advanced-care therapies. Ostomy Wound Manage 55:28–38

Soni N, Prakash S (2012) Efficacy of fungus mediated silver and gold nanoparticles against *Aedes aegypti* larvae. Parasitol Res 110:175–184

Sundaravadivelan C, Padmanabhan MN (2014) Effect of mycosynthesized silver nanoparticles from filtrate of *Trichoderma harzianum* against larvae and pupa of dengue vector *Aedes aegypti* L. Environ Sci Pollut Res 21(6):4624–4633

Thakker JN, Dalwadi P, Dhandhukia PC (2013) Biosynthesis of gold nanoparticles using *Fusarium oxysporum* f. Sp. *cubense* JT1, a plant pathogenic fungus. Biotechnology 2013:1–5

Vardakas KZ, Horianopoulou M, Falagas ME (2008) Factors associated with treatment failure in patients with diabetic foot infections: an analysis of data from randomized controlled trials. Diab Res Clin Pract 80:344–351

Vazquez-Muñoz R, Avalos Borja M, Castro-Longoria E (2014) Ultrastructural analysis of *Candida albicans* when exposed to silver nanoparticles. PLoS One 9(10):e108876

Verma VC, Kharwar RN, Gange AC (2010) Biosynthesis of antimicrobial silver nanoparticles by the endophytic fungus *Aspergillus clavatus*. Nanomedicine 5:33–40

Vigneshwaran N, Kathe AA, Varadarajan PV, Nachane RP, Balasubramanya RH (2007) Silver-protein (core-shell) nanoparticle production using spent mushroom substrate. Langmuir 23:7113–7117

Wang L, Liu CC, Wang YY, Xu H, Su H, Cheng X (2016) Antibacterial activities of the novel silver nanoparticles biosynthesized using *Cordyceps militaris* extract. Curr Appl Phys 16:969–973

Weinstein RA, Gaynes R, Edwards JR (2005) Overview of nosocomial infections caused by gram-negative bacilli. Clin Infect Dis 41:848–854

Williams R (2001) Resistance as a worldwide problem. In: Lewis K, Salyers AA, Taber HW, Wax RG (eds) Bacterial resistance to antimicrobials. Marcel Dekker, Inc., New York, pp 232–238

Xue BJ, He D, Gai S, Wang DY, Yokoyama K, Wang L (2016) Biosynthesis of silver nanoparticles by the fungus *Arthroderma fulvum* and its antifungal activity against genera of *Candida*, *Aspergillus* and *Fusarium*. Int J Nanomedicine 11:1899–1906

Yazdanpanah L, Nasiri M, Adarvishi S (2015) Literature review on the management of diabetic foot ulcer. World J Diabetes. 15 6(1):37

Yehia RS, Al-Sheikh H (2014) Biosynthesis and characterization of silver nanoparticles produced by *Pleurotus ostreatus* and their anticandidal and anticancer activities. World J Microbiol Biotechnol 30:2797–2803

Zhang Z, Yang M, Huang M, Hu Y, Xie J (2007) Study on germicidal efficacy and toxicity of compound disinfectant gel of nanometer silver and chlorhexidine acetate. Chin J Health Lab Technol 17:1403–1406

Chapter 13
Nanobiotechnology Applications in Special Reference to Fungi

Safiye Elif Korcan and Muhsin Konuk

Abstract Nanobiotechnology is placed in the intersection point of nanobiology, biotechnology, nanotechnology, and biology. This technique approach provides an angle to scientists to imagine new systematic gates to study on. From the point of biological sciences, it is an inspiring area for the studies which has not been created. The fungi can synthesize nanoparticles both inside and outside of their cells. In extracellular synthesis, after growing and obtaining the biomass these cells are incubated in the presence of metal salt solutions. The synthesis of nanobioparticles can be observed easily by looking at the color changes in the cultures. After completing the synthesis, nanoparticles were then subjected to centrifuge in high speed and density gradient. Then they were collected by washing with water or organic solvents like EtOH/MeOH. The main focus of this review is to introduce the application of fungi in the synthesis of nanoparticles biologically.

13.1 Introduction

The word "nano" is used to indicate the dimension of less than 100 nm. A nanoparticle (NP) (nanopowder, nanocluster, or nanocrystal) is an ultramicroscopic particle. Nanoparticles (NPs) have different physical, chemical, electronic, electrical, mechanical, magnetic, thermal, dielectric, optical, and biological characters. Developments of this area present great potential of various sectors like energy, environment, agriculture, and healthcare. Hence, it has been building great expectations not only in the academia but also among the investors, governments, and industries (Jain et al. 2011).

S.E. Korcan
Uşak University, Vocational School of Health Services, Department of Medical Services and Techniques, Uşak, Turkey
e-mail: safiye.korcan@usak.edu.tr

M. Konuk (✉)
Üsküdar University, School of Engineering and Natural Sciences, Molecular Biology and Genetics Department, 34662 Istanbul, Turkey
e-mail: mkonuk@gmail.com

R. Prasad (ed.), *Fungal Nanotechnology*, Fungal Biology,
DOI 10.1007/978-3-319-68424-6_13

Table 13.1 Mechanism of nanoparticle biosynthesis using different sources

Biological systems	Possible mechanism
Plant	Secondary metabolites (alkaloids, flavonoids, saponins, steroids, tannins, and other nutritional compounds) act as reducing and stabilizing agents
Algae	Polysaccharides have hydroxyl groups and other functionalities that can play important roles in both the reduction and the stabilization of nanoparticles
Fungi	Reducing enzyme intracellularly or extracellularly and the procedure of biomimetic mineralization
Yeast	Membrane bound (as well as cytosolic) oxidoreductases and quinones
Bacteria	The microbial cell reduces metal ions by the use of specific reducing enzymes like NADH-dependent reductase or nitrate-dependent reductase
Virus	Tobacco mosaic virus (TMV) was used as template for the synthesis of iron oxides by oxidative hydrolysis, co-crystallization of CdS and PbS, and the synthesis of SiO2 by sol-gel condensation. It happened with the help of external groups of glutamate and aspartate on the external surface of the virus. Self-assembled viral capsids of genetically engineered viruses were exploited as biological templates for the assembly of quantum dot nanowires

Moghaddam et al. (2015); Yen and Mashitah (2012)

There are diverse techniques to synthesize different kinds of NPs (Xiangqian et al. 2011). Physical and chemical ones are popular but the use of toxic chemicals greatly limits their biomedical applications. For this reason, improving of reliable, nontoxic, and environmentally friendly methods for synthesis of NPs is very important (Prasad 2014; Prasad et al. 2016, 2017). Research in biotechnology has revealed that there are reliable, eco-friendly processes for synthesis of novel nanomaterials. Biological synthesis of nanoparticles using various biological systems such as yeast, bacteria, fungi, algae, and plant extract has also been in our knowledge (Yen and Mashitah 2012; Prasad et al. 2016).

13.2 Nanoparticle Synthesis Using Microorganisms

In the last decade, the application of green nanotechnology has been investigated as an alternative way to chemical and physical techniques. Green synthesis of nanoparticles can be done by polysaccharide method, tollens method, irradiation method, biological methods, and polyoxometalates method (Sharma et al. 2009). As shown in Table 13.1, biological syntheses of NPs in different biological systems have been reported (Yen and Mashitah 2012). As it is well known, many biological systems accumulate inorganic material inside or outside of the cell to form NPs. Many microbial species can produce metal NPs (gold, silver, goldsilver alloy, selenium, tellurium, platinum, palladium, silica, titania, zirconia, etc.). This kind of syntheses of NPs brings together the nanotechnology and biotechnology.

13.2.1 Advantages of Biological Synthesis of Nanoparticles

(i) Routine methods for synthesing of the metal nanomaterials often need to use of organic solvents and/or high-energy input. In opposition, microbes have evolved to possess molecular machineries to detoxify heavy metals, mainly by operating metal-binding peptides (Park et al. 2016).
(ii) Biological methods for nanoparticle synthesis would help avoiding many of the detrimental features by enabling synthesis at mild pH, pressure and temperature, and at a substantially lower cost (Jain et al. 2011).
(iii) Biological process is also an environmentally friendly way, because, both production and remediation of NPs can be achieved at the same time.
(iv) Diverse NPs, including those that have never been chemically synthesized, can be synthesized biologically (Park et al. 2016).

13.2.2 Advances of Fungal Synthesis of Nanoparticles

The fungi such as *Fusarium oxysporum*, *Colletotrichum sp.* (Shankar et al. 2003), *Trichothecium sp.*, *Trichoderma asperellum*, *T. viride*, (Ahmad et al. 2005; Fayaz et al. 2010), *Phanerochaete chrysosporium* (Fayaz et al. 2006), *F. solani* (Ingle et al. 2009), *F. semitectum* (Basavaraja et al. 2008), *A. fumigatus* (Bhainsa and D'Souza 2006), *Coriolus versicolor* (Bhainsa et al. 2009), *Aspergillus niger* (Gade et al. 2008), *Phoma glomerata* (Birla et al. 2009), *Penicillium brevicompactum* (Shaligram et al. 2009), *Cladosporium cladosporioides* (Balaji et al. 2009), *Penicillium fellutanum* (Kathiresan et al. 2009), and *Volvariella volvaceae* (Philip 2009) have been investigated for NPs synthesis. Potential fungal isolates used for the biosynthesis of nanoparticles were given in Table 13.2.

13.2.3 Fungi Are More Advantageous Compared to Other Microorganisms

(a) Mycelial mesh of fungi can flow pressure. Aggregation and other conditions in bioreactors/chambers might be also compared to plant and bacteria.
(b) These are critical to grow and easy to both handle and manufacture. The extracellular reductive protein secretions are high and can be easily managed.
(c) Since the nanoparticles precipitated outside the cell are devoid of unnecessary cellular components, they might be directly used in various applications (Narayanan and Sakthivel 2010).
(d) Since fungi have the advantage of producing very high secreted proteins, this feature might increase nanoparticle synthesis grade.

Table 13.2 Potential fungal isolates used for the biosynthesis of nanoparticles

Fungi	NPs	Shape and location	Min and max particle size (nm)	References
Fusarium oxysporum	Pt	Rectangular, triangular, hexagonal,square,spherical, and aggregates	70–180	Govender et al. (2009); Moghaddam et al. (2015)
	Cd	Spherical, extracellular	9–15	Kumar et al. (2007a, b); Moghaddam et al. (2015)
	Ag	Aggregates, spherical, extracellular	5–50	Ahmad et al. (2003); Kumar et al. (2007a, b); Moghaddam et al. (2015)
	Au	Triangular, spherical, extracellular, or intracellular	2–50	Mandal et al. (2006); Zhang et al. (2011); Moghaddam et al. (2015); Khandel and Kumar (2016)
	$PbCO_3$, $CdCO_3$	Spherical, extracellular	120–200	Sanyal et al. (2005); Li et al. (2011)
	$SrCO_3$	Needlelike, extracellular	10–50	Rautaray et al. (2004); Li et al. (2011)
	CdSe	Spherical, extracellular	9–15	Kumar et al. (2007a, b); Narayanan and Sakthivel (2010)
	CdS	Spherical, extracellular	5–20	Ahmad et al. (2002); Salahuddin and Azamal (2016)
	TiO_2	Spherical, extracellular	6–13	Bansal et al. (2005); Salahuddin and Azamal (2016)
	$BaTiO_3$	Spherical, extracellular	4–5	Bansal et al. (2006); Narayanan and Sakthivel (2010)

(continued)

Table 13.2 (continued)

Fungi	NPs	Shape and location	Min and max particle size (nm)	References
	ZrO_2	Spherical,	3–11	Bansal et al. (2004); Salahuddin and Azamal (2016)
	Si	Quasi-spherical	5–15	Bansal et al. (2005); Narayanan and Sakthivel (2010)
	Bi_2O_3	Quasi	5–8	Uddin et al. (2008); Narayanan and Sakthivel (2010)
	BT	Extracellular	4–5	Bansal et al. (2006); Khandel and Kumar (2016)
	Fe_3O_4	Irregular, quasi-spherical	20–50	Bharde et al. (2006); Khandel and Kumar (2016)
Fusarium oxysporum f. sp. *lycopersici*	Pt	Hexagonal, pentagonal, circular, squares, rectangles Extra- and intracellular	10–100	Riddin et al. (2006); Govender et al. (2009); Moghaddam et al. (2015); Narayanan and Sakthivel (2010)
Fusarium spp.	Zn	Alteration intracellular	100–200	Velmurugan et al. (2010); Moghaddam et al. (2015)
Fusarium solani	Ag	Spherical, extracellular	5–35	Maliszewska et al. (2009a, b); Khandel and Kumar (2016)
Fusarium culmorum	Ag, Au, Pb, Cu	Spherical, extracellular	5–10	Bharde et al. (2006); Khandel and Kumar (2016)
Aspergillus clavitus	Ag	Extracellular	550–650	Saravanan and Nanda (2010); Moghaddam et al. (2015)
	Au	Triangular, spherical and hexagonal, extracellular	24.4	Verma et al. (2011)

(continued)

Table 13.2 (continued)

Fungi	NPs	Shape and location	Min and max particle size (nm)	References
Aspergillus fumigatus	ZnO	Spherical and hexagonal, extracellular	1.2–6.8	Raliya (2013); Moghaddam et al. (2015)
	Ag, Ag-Au	Mostly spherical. Extracellular	5–25/15->120	Bhainsa and D'Souza (2006)
Aspergillus oryzae TFR9	$FeCl_3$	Spherical	10–24.6	Binupriya et al. (2010a, b); Raliya (2013); Moghaddam et al. (2015); Siddiqi and Azamal (2016)
Aspergillus oryzae	Ag, Zn, Au	Spherical, extracellular	2.78–5.76	Khandel and Kumar (2016)
Aspergillus oryzae var. viridis	Au	Various shapes Mycelial surface	10–60	Binupriya et al. (2010a, b); Siddiqi and Azamal (2016)
Aspergillus tubingensis	$Ca_3P_2O_8$	Spherical, extracellular	28.2	Tarafdar et al. (2012); Siddiqi and Azamal (2016)
Aspergillus niger	Au	Nanowalls, spiral plates, polydispersed or spherical,	12.8–20	Xie et al. (2007); Bhambure et al. (2009)
	Ag	Spherical, extracellular	3–30	Alani et al. (2012); Moghaddam et al. (2015)
Aspergillus flavus	Ag	Spherical, cell wall surface	8.92–17	Vigneshwaran et al. (2007a, b); Moghaddam et al. (2015)
	TiO_2	Extracellular	12–74	Vigneshwaran et al. (2007a, b); Rajakumar et al. (2012); Raliya et al. (2015); Moghaddam et al. (2015)
Aspergillus clavitus	Ag	Extracellular	100–200	Saravanan and Nanda (2010)
Aspergillus terreus	Ag, Au-Ag	Spherical, extracellular	1–20	Khandel and Kumar (2016)

(continued)

Table 13.2 (continued)

Fungi	NPs	Shape and location	Min and max particle size (nm)	References
A. sydowii	Au	Spherical, extracellular	8.7–15.6	Vala (2015); Siddiqi and Azamal (2016)
A. terreus	Ag	Spherical, extracellular	1–20	Li et al. (2012); Siddiqi and Azamal (2016)
Aspergillus versicolor mycelia	Hg	Alteration. Surface of mycelia	20.5 ± 1.82	Das et al. (2008); Moghaddam et al. (2015)
Alternaria alternata	Ag, Cd	Spherical, extracellular	20–60	Gajbhiye et al. (2009); Khandel and Kumar (2016)
Rhizopus oryzae	Au	Nanocrystalline or triangular, hexagonal, pentagonal, spheroidal, sea urchin-like, 2D nanowires, nanorods. Cell surface	Various 10	Gericke and Pinches (2006); Das et al. (2009); Maliszewska et al. (2009a, b); Das et al. (2010); Moghaddam et al. (2015)
Rhizopus stolonifer	Au	Irregularly (uniform)	1–5	Binupriya et al. (2010a, b); Sarkar et al. (2012); Moghaddam et al. (2015)
	Ag	Quasi-spherical	25–30	
Rhizopus nigricans	Ag	Spherical, extracellular	7–20	Mohammadian et al. (2007); Khandel and Kumar (2016)
Phanerochaete chysosporium	Ag	Spherical, pyramidal, extracellular	50–200	Sanghi and Verma (2009); Khandel and Kumar (2016)
	Au	Spherical, extracellular	10–100	Philip (2009); Moghaddam et al. (2015)
Phyllanthus amarus	Ag	Spherical, extracellular	30	

(continued)

Table 13.2 (continued)

Fungi	NPs	Shape and location	Min and max particle size (nm)	References
Pleurotus sajor-caju	Au, Ag	Spherical, extracellular	20–40	Husseiny et al. (2007); Khandel and Kumar (2016)
				Vigneshwaran and Kathe (2007)
Penicillium fellutanum	Ag	Mostly spherical, extracellular	5–25	Kathiresan et al. (2009); Khandel and Kumar (2016)
Penicillium strain J3	Ag	Mostly spherical	10–100	Maliszewska et al. (2009a, b); Moghaddam et al. (2015)
Penicillium brevicompactum WA2315 (139)	Ag	Spherical, extracellular	58.35 ± 17.88	Shaligram et al. (2009); Khandel and Kumar (2016)
Penicillium brevicompactum	Au	Spherical, triangular and hexagonal Extracellular	10–60	Selvakannan et al. (2004); Khandel and Kumar (2016)
Penicillium citrinum	Ag	Spherical. Extracellular	5–25	Kathiresan et al. (2009); Khandel and Kumar (2016)
P. fellutanum	Ag	Spherical	5–25	Kathiresan et al. (2009); Siddiqi and Azamal (2016)
P. nagiovense AJ12	Ag	Spherical cell-free filtrate	25 ± 2.8	Maliszewska et al. (2014); Siddiqi and Azamal (2016)
P. rugulosum	Au	Spherical, triangular, hexagonal	20–80	Mishra et al. (2012); Siddiqi and Azamal (2016)
Penicillium sp.	Au	Spherical cell filtrate	30–50	Du et al. (2011); Siddiqi and Azamal (2016)

(continued)

Table 13.2 (continued)

Fungi	NPs	Shape and location	Min and max particle size (nm)	References
Trichoderma viride	Ag	Spherical, rod-like. Extracellular	2–100	Mukherjee et al. (2008); Fayaz et al. (2009a, b); Fayaz et al. (2010a, b); Moghaddam et al. (2015)
Trichoderma asperellum	Ag	Nanocrystalline or spherical. Extracellular	13–18	Mukherjee et al. (2008); Moghaddam et al. (2015)
Trichoderma reesei	Ag	Extracellular	5–50	Vahabi et al. (2011)
Trichoderma Koningii	Au	Small spheres to polygons. Cell-free filtrate	10–40	Maliszewska et al. (2009a, b); Maliszewska (2013); Siddiqi and Azamal (2016)
Trichoderma harzianum	Cu, Ag	Spherical. Extracellular	20–35	Gajbhiye et al. (2009); Khandel and Kumar (2016)
Tricholoma crassum	Au	Spherical. Extracellular	8.62–9.12	Sawle et al. (2008); Khandel and Kumar (2016)
Pleurotus sajor-caju	Ag	Spherical extracellular	30.5	Vigneshwaran and Kathe (2007)
Volvariella volvaceae	Au-Ag	Triangular,spherical, hexagonal extracellular,	20–150	Philip (2009); Thakkar et al. (2010); Moghaddam et al. (2015)
Cladosporium cladosporioides	Ag	Mostly spherical or hexagonal. Extracellular	10–100	Balaji et al. (2009); Khandel and Kumar (2016)
Cylindrocladium floridanum	Au	Spherical. Extracellular	19.5	Zhang et al. (2012); Khandel and Kumar (2016)
Cochliobolus lunatus	Ag	Spherical. Extracellular	5–10	Khandel and Kumar (2016)

(continued)

Table 13.2 (continued)

Fungi	NPs	Shape and location	Min and max particle size (nm)	References
Cochlibolus lunatus	Cu, Al	Quasi- spherical. Extracellular	3–21	Salunkhe et al. (2011); Raheman et al. (2011)
Hypocrea lixii	Cu	Spherical. Extracellular	24.5	Deplanche et al. (2010); Khandel and Kumar (2016)
Phoma sorghina	Ag	Rod shaped. Extracellular	120–160 × 30–40	Raheman et al. (2011)
Pestalotia sp.	Ag	Spherical extracellular or intracellular	10–40	Raheman et al. (2011)
Coriolus versicolor	Ag	Extra- and intracellular. Spherical	25–491	Sanghi and Verma (2009); Moghaddam et al. (2015)
Verticillium sp.	Fe_3O_4	Extracellular. Cubo-octahedral, quasi-spherical	20–400	Bharde et al. (2006); Moghaddam et al. (2015)
	Au	Spherical. Cell wall, cytoplasmic membrane amd ıntracellular	20	Mukherjee et al. (2001); Ramanathan et al. (2013)
Verticillium luteoalbum	Ag	Spherical. Extracellular	12–22	Bawaskar et al. (2010); Fayaz et al. (2009a, b); Khandel and Kumar (2016)
	Au	Spherical. Extracellular	12–15	
Yarrowia lipolytica	Au	Hexagonal, triangular, extracellular	15	Agnihotri et al. (2009); Pimprikar et al. (2009)

(e) Fungal mycelia provide higher surface area than bacteria and this advantage could be used to support the interaction of metal ions and fungal agents. This is enhancing the conversion of ions to metallic nanoparticles.

(f) Fungi also have the advantage to ease the downstream processing when extracellular nanoparticles are produced.

(g) Scalability is another factor for consideration of commercial production of nanoparticles. This gives fungi the edge as the chassis of choice for long-term development as they might be easily used in large-scale reactors (Pantidos and Horsfall 2014).

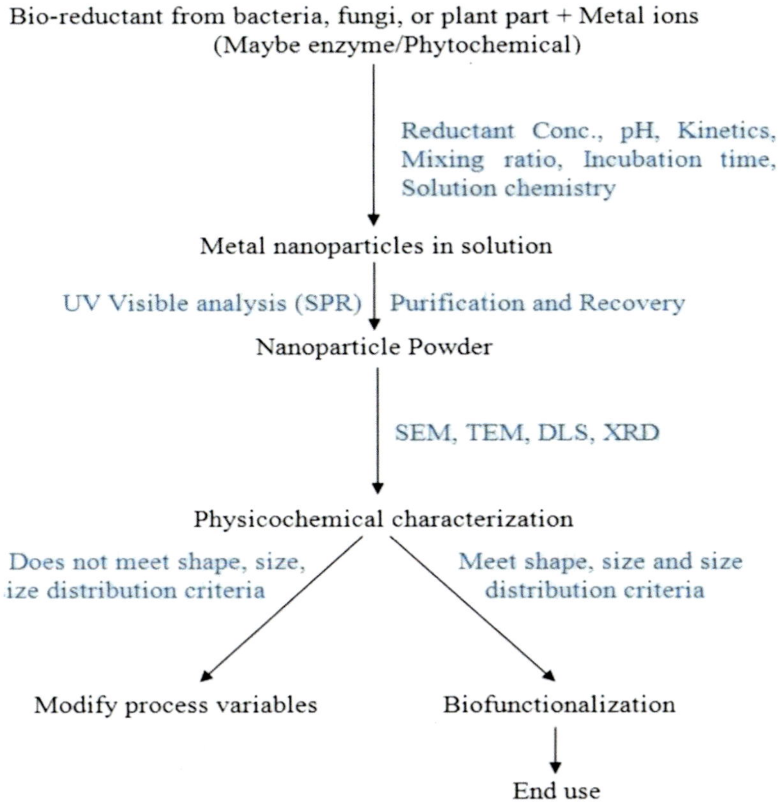

Fig. 13.1 Generalized flow chart for the biosynthesis of metallic nanoparticles (Rath et al. 2014; Punjabi et al. 2015)

13.3 Biosynthesis of Nanoparticles by Fungi

Researchers paid attention in recent years that the novel field of nano-biosynthesis of metal nanoparticles called "myconanotechnology." This new field is at the interference of nanotechnology and mycology combination which is interesting as a new applied science with a substantial potential due to the wide range of diversity of fungi. The fungal systems have already been used for the biosynthesis of metal nanoparticles of silver, gold, zirconium, silica, titanium, iron (magnetite), and platinum as well as ultrafine oxide nanoparticles, such as Sb_2O_3 and TiO_2. A generalized flow chart for the biosynthesis of metallic nanoparticles is shown in Fig. 13.1.

As happens in all cells, microbial cells also need metal ions primerly as cofactors. Metal/metal ions can interact with fungi in various ways and rely on the type of metal, organism, and environment. They accomplish toxic effects in some ways, like inhibiting the enzymes. Microorganisms have the ability to survive at high concentrations of toxic metals. The adaptation of fungi exposed to heavy metal ions has been examined to increase the tolerance of fungi. Therefore, microbial cells have evolved the ability to manage proper metal-protein interactions (Tottey et al. 2005; Kang et al. 2008; Anahid et al. 2011; Park et al. 2016). Microorganisms have diverse mechanisms of developing nanoparticles. Silver nanoparticle synthesis was suggested as a defensive mechanism of cells for silver. Ahmad et al. (2003) reported that NADH-dependent enzymes are responsible for the biosynthesis of nanoparticles. The reduction mechanism seems to be initiated by electron transfer from the NADH by NADH-dependent reductase as electron carrier.

Two mechanisms have been suggested for heavy metal tolerance/detoxication in fungi:

1. Extracellular (chelation and cell wall binding) separation
2. Intracellular physical separation of metal by binding to ligands (peptides or others) to prevent them from metal sensitive cellular targets.

During the intracellular synthesis of gold nanoparticles (A), the gold metal ions firstly bind on the fungal cell surface, through electrostatic interaction force which is generated due to opposite charges present on the metal ion surface and fungal cell surface. After that, absorbed metal ions are reduced by enzymes of the fungal cell wall. This is the result of the positively charged groups of these enzymes and this leads to the aggregation and formation of metal nanoparticles. In case of extracellular synthesis of silver nanoparticles (B) due to the nitrate reductase presents in the cell of fungi. This enzyme reduces the silver metal ions into silver nanoparticles. This finally results to the formation of highly stable silver nanoparticles (Fig. 13.2) (Khandel and Kumar 2016).

Biogenic synthesis of metal nanoparticles engages bioreduction of metal salts to elemental metals. This might be stabilizing the organic molecules present in the microorganisms such as fungi and bacteria. The other way of producing metal nanoparticles is biosorption. In this way, metal ions in the aqueous medium are stuck to the organisms' cell wall surface (Siddiqi and Azamal 2016).

Extracellular mechanisms are mainly intimated in the avoidance of metal entry. In this mechanism, different organic molecules, which do not belong to the cell wall matrix, are excreted by the fungal cell to chelate metal ions. This binding is called biosorption. In general, surface of the cell surface is negatively charged due to the presence of several anionic structures, such as glucan and chitin. This feature gives microorganisms the ability to bind metal cations (Anahid et al. 2011). Chelating agents can be organic or inorganic compounds and capable to bind to metal ions to forming complex ringlike structure called "chelates." Chelating agents have "ligand"-binding atoms which form either two or one and one coordinate or two coordinate covalent linkages in the case of bidentate chelates. Mainly, S, N, and O atoms function as ligand atoms in the form of chemical groups like –SH, −S-S, −NH_2, =NH, −OH, $-OPO_3H$, or >C=O. Bidentate or multidentate ligands form ring structures that include the metal ion and the two ligand atoms attached to the metal (Andersen 1999).

Siderophores are small, high-affinity iron-chelating compounds secreted by microorganisms such as bacteria and fungi. It was reported that most fungi produce

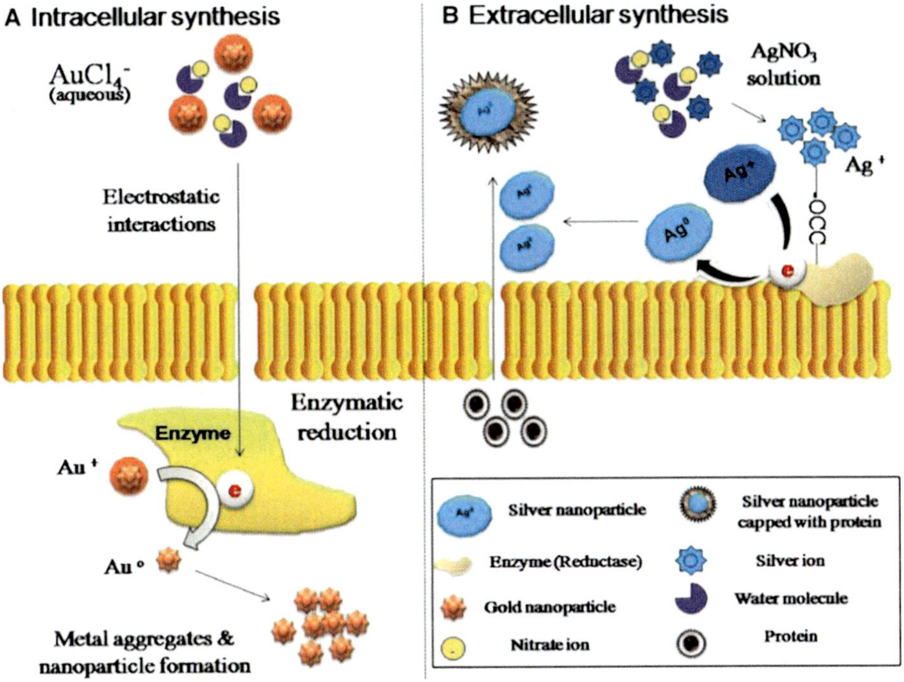

Fig. 13.2 Mechanism of intracellular and extracellular synthesis of gold (Au) and silver (Ag) nanoparticles through fungi (Khandel and Kumar 2016)

hydroxamate-type siderophores, but there is only little information regarding the production of catecholate-type compounds by fungi (Gadd 1999; Renshaw et al. 2002; Haselwandter and Winkelmann 2002). However, the output of catecholate-type chelating compounds has recently been depicted in wood-rotting fungi (Arantes and Milagres 2006). Whether these compounds are true or not is unknown, since siderophores are debatable due to their role in iron transport (Renshaw et al. 2002).

The presence of unspecific metal-chelating compounds such as organic acids was also examined in the fungal cultures as the low pH of the culture filtrates proposes the production of these acids. Oxalic, citric, and succinic acids are common metabolites produced by several mycorrhizal fungi. Their production is associated with the solubilization of insoluble metal-containing compounds (Fomina et al. 2005). It is known that depending on the concentration of these organic acids, they can react with CAS reagent in the same form as true siderophores and cause color changes in the mixtures.

The biosynthesis of metal nanoparticles by the fungi has also been reported. In the colorimetric examinations, the filtrate color changes from almost yellow to brown. This is a clear indication of the silver nanoparticles production in the reaction mixture (Juraifani and Ghazwani 2015). This result has also indicated that organic acids can play an important role in the transport and metabolism of metals in some microorganisms (Carson et al. 1992; Machuca et al. 2007).

Chelating compounds was determined by the Schwyn and Neilands (1987) spectrophotometrically using the Chrome Azurol S (CAS) reagent containing Fe(III) Absorbance at 630 nm after 1 h of incubation at room temperature. Percentages of compounds were calculated by subtracting the sample absorbance from the reference; and values >10% were considered as positive. They also investigated and developed the measurement in solid medium using the CAS agar-plate assay. This also depended on the color during the incubation period.

Biosynthesis of metal nanoparticles involves in bioreduction of metal salts to elemental metal. This might stabilize the organic compounds already present in the microorganisms. Sneha et al. (2010) exhibited the gold or silver ions were first captured on the surface of the fungal cells via electrostatic interaction between the ions and negatively charged cell wall from the carboxylate groups in the enzymes. This condition clearly indicates that metal ions are first sticks on the surface or inside of the microbial cells then reduced to nanoparticles in the presence of the related enzymes (Benzerara et al. 2010; Li et al. 2011). The reduction process takes place on the surface by the enzymes found in the cell wall (Mukherjee et al. 2001). Microorganisms affect the NP formation in two distinctive ways. First way, they could modify the composition of the solution in order that the solution becomes more supersaturated than its previous phase. Second way, microbes could impact the mineral formation through the organic polymers production (Benzerara et al. 2010; Li et al. 2011).

Some phenolic compounds such as naphthoquinone and anthraquinones show excellent redox properties and can act as electron shuttle in silver reduction. Specific extracellular enzymes act on a specific metal (Medentsev and Alimenko 1998; Bell et al. 2003; Siddiqi and Azamal 2016). For instance, nitrate reductase is essential for ferric ion reduction. Nitrate reductase system might be responsible for the bioreduction and formation of silver nanoparticles (Kumar et al. 2003). Similarly, a number of studies with *Fusarium oxysporum* demonstrated that the reduction of silver ions happens in the presence of a nitrate-dependent reductase and a shuttle quinone for extracellular process (Durán et al. 2005). These findings suggested that metal ion reduction needs not only the enzyme but also an electron shuttle (Durán et al. 2005).

NADH and NADH-dependent nitrate reductases are important factors in the biosynthesis of metal NPs. This enzymes catalyze NAD(P)H reduction of nitrate to nitrite. Eukaryotic assimilatory nitrate reductase (NR) catalyzes the following reaction:

$$\mathrm{NO^{-3} + NADH \rightarrow NO^{-2} + NAD^{++}OH^{-}}$$
$$\Delta G = -34.2\,\mathrm{kcal/mol}\left(-143\,\mathrm{kJ/mol}\right); \Delta E = 0.74\,\mathrm{V}$$

Two forms of the enzyme, NAD(P)H-bispecific forms (EC 1.6.6.2) and NADPH-specific forms (EC 1.6.6.3), are found in fungi. Campbell (1999) suggested that NR could also contribute to iron reduction in vivo since it catalyzes NADH ferric citrate reduction. *Fusarium oxysporum* MT 811 is able to reduce nitrates and nitrites to N_2. Fungal NR is similar to the bacterial enzyme that contains copper in their active site. The reduction of NO to N_2 is catalyzed by cytosolic and mitochondrial NORases (cytochrome P450). Its synthesis is specifically induced by nitrate and nitrite but repressed under aeration (Takaya et al. 1999).

Morozkina et al. (2005) reported that NRases from *F. oxysporum* mycelia grown aerobically and anaerobically differ in molecular weight, activity in several mineral sources of nitrogen, and optimum temperature. This shows that NRase probably exists in two different forms which function under both aerobic and anaerobic conditions. NRase of *F. oxysporum* grown is inhibited by ammonium ions under aerobic conditions. NRase from the anaerobically grown mycelium had low sensitivity to ammonium ions. *F. oxysporum* has also been shown to produce cadmium sulfide (CdS), lead sulfide (PbS), zinc sulfide (ZnS), and molybdenum sulfide (MoS) nanoparticles, when the appropriate salt is added to the growth medium (Ahmad et al. 2002).

When NR assay was carried out by the reaction of nitrite with 2,3-diaminonaphthalene, it could initiate NP formation by many fungi including *Penicillium* species. However, the exact mechanism of the formation of nanoparticles is yet to be elucidated. Some *Aspergillus flavus* proteins are responsible for synthesis of silver nanoparticles. It was reported that the synthesis procedure takes place in two steps. In the first one, reduction process of bulk silver ions into silver nanoparticles occurs, and in the next step, synthesized nanoparticles are capped. The protein-nanoparticle interactions could play very meaningful role such as providing stability to nanoparticles (Fig. 13.3) (Jain et al. 2011). However, this interaction between protein and nanoparticles is still not completely understood. Understanding the protein-nanoparticle interactions would lead us to form future "nano-factories."

Intracellular metal trafficking systems (IMTS) work to reduce the metal loading in the cytosol. In the IMTS, metal transport proteins might involve in metal tolerance. This could be either by expelling toxic metal ions from the cytosol out or letting the metals into vacuolar systems (Anahid et al. 2011). Some microorganisms could accumulate and detoxify heavy metals owing to various reductase enzymes. These enzymes are able to reduce metal salts to metal nanoparticles with a narrow size and less polydispersity. The size of the NPs is related to their nucleating activities. In accordance to the location where nanoparticles are formed, they can be classified into intracellular and extracellular NPs (Mann 2001). *Trichothecium* sp., *Verticillium luteoalbum*, and *Phoma* sp. have been explored for intracellular gold and silver nanoparticle synthesis. Vigneshwaran et al. (2007a, b) reported the accumulation of silver nanoparticles on the surface of its cell wall in *Aspergillus flavus*. The intracellular formation and accumulation of NPs are composed of transporting metal ions into the microbial cell in the presence of enzymes (Zhang et al. 2011). An intracellular synthesis of NPs needs additional steps such as ultrasound treatment or usage of suitable detergents to release the synthesized nanoparticles (Babu and Gunasekaran 2009; Das et al. 2014).

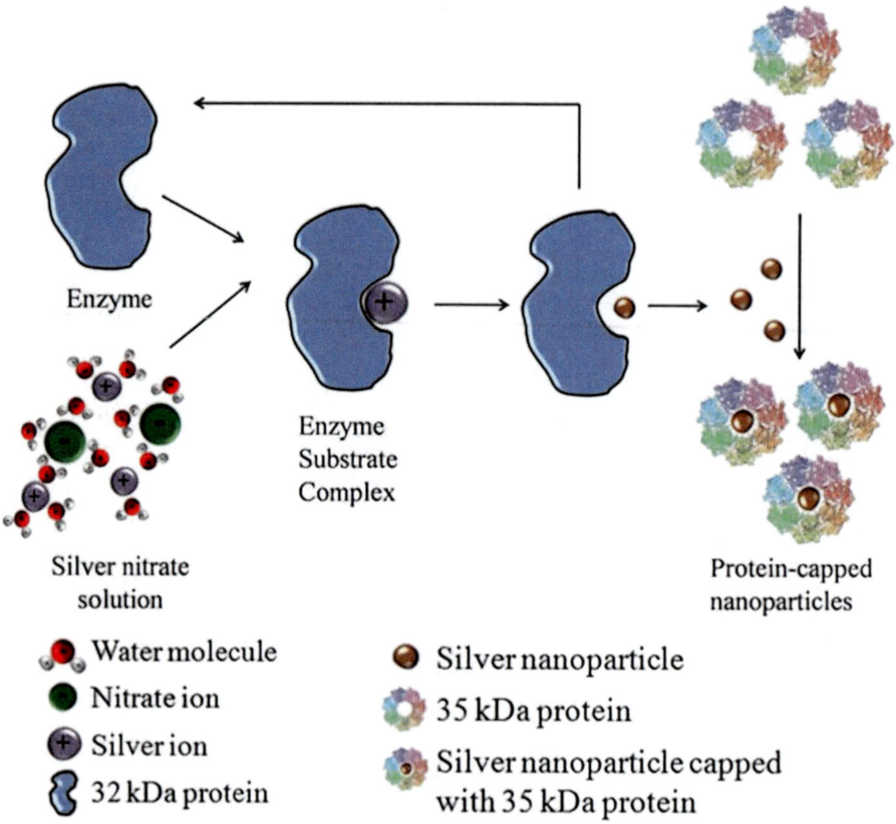

Fig. 13.3 Mechanism showing the role of extracellular proteins in the synthesis of silver nanoparticles (Jain et al. 2011)

Extracellular biosynthesis is cheap and it requires simpler processes. This favors large-scale production of NPs to examine its potential applications. Because of this possible advantage, many studies have been focussed on synthesis of metal nanoparticles outside of the cells (Durán et al. 2005; Das et al. 2014).

13.4 Factors Affecting Biosynthesis of Metal Nanoparticles

Major parameters (include temperature, pH, the presence of specific enzymes, type of biomass, exposure time to substrate, and the substrate concentration) affect the physical and chemical characters of nanoparticles.

pH is an important factor for shape of nanoparticles. Gericke and Pinches (2006) demonstrated the change in the shape of NPs with the variation of pH. Similarly, Davis and Ogden (1997) discovered that reduction of metal ions were highly sensitive to pH. Dhillon et al. (2012) reported that the movement of ions and activity of microbial biomass were controlled by variation in temperature. It could be also suggested that temperature plays an important role on the growth of organism as well as on metal uptake by the surrounding environment. In addition to temperature and pH, concentration of metal ions and type of enzyme also influence the synthesis of metal nanoparticles. The concentration of reactants decides the rate of reaction and also affects the size and shape of the synthesized particles. According to the study carried out by Gericke and Pinches (2006), synthesis of nanoparticles at different time intervals and their influence on synthesis process were also studied. It was found that incubation time increases the shape and size of nanoparticles. It has been also reported that with increase in the incubation time, the synthesis of nanoparticles also increases (Khandel and Kumar 2016).

13.5 Characterization of Metal Nanoparticles

After the biosynthesis of metal nanoparticles, characterizations of the nanoparticles are also an important step for the identification (size, shape, chemical composition, surface area, and dispersity). For the characterization of nanomaterials, different techniques are employed. These techniques are divided into two categories.

13.5.1 Determination of the Size, Shape, and Conformity of the Nanoparticles Synthesized

This includes mainly X-ray Diffraction (XRD), both Scanning electron microscopy (SEM), and Transmission electron microscopy (TEM), Dynamic light scattering (DLS), and Atomic force microscopy (AFM) analysis.

13.5.2 Functional Group Identification of Synthesized Nanoparticles

Involves in UV-visible spectroscopic analysis, energy dispersive spectroscopy (EDS), and Fourier transforms infrared spectroscopy (FTIR) analysis techniques (Khandel and Kumar 2016).

Table 13.3 Application of fungus-mediated synthesis of metal nanoparticles

NPs	Fungi	Application
Ag	*Alternaria alternata*	Enhancement in antifungal activity of fluconazole against *Phoma glomerata* and water quality monitoring
	Aspergillus clavatus	Antimicrobial activity
	A. niger	Antibacterial activity. Wound healing activity
	Aspergillus sp.	Antimicrobial activity
	Aspergillus tubingensis	44% inhibition of Syncitial virus infection
	Colletotrichum gloeosporioides	Antifungal activity
	Fusarium acuminatum	Antibacterial activity
	F. oxysporum	Textile fabrics
	F. solani	Textile fabrics
	Lecanicillium lecanii	Textile fabrics
	Macrophomina phaseolina	Antimicrobial properties against multidrug-resistant bacteria
	Neurospora oryzae	Only 1–10 nm nanoparticles attached to virus restraining virus from attaching to host cells. HIV
	Penicillium oxalicum	Catalytic activity
	Penicillium sp.	Antibacterial activity against MDR *E. coli* and *S. Aureus*
	Phytophthora infestans	Antimicrobial activity
	Pleurotus ostreatus	Antimicrobial activity
	Raffaelea sp.	Antifungal activity
	Trichoderma crassum	Antimicrobial activity
	T. viride	Vegetable and fruit preservation
Au	*Aspergillus japonicus AJP01*	Catalytic activity
	A. niger	Toxic to mosquito larvae
	Rhizopus oryzae	Water hygiene management
Cds	*Saccharomyces pombe*	Electric diode
	F. oxysporum	Live cell imaging and diagnostics
Carbon nanotubes sensors	*F. oxysporum*	Developed for glucose, ethanol, sulfides, and sequence-specific DNA analysis
Carbon nanotubes with enzymes	*Phoma glomerata*	Establish a fast electron transfer from the active site of the enzyme through the CNT to an electrode, in many cases enhancing the electrochemical activity of the biomolecules

Siddiqi and Husen (2016)

13.6 Applications of Nanobioparticles in Fungi

Owing to their great properties, nanoparticles have significant application in many fields such as cosmetics, catalysts, lubricants, fuel additives, paints, agrochemicals, food packaging, textile engineering, electronics, optics, environmental sensing, nanomedicine, drug and gene delivery agents, biodetection of pathogens, tumor destruction via heating (hyperthermia), magnetic resonance imaging, and phagokinetic studies (Prasad et al. 2014, 2016, 2017; Aziz et al. 2016; Siddiqi and Azamal 2016). Fungus-mediated synthesis of metal nanoparticles is getting much attention due to their extensive application in various sectors (Table 13.3).

13.7 Conclusion

The synthesis of functional nanoparticles by using microorganisms has taken a big concern in recent years. Microorganisms could alter the oxidation state of the metals. These microbial processes provide us new opportunities to synthesize metal nanomaterials biologically. On the contrary to chemical and physical methods, microbic synthesis of nanomaterials can be achieved under optimal environmental conditions in aquatic media. This approach became one of the sustainable development tools for the green bionanotechnological researches.

The mechanism of the biosynthesis has not been clear yet, but we can easily say that it is an enzyme-dependent occurrence for microorganisms. For this reason, we need to determine and characterize the specific enzymes and enlighten the pathways involved in these processes. It is strongly believed in that this area of science is very promising for the future of medicine and other health sciences.

References

Agnihotri M, Joshi S, Ravikumar A, Zinjarde S, Kulkarni S (2009) Biosynthesis of gold nanoparticles by the tropical marine yeast Yarrowia lipolytica NCIM 3589. Mater Lett 63:1231–1234

Ahmad A, Mukherjee P, Mandal D, Senapati S, Khan MI, Kumar R, Sastry M (2002) Enzyme mediated extracellular synthesis of CdS nanoparticles by the fungus, *Fusarium oxysporum*. J Am Chem Soc 124(41):12108–12109

Ahmad A, Mukherjee P, Senapati S, Mandal DM, Khan IM, Kumar R, Sastry M (2003) Extracellular biosynthesis of silver nanoparticles using the fungus *Fusarium oxysporum*. Colloids Surf B: Biointerfaces 28:313–318

Ahmad A, Senapati S, Khan MI, Kumar R, Sastry M (2005) Extra/intracellular biosynthesis of gold nanoparticles by an alkalotolerant fungus, trichothecium sp. J Biomed Nanotechnol 1(1):47–53

Alani F, Moo-Young M, Anderson W (2012) Biosynthesis of silver nanoparticles by a new strain of Streptomyces sp. compared with *Aspergillus fumigatus* (2012). World J Microbiol Biotechnol 28:1081–1086

Anahid S, Yaghmaei S, Ghobadinejad Z (2011) Heavy metal tolerance of fungi. Scientia Iranica C 18(3):502–508

Andersen O (1999) Principles and recent developments in chelation treatment of metal intoxication. Chem Rev 99:2683–2710

Arantes V, Milagres AMF (2006) Evaluation of different carbon sources for production of iron-reducing compounds by *Wolfiporia cocos* and *Perenniporia medulla*-panis. Process Biochem 41:887–891

Aziz N, Pandey R, Barman I, Prasad R (2016) Leveraging the attributes of *Mucor hiemalis*-derived silver nanoparticles for a synergistic broad-spectrum antimicrobial platform. Front Microbiol 7:1984. https://doi.org/10.3389/fmicb.2016.01984

Babu GMM, Gunasekaran P (2009) Production and structural characterization of crystalline silver nanoparticles from *Bacillus cereus* isolate. Coll Surf B 74:191–195

Balaji DS, Basavaraja S, Deshpande R, Mahesh DB, Prabhakar BK, Venkataraman A (2009) Extracellular biosynthesis of functionalized silver nanoparticles by strains of *Cladosporium cladosporioides* fungus. Colloids Surf B: Biointerfaces 6(1):88–92

Bansal V, Rautaray D, Ahmad A, Sastry M (2004) Biosynthesis of zirconia nanoparticles using the fungus *Fusarium oxysporum*. J Mater Chem 14:3303–3305

Bansal V, Rautaray D, Bharde A, Ahire K, Sanyal A, Ahmad A (2005) Fungus-mediated biosynthesis of silica and titania particles. J Mater Chem 15:2583

Bansal V, Poddar P, Ahmad A, Sastry MJ (2006) Room-temperature biosynthesis of ferroelectric barium titanate nanoparticles. Am Chem Soc 128:11958

Basavaraja S, Balaji SD, Lagashetty A, Rajasab AH, Venkataraman A (2008) Extracellular biosynthesis of silver nanoparticles using the fungus *Fusarium semitectum*. Mater Res Bull 43:1164–1170

Bawaskar M, Gaikwad S, Ingle A, Rathod D, Gade A, Duran N, Marcato P, Rai M (2010) A new report on mycosynthesis of silver nanoparticles by *Fusarium culmorum*. Curr Nanosci 6(4):376–380

Bell AA, Wheeler MH, Liu J, Stipanovic RD, Puckhaber LS, Orta H (2003) Fabrication of metal nanoparticles from fungi and metal salts: scope and application. Pest Manag Sci 59:736–747

Benzerara K, Miot J, Morin G, Ona-Nguema G, Skouri-Panet F, Férard C (2010) Significance, mechanisms and environmental implications of microbial biomineralization. Compt Rendus Geosci 343(2–3):160–167

Bhainsa KC, D'Souza SF (2006) Extracellular biosynthesis of silver nanoparticles using the fungus *Aspergillus fumigatus*. Coll Surf B Biointerfaces 47:160–164

Bhambure R, Bule M, Shaligram N, Kamat M, Singhal R (2009) Extracellular biosynthesis of gold nanoparticles using *Aspergillus niger* its characterization and stability. Chem Eng Technol 32:1036–1041

Bharde A, Rautaray D, Bansal V, Ahmad A, Sarkar I, Yusuf SM, Sanyal M, Sastry M (2006) Extracellular biosynthesis of magnetite using fungi. Small 2:135–141

Binupriya AR, Sathishkumar M, Vijayaraghavan K, Yun SI (2010a) Bioreduction of trivalent aurum to nanocrystalline gold particles by active and inactive cells and cell free extract of *Aspergillus oryzae* var. viridis. J Hazard Mater 177:539–545

Binupriya AR, Sathishkumar M, Yun SI (2010b) Biocrystallization of silver and gold ions by inactive cell filtrate of *Rhizopus stolonifer*. Colloids Surf B: Biointerfaces 79:531–534

Birla SS, Tiwari VV, Gade AK, Ingle AP, Yadav AP, Rai MK (2009) Fabrication of silver nanoparticles by *Phoma glomerata* and its combined effect against *Escherichia coli, Pseudomonas aeruginosa* and *Staphylococcus aureus*. Appl Microbiol 48(2):173–179

Campbell WH (1999) Nitrate reductase structure, function and regulation: bridging the gap between biochemistry and physiology. Annu Rev Plant Physiol Plant Mol Biol 50:277–303

Carson KC, Holliday S, Glenn AR, Dilworth MJ (1992) Siderophore and organic acid production in root nodule bacteria. Arch Microbiol 157:264–271

Das S, Das A, Guha A (2008) Adsorption behavior of mercury on functionalized *Aspergillus versicolor* mycelia: atomic force microscopic study. Langmuir 25:360–366

Das SK, Das AR, Guha AK (2009) Gold nanoparticles: microbial synthesis and application in water hygiene management. Langmuir 25:8192–8199

Das SK, Das AR, Guha K (2010) Microbial synthesis of multishaped gold nanostructures. Small 6:1012–1021

Das VL, Thomas R, Varghese RT, Soniya EV, Jyothis M, Radhakrishnan EK (2014) Extracellular synthesis of silver nanoparticles by the Bacillus strain CS 11 isolated from industrialized area. 3 Biotech 4(2):121–126

Davis MW, Ogden WC (1997) Free resources and advanced alignment for cross-language text retrieval. In: TREC, p 385–395

Deplanche K, Caldelari I, Mikheenko IP, Sargent F, Macaskie LE (2010) Involvement of hydrogenases in the formation of highly catalytic Pd (0) nanoparticles by bioreduction of Pd (II) using *Escherichia coli* mutant strains. Microbiology 156:2630–2640

Dhillon GS, Brar SK, Kour S, Verma M (2012) Green approach for nanoparticle biosynthesis by fungi: current trends and applications. Crit Rev Biotechnol 32:49–73

Du L, Xian L, Feng JX (2011) Rapid extra–/intracellular biosynthesis of gold nanoparticles by the fungus Penicillium sp. J Nanopart Res 13:921–930

Durán N, Marcato PD, Alves OL, De Souza GIH, Esposito E (2005) Mechanistic aspects of biosynthesis of silver nanoparticles by several *Fusarium oxysporum* strains. J Nanobiotechnol 3(8):4–7

Fayaz N, Kathe AA, Varadarajan PV, Nachane RP, Balasubramanya RH (2006) Biomimetics of silver nanoparticles by white rot fungus, *Phanerochaete chrysosporium*. Coll Surf B: Interf 53(1):55–59

Fayaz AM, Balaji K, Girilal M, Kalaichelvan PT, Venkatesan R (2009a) Mycobased synthesis of silver nanoparticles and their incorporation into sodium alginate films for vegetable and fruit preservation. J Agric Food Chem 57:6246–6252

Fayaz AM, Balaji K, Kalaichelvan PT, Venkatesan R (2009b) Fungal based synthesis of silver nanoparticles-an effect of temperature on the size of particles. Colloids Surf B Biointerfaces 74:123–126

Fayaz AM, Balaji K, Girilal M, Yadav R, Kalaichelvan PT, Venketesan R (2010) Biogenic synthesis of silver nanoparticles and their synergistic effect with antibiotics: a study against Gram-positive and Gram-negative bacteria. Nanomed-Nanotechnol 6:103–109

Fayaz M, Tiwary CS, Kalaichelvan PT, Venkatesan R (2010b) Blue orange light emission from biogenic synthesized silver nanoparticles using *Trichoderma viride*. Colloids Surf B: Biointerfaces 75:175–178

Fomina MA, Alexander IJ, Colpaert JV, Gadd GM (2005) Solubilization of toxic metal minerals and metal tolerance of mycorrhizal fungi. Soil Biol Biochem 37:851–866

Gadd GM (1999) Fungal production of citric and oxalic acid: importance in metal speciation, physiology and biogeochemical process. Adv Microb Physiol 11:47–91

Gade AK, Bonde PP, Ingle AP, Marcato P, Duran N, Rai MK (2008) Exploitation of aspergillus niger for synthesis of silver nanoparticles. J Biobased Mater Bioenergy 2:1–5

Gajbhiye MB, Kesharwani JG, Ingle AP, Gade AK, Rai MK (2009) Fungus mediated synthesis of silver nanoparticles and its activity against pathogenic fungi in combination of fluconazole. Nanomedicine: NBM 5:282–286

Gericke M, Pinches A (2006) Biological synthesis of metal nanoparticles. Hydrometallurgy 83:132–140

Govender Y, Riddin T, Gericke M, Whiteley CG (2009) Bioreduction of platinum salts into nanoparticles: a mechanistic perspective. Biotechnol Lett 31:95–100

Haselwandter K, Winkelmann G (2002) Ferricrocin an ectomycorrhizal siderophore of *Cenococcum geophilum*. Biometals 15:73–77

Husseiny MI, El-Aziz MA, Badr Y, Mahmoud MA (2007) Biosynthesis of gold nanoparticles using *Pseudomonas aeruginosa*. Spectrochim Acta A Mol Biomol Spectrosc 67:1003–1006

Ingle A, Gade A, Bawaskar M, Rai M (2009) *Fusarium solani*: a novel biological agent for the extracellular synthesis of silver nanoparticles. J Nanopart Res 11:2079–2085

Jain N, Bhargava A, Majumdar S, Tarafdar JC, Panwar J (2011) Extracellular biosynthesis and characterization of silver nanoparticles using *Aspergillus flavus* NJP08: a mechanism perspective. Nanoscale 3:635–641

Juraifani AAAA, Ghazwani AA (2015) Biosynthesis of silver nanoparticles by *Aspergillus niger, Fusarium oxysporum* and *Alternaria solani*. Afr J Biotechnol 14(26):2170–2174

Kang SH, Bozhilov KN, Myung NV, Mulchandani A, Chen W (2008) Microbial synthesis of CdS nanocrystals in genetically engineered *E. coli*. Angew Chem Int Ed 120:5186–5189

Kathiresan K, Manivanan S, Nabeel MA, Dhivya B (2009) Studies on silver nanoparticles synthesized by a marine fungus, *Penicillium fellutanum* isolated from coastal mangrove sediment. Colloids Surf B: Biointerfaces 71:133–137

Khandel P, Kumar SS (2016) Microbes mediated synthesis of metal nanoparticles: current status and future prospects. Dig J Nanomater Biostruct 6(1):1–24

Kumar A, Mandal S, Selvakannan PR, Pasricha R, Mandal AB, Sastry M (2003) Investigation into the interaction between surface-bound alkyl amines and gold nanoparticles. Langmuir 19:6277–6282

Kumar AS, Abyaneh MK, Gosavi SW, Kulkarni SK, Pasricha R, Ahmad A, Khan MI (2007a) Nitrate reductase-mediated synthesis of silver nanoparticles from $AgNO_3$. Biotechnol Lett 29(3):439–445

Kumar SA, Ansary AA, Ahmad A, Khan MI (2007b) Extracellular biosynthesis of CdSe quantum dots by the fungus *Fusarium oxysporum*. J Biomed Nanotechnol 3:190–194

Li XX, Chen Z-S, Chen G (2011) Biosynthesis of nanoparticles by microorganisms and their applications. J Nanomater 2011:1–16. https://doi.org/10.1155/2011/270974

Li G, He D, Qian Y, Guan B, Gao S, Cui Y, Yokoyama K, Wang L (2012) Fungus-mediated green synthesis of silver nanoparticles using *Aspergillus terreus*. Int J Mol Sci 13:466–476

Machuca A, Pereira G, Aguiar A, Milagres AMF (2007) Metal-chelating compounds produced by ectomycorrhizal fungi collected from pine plantations. Lett Appl Microbiol 44:7–12

Maliszewska I (2013) Microbial mediated synthesis of gold nanoparticles: preparation, characterization and cytotoxicity studies. Dig J Nanomater Bios 8:1123–1131

Maliszewska I, Juraszek A, Bielska K (2014) Green synthesis and characterization of silver nanoparticles using ascomycota fungi Penicillium nalgiovense AJ12. J Clust Sci 25:989–1004

Maliszewska I, Aniszkiewicz Ł, Sadowski Z (2009a) Biological synthesis of gold nanostructures using the extract of *Trichoderma koningii*. Acta Phys Polon A 116:163–165

Maliszewska M, Szewczy K, Waszak K (2009b) Biological synthesis of silver nanoparticles. J Phys: Conf Ser 146(1):012025

Mandal D, Bolander ME, Mukhopadhyay D, Sarkar G, Mukherjee P (2006) The use of microorganisms for the formation of metal nanoparticles and their application. Appl Microbiol Biotechnol 69:485–492

Mann S (2001) Biomineralization: principles and concepts in bioinorganic materials chemistry. Oxford University Press, Oxford

Medentsev AG, Alimenko VK (1998) Naphthoquinone metabolites of the fungi. Photochemistry 47:935–959

Mishra A, Tripathy SK, Yuna SI (2012) Fungus mediated synthesis of gold nanoparticles and their conjugation with genomic DNA isolated from *Escherichia coli* and *Staphylococcus aureus*. Process Biochem 47:701–711

Moghaddam A B, Farideh N, Moniri M, Tahir P Md. Azizi S, Mohamad R (2015) Nanoparticles biosynthesized by fungi and yeast: a review of their preparation, properties, and medical applications. Molecules 20:16540–16565

Mohammadian M, Shojaosadati SA, Habibi Rezaee M (2007) *Fusarium oxysporum* mediates photogeneration of silver nanoparticles. Scientia Iranica 14(4):323–326

Morozkina EV, Kurakov AV, Nosikov AN, Sapova EV, L'vov NP (2005) Properties of nitrate reductase from *Fusarium oxysporum* 11dn1 fungi grown under anaerobic conditions. Appl Biochem Microbiol 41(3):254–258

Mukherjee P, Ahmad A, Mandal D, Senapati S, Sainkar SR, Khan MI, Ramani R, Parischa R, Ajaykumar PV, Alam M, Sastry M, Kumar R (2001) Bioreduction of AuCl4-ions by the fungus, Verticillium sp. and surface trapping of the gold. Angew Chem Int Ed Eng 40(19):3585–3588

Mukherjee P, Roy M, Mandal BP, Dey GK, Mukherjee PK, Ghatak J, Tyagi AK, Kale SP (2008) Green synthesis of highly stabilized nanocrystalline silver particles by a non-pathogenic and agriculturally important fungus *T. asperellum*. Nanotechnology 19:1–7

Narayanan KB, Sakthivel N (2010) Biological synthesis of metal nanoparticles by microbes. Adv Colloid Interf Sci 156:1–13

Pantidos N, Horsfall LE (2014) Biological synthesis of metallic nanoparticles by bacteria, fungi and plants. Nanomed Nanotechnol 5(5):1–10

Park JT, Lee GK, Lee YS (2016) Advances in microbial biosynthesis of metal nanoparticles. Appl Microbiol Biotechnol 100:521–534. https://doi.org/10.1007/s00253-015-6904-7

Philip D (2009) Biosynthesis of Au, Ag and Au-Ag nanoparticles using edible mushroom extract. Spectrochim Acta Part A 73:374–381

Pimprikar PS, Joshi SS, Kumar AR, Zinjarde SS, Kulkarni SK (2009) Influence of biomass and gold salt concentration on nanoparticle synthesis by the tropical marine yeast *Yarrowia lipolytica* NCIM 3589. Colloids Surf B: Biointerfaces 74:309–316

Prasad R (2014) Synthesis of silver nanoparticles in photosynthetic plants. J Nanoparticles, Article ID 963961, https://doi.org/10.1155/2014/963961

Prasad R, Kumar V, Prasad KS (2014) Nanotechnology in sustainable agriculture: present concerns and future aspects. Afr J Biotechnol 13(6):705–713

Prasad R, Pandey R, Barman I (2016) Engineering tailored nanoparticles with microbes: quo vadis. WIREs Nanomedicine Nanobiotechnology 8:316–330. https://doi.org/10.1002/wnan.1363

Prasad R, Bhattacharyya A, Nguyen QD (2017) Nanotechnology in sustainable agriculture: recent developments, challenges, and perspectives. Front Microbiol 8:1014. https://doi.org/10.3389/fmicb.2017.01014

Punjabi K, Choudhary P, Samant L, Mukherjee S, Vaidya S, Chowdhary A (2015) Biosynthesis of nanoparticles: a review. Int J Pharm Sci Rev Res 30(1):219–226

Raheman F, Deshmukh S, Ingle A, Gade A, Rai M (2011) Silver nanoparticles: novel antimicrobial agent synthesized from an endophytic fungus Pestalotia sp. isolated from leaves of *Syzygium cumini* (L). Nano Biomed Eng 3(3):174–178

Rajakumar G, Rahuman A, Roopan SM, Khanna VG, Elango G, Kamaraj C, Zahir AA, Velayutham K (2012) Fungus-mediated biosynthesis and characterization of TiO_2 nanoparticles and their activity against pathogenic bacteria. Spectrochim Acta A Mol Biomol Spectrosc 91:23–29

Raliya R (2013) Rapid, low-cost, and ecofriendly approach her for iron nanoparticle synthesis using *Aspergillus oryzae* TFR9. J Nanoparticles 2013:1–4

Raliya R, Biswas P, Tarafdar JC (2015) TiO_2 nanoparticle biosynthesis and its physiological effect on mung bean (*Vigna radiata* L.) Biotechnol Rep 5:22–26

Ramanathan R, Field MR, O'Mullane AP, Smooker PM, Bhargava SK, Bansal V (2013) Aqueous phase synthesis of copper nanoparticles: a link between heavy metal resistance and nanoparticle synthesis ability in bacterial systems. Nanoscale 5:2300–2306

Rath M, Panda SS, Dhal NK (2014) Synthesis of silver nano particles from plant extract and its application in cancer treatment: a review. Int J Pl An Env Sci 4(3):137–145

Rautaray D, Sanyal A, Adyanthaya SD, Ahmad A, Sastry M (2004) Biological synthesis of strontium carbonate crystals using the fungus *Fusarium oxysporum*. Langmuir 20(16):6827–6833

Renshaw JC, Robson GD, Trinci APJ, Wiebe MG, Livens FR, Collison D, Taylor RJ (2002) Fungal siderophores: structures, functions and applications. Mycol Res 106:1123–1142

Riddin TL, Gericke M, Whiteley CG (2006) Analysis of the inter- and extracellular formation of platinum nanoparticles by Fusarium oxysporum f. sp. lycopersici using response surface methodology. Nanotechnology 17:3482

Salahuddin KS, Azamal H (2016) Fabrication of metal nanoparticles from fungi and metal salts: scope and application. Nanoscale Res Lett 11(98):2–15

Salunkhe RB, Patil SV, Patil CD, Salunke BK (2011) Larvicidal potential of silver nanoparticles synthesized using fungus *Cochliobolus lunatus* against *Aedes aegypti* (Linnaeus, 1762) and Anopheles stephensi Liston (Diptera; Culicidae). Parasitol Res 109(3):823–831

Sanghi R, Verma P (2009) Biomimetic synthesis and characterisation of protein capped silver nanoparticles. Bioresour Technol 100:501–504

Sanyal A, Rautaray D, Bansal V, Ahmad A, Sastry M (2005) Heavy-metal remediation by a fungus as a means of production of lead and cadmium carbonate crystals. Langmuir 21(16):7220–7224

Saravanan M, Nanda A (2010) Extracellular synthesis of silver bionanoparticles from *Aspergillus clavatus* and its antimicrobial activity against MRSA and MRSE. Colloids Surf B: Biointerfaces 77(2):214–218

Sarkar J, Ray S, Chattopadhyay D, Laskar A, Acharya K (2012) Mycogenesis of gold nanoparticles using a phytopathogen *Alternaria alternata*. Bioprocess Biosyst Eng 35:637–643

Sawle BD, Salimath B, Deshpande R, Bedre MD, Prabhakar BK, Venkataraman A (2008) Biosynthesis and stabilization of Au and Au-Ag alloy nanoparticles by fungus, *Fusarium semitectum*. Sci Technol Adv Mater 9(3):1–6

Schwyn B, Neilands J (1987) Universal chemical assay for the detection and determination of siderophores. Anal Biochem 160:46–56

Selvakannan PR, Mandal S, Phadtare S, Gole A, Pasricha R, Adyanthaya SD, Sastry M (2004) Water-dispersible tryptophan-protected gold nanoparticles prepared by the spontaneous reduction of aqueous chloroaurate ions by the amino acid. J Colloid Interface Sci 269:97–102

Shaligram NS, Bule M, Bhambure R, Singhal RS, Singh SK, Szakacs G, Pandey A (2009) Biosynthesis of silver nanoparticles using aqueous extract from the compactin producing fungal strain. Process Biochem 44(8):939–943

Shankar SS, Ahmad A, Pasricha R, Sastry M (2003) Bioreduction of chloroaurate ions by geranium leaves and its endophytic fungus yields gold nanoparticles of different shapes. J Mater Chem 13(7):1822–1826

Sharma VK, Yngard R, Lin Y (2009) Silver nanoparticles: green synthesis and their antimicrobial activities. Adv Colloid Interf Sci 145:83–96

Siddiqi KS, Husen A (2016) Fabrication of Metal Nanoparticles from Fungi and Metal Salts: Scope and Application. Nanoscale Res Lett 11(98):1–15

Sneha K, Sathishkumar M, Mao J, Kwak SI, Yun SY (2010) Corynebacterium glutamicum-mediated crystallization of silver ions through sorption and reduction processes. Chem Eng J 162(3):989–996

Park TJ, Lee KG, 2 & Lee SY (2016) Advances in microbial biosynthesis of metal nanoparticles, Appl Microbiol Biotechnol 100(2):521–534

Takaya N, Suzuki S, Kuwazaki S, Shoun H, Maruo F, Yamaguchi M, Takeo K (1999) Cytochrome p450nor, a novel class of mitochondrial cytochrome P450 involved in nitrate respiration in the fungus *Fusarium oxysporum*. Arch Biochem Biophys 372(2):340–346

Tarafdar JC, Raliya R, Rathore I (2012) Microbial synthesis of phosphorous nanoparticle from tricalcium phosphate using *Aspergillus tubingensis* TFR-5. J Bionanosci 6:84–89

Thakkar KN, Mhatre SS, Parikh RY (2010) Biological synthesis of metallic nanoparticles. Nanomedicine 6:257–262

Tottey S, Harvie DR, Robinson NJ (2005) Understanding how cells allocate metals using metal sensors and metallochaperones. Acc Chem Res 38:775–783

Uddin I, Adyanthaya S, Syed A, Selvaraj K, Ahmad A, Poddar P (2008) Structure and microbial synthesis of sub-10 nm Bi_2O_3 nanocrystals. J Nanosci Nanotechnol 8:3909

Vahabi K, Mansoori GA, Karimi S (2011) Biosynthesis of silver nanoparticles by fungus *Trichoderma reesei* (a route for large-scale production of AgNPs). Insci J 1:65–79

Vala AK (2015) Exploration on green synthesis of gold nanoparticles by a marine-derived fungus *Aspergillus sydowii*. Environ Prog Sustain Energy 34:194–197

Velmurugan P, Shim J, You Y, Choi S, Kamala-Kannan S, Lee KJ, Kim HJ, Oh BT (2010) Removal of zinc by live, dead, and dried biomass of *Fusarium* spp. isolated from the abandoned-metal mine in South Korea and its perspective of producing nanocrystals. J Hazard Mater 182:317–324

Verma VC, Singh SK, Solanki R, Prakash S (2011) Biofabrication of anisotropic gold nanotriangles using extract of endophytic *aspergillus clavatus* as a dual functional reductant and stabilizer. Nanoscale Res Lett 6:16–22

Vigneshwaran N, Kathe A (2007) Silver-protein (core-shell) nanoparticle production using spent mushroom substrate. Langmuir 23:7113–7117

Vigneshwaran N, Ashtaputre NM, Varadarajan PV, Nachane RP, Paralikar KM, Balasubramanya R (2007a) Biological synthesis of silver nanoparticles using the fungus *Aspergillus flavus*. Mater Lett 61:1413–1418

Vigneshwaran N, Kathe AA, Varadarajan PV, Nachane RP (2007b) Silver-protein (core-shell) nanoparticle production using spent mushroom substrate. Langmuir 23(13):7113–7117

Xiangqian L, Huizhong X, Chenz CG (2011) Biosynthesis of nanoparticles by microorganisms and their applications. J Nanomater 32:16

Xie J, Lee JY, Wang DI, Ting YP (2007) High-yield synthesis of complex gold nanostructures in a fungal system. J Phys Chem C 111:16858–16865

Yen SC, Mashitah MD (2012) Characterization of Ag nanoparticles produced by white-rot fungi and its in vitro antimicrobial activities. Int Arab J Antimicrob Agents 2:1–8

Zhang X, He X, Wang K, Yang X (2011) Different active biomolecules involved in biosynthesis of gold nanoparticles by three fungus species. J Biomed Nanotechnol 7:245–254

Zhang H, Smith JA, Oyanedel-Craver V (2012) The effect of natural water conditions on the antibacterial performance and stability of silver nanoparticles capped with different polymers. Water Res 46:691–699

Index

R. Prasad (ed.), *Fungal Nanotechnology*, Fungal Biology,
DOI 10.1007/978-3-319-68424-6